Governing Climate

Governing Climate

HOW SCIENCE AND POLITICS HAVE
SHAPED OUR ENVIRONMENTAL FUTURE

Zeke Baker

UNIVERSITY OF CALIFORNIA PRESS

University of California Press
Oakland, California

© 2024 by Zeke Baker

Library of Congress Cataloging-in-Publication Data

Names: Baker, Zeke, author.
Title: Governing climate : how science and politics have shaped our
    environmental future / Zeke Baker.
Description: Oakland, California : University of California Press,
    [2024] | Includes bibliographical references and index.
Identifiers: LCCN 2024017610 (print) | LCCN 2024017611 (ebook) |
    ISBN 9780520401297 (cloth) | ISBN 9780520401303 (paperback) |
    ISBN 9780520401327 (epub)
Subjects: LCSH: Climatic changes—Political aspects—United States. |
    Climatic changes—Government policy—United States. |
    Environmental policy—United States. | Environmental protection—
    United States.
Classification: LCC QC903.2.U6 .B36 2024 (print) |
    LCC QC903.2.U6 (ebook) | DDC 363.7/05610973—dc23/eng/20240509
LC record available at https://lccn.loc.gov/2024017610
LC ebook record available at https://lccn.loc.gov/2024017611

33 32 31 30 29 28 27 26 25 24
10 9 8 7 6 5 4 3 2 1

For Sebastian, Raven, and Jasper
Instructors in joy, balance, and the promise
of new life—the weight of the future notwithstanding

Contents

Illustrations

TABLES

Acknowledgments

I acknowledge financial support and employment from the University of California–Davis Office of the Provost, the University of California–Davis Department of Sociology, the University of Oklahoma Cooperative Institute for Mesoscale Meteorological Studies, the National Oceanic and Atmospheric Administration, and the Sonoma State University Office of the Provost and School of Social Sciences. The resources, collegiality, joy, and rigorous atmosphere that characterize these institutions facilitated the work that resulted in this book.

The nuts and bolts of this project are indebted to professional associations that have formed critical sites of creativity, growth, exposure, and debate since 2013. Chapters, ideas, and concepts in the book have benefited from regular scrutiny and engagement provided by the following organizations: the Social Science History Association (particularly the Education/Knowledge/Science Network), the American Sociological Association (Environmental Sociology Section, Science, Knowledge and Technology Section, and Comparative-Historical Section), the Society for the Social Studies of Science (4S), the History of Science Society, the American Geophysical Union, the American Meteorological Society, and the International Commission on the History of Meteorology.

The University of California–Davis formed the intellectual community that helped to cultivate this project from its inception up through the completion of a PhD in 2019. I acknowledge the fertile grounds of the Social Inquiry Workshop, the Power and Inequality Workshop, the Environments and Societies Workshop, and countless opportunities and critical conversations within the Sociology Department. I acknowledge the learning community with my students, particularly research assistants Shaina Prasad, Joselyne Quiroz, and Alison Barnard. I would not have succeeded in this book without my UC Davis peer-colleagues, especially Kelsey Meagher, Joanna Hale, Sam Haraway, and Angela Carter. For his consistent ecological perspective, I acknowledge Stephen E. Fick. I acknowledge Jessica Gold and Zosia Wlodarczyk for being light in dark places.

The members of my dissertation committee deserve special recognition. I acknowledge the long, tough, and beautiful conversations with Patrick Carroll, who taught me to dig into the technical bases of knowledge-making and state formation processes. No other individual has so deeply changed my intellectual thought patterns. I also acknowledge the geographical perspective, skepticism, and kindness of Diana K. Davis, who would school me in political ecology, ask the hardest questions, and then invite me over for dinner and let me pick her mulberries. Utmost respect is reserved for John R. Hall and Stephanie Lee Mudge, whose attention, immense work, and careful notes—always critical, but with support in the wings—I cherish to this day. I could have made it nowhere without you, Stephanie and John.

Opportunities to work with geographers and meteorologists were important to the maturation of this project. I acknowledge the leadership and perspective granted especially by Julia Ekstrom (UC Davis/California Department of Water Resources) and Eugene Petrescu (U.S. National Weather Service). I acknowledge mind-changing experiences in the Pribilof Islands, Alaska, facilitated by the National Weather Service–Alaska Region and the Aleut Community of Saint Paul Island.

Although I will miss important people, I acknowledge feedback and engagement on parts of this work from the following individuals: Allison Ford, James Dean, Aaron Panofsky, Claire Decoteau, David M. McCourt, Sam Randalls, Martin Mahony, Deborah Coen, Jamie Pietruska, Tom Beamish, Ming-Cheng Lo, Laura Grindstaff, Luis Guarnizo, Louis

Warren, Marion Fourcade, Jordan Besek, Sergio Sismondo, Anthony Chen, Scott Frickel, Stephen Zehr, and all the commentators and peer reviewers who volunteered their time and effort to engage this project at various stages of its development. Special thanks are reserved for my reviewers at the University of California Press and especially my editor, Naja Pulliam Collins, for guiding this work with patience, red ink, and care.

I acknowledge the nudging of Kirsten Carton to pursue a path in sociology. I credit Noah Toly for provoking my decision to write a book in the environmental social sciences and showing me how to treat intellectual life as a deeply vocational practice. Thanks to Isaac, Marina, Priscilla, Alice, Dan, Luke, Shelby, Andrew, Sarah, Amanda, Matthew, Nick, Angela, and Rebekah most of all, for their companionship over the journey. I acknowledge inspiration to pursue higher education from my parents, Jim and Liz Baker, who raised me in Kyiv, Ukraine, with a burning yet compassionate curiosity about the vexed trajectories of the modern world. Finally, I credit Lyda Grigorivna Zubavienko for demonstrating the art of perseverance in life and learning.

Introduction

In the late 1950s Roger Revelle and Hans Suess, among other select re-searchers, started publishing research about carbon dioxide accumulation in Earth's atmosphere and a plausible relationship to global warming. Armed with new scientific facts, these researchers also had prestigious standing in scientific and government institutions. Public officials in the United States soon took them seriously. They had credibility and public visibility, granting them a fair amount of power to shape scientific re-search about climate and its role in government. But they ultimately failed to shape politics and a society that could confront the problems their sci-ence eventually came to raise in louder and clearer terms—namely the problems of global warming and climate change. What happened?

One way to answer this question is to point to partisan interests and economic elites, particularly in the United States, who made crooked deals as a deliberate means to misinform the public about climate change and protect fossil fuel interests from government regulation. This is an important, well-documented answer. It is also a familiar one. Those who believe climate change is happening and are concerned about it probably have some knowledge about the history of the climate change issue, fea-turing climate scientists, public figures like former vice president Al Gore;

1

"denialists" closely associated with the Republican Party; and decades of warnings about climate change that public officials have, generally speaking, yet to heed. A *New York Times* investigative essay turned-book, written in 2018 and ultimately titled *Losing Earth: The Decade We Could Have Stopped Climate*, has helped to refine this story. In it, author Nathaniel Rich reconstructs the period in the 1970s during which scientists, primarily working in the United States, pronounced that global warming was caused by humans' burning of fossil fuels and that something serious must be done to reduce the emission of carbon dioxide and other planet-warming gasses. That was *half a century* ago. The scathing narrative threaded through Rich's account is that fossil fuel industries have aligned with conservative politicians in the U.S. to systematically influence, thwart, and finally deny scientific knowledge. Naomi Oreskes and Erik M. Conway (2010) call these culprits of misinformation "merchants of doubt" and show how they have been connected to a range of organizations, donors, lobbyists, political operatives, and elected officials that have held sway in wealthy countries, especially the United States, with significant implications for national and international climate policymaking in recent decades.

Although what sociologist Robert Brulle has labeled the "climate change counter-movement" that has attacked climate science can explain a lack of public trust in climate science and some of the polarization among elected officials regarding climate policy, it cannot explain any of the original authority that climate science may have possessed. That would be to put the cart before the horse. Moreover, politically organized denial of climate science did not begin to get underway until the 1970s and 1980s. Therefore, it cannot account for what we can call the politics of climate knowledge before and during the mid-twentieth century, periods I argue were central to shaping climate science in relation to government.

Another way to consider "what happened" is to recognize that climate science was already long related to politics and government, only scientists and others were striking new nerves during the turbulent decades of the 1960s and 1970s—the decades that saw sweeping social movements, the rise of environmentalism, and Cold War "big science" come of age. Yet was it scientific discoveries themselves that struck these nerves? The idea that what historian of science Spencer Weart (2008) calls "the discovery of global warming" is the reason behind subsequent climate politics—a kind of scientific first mover—is a common perspective. That carbon dioxide accumulation might raise the global temperature because of the burning of fossil fuels is *inherently* troubling, is it not? Such a claim, it turns out, is anachronistic. Even Roger Revelle, the "father of the greenhouse effect," wrote in his landmark climate change studies in the late 1950s that global warming could likely provide economic benefits to world society or else be solved through technical interventions. Early climate scientists were not typically "environmentalists." Environmental protection was not an initial basis for either scientific or political claims regarding global climate change.

My specific argument regarding the rise of climate science and its implications for society and government in the 1950s and 1960s is that an interdisciplinary scientific elite effectively reshaped existing scientific institutions—including scientific disciplines, research centers, and paradigms of how to do good climate science. Further, they did so by virtue of their capacity to tie problems of atmospheric research to problems of war and state-making in the post–World War II period. The character and degree of overlap between science and government explain the discovery and significance of global warming, not the other way around.

As important as this perspective may be to explaining the rise of climate science and politics in the latter half of the twentieth century, it remains but a piece of the much larger story I believe needs to be told. This larger story is that climate, science, and government have long been intertwined in ways that have durably shaped how society relates to nature. In this view, the mid-twentieth century is not an origin story, nor is it simply a historical backdrop to the allegedly *recent* and *contemporary* problem of climate change. Partisan debates and policy decisions regarding how to mitigate global warming may be new. The International Panel on Climate Change (IPCC) and UN Framework Convention on Climate Change (UNFCCC) are indeed less than forty years old. But climate politics, if conceived broadly, is centuries old. And this deeper story matters. It matters not only by helping to explain where contemporary climate science came from, but also because if we can learn to see how climate, science, and government have shaped one another over time, then we can gain some understanding of how they may do so in the future. And although the past cannot be changed, the future remains a more open and potent site of struggle. I believe that struggle is best approached with a historical perspective in mind.

To take a wider historical view of climate politics, in this book I treat science, government, and the climate itself as a configuration—a constellation of actors, institutions, things, events, patterns, and ideas that make up a sort of whole. This approach first means not treating climate simply as a physical entity out there in nature that can be observed, known, and represented objectively. It also means not thinking about science in temptingly idealistic terms, that is, as having raw access to matters of fact and truth about the world. Further, it means considering politics and ideology as factors that help create scientific knowledge (rather than simply getting in the way).

If we treat climate science—including scientists themselves, the technologies employed in the process of making knowledge, and the parts of society that are touched by climate science—as embedded in such a configuration, then we can start to see new aspects of the "what happened?" question. First, as already indicated, this approach makes into a puzzle the original authority that climate science garnered in the mid-twentieth century. Was discovery sufficient, or were other factors significant? Second, it raises questions about how the social, technical, political, and economic context in which climate change emerged as an issue for science mattered. To do so necessarily involves looking backward, to what came before. Did the emergence of climate science displace other ways of knowing? Other *configurations* of science, politics, and climate? Did it rearrange other social institutions? When looking backward, it becomes possible to see how the specific configuration of the post–World War II decades was indeed unique, forming an amazing context for the transformation of science as an institution. It is within such a context that global warming gained ground as an issue. But pushed further, situating climate science in its historical circumstances does something more: it can locate the recent period in a larger set of historical patterns and trajectories, including those that remain with us and, as I explore, may be decisive for how society deals with climate change today.

One upshot of understanding the domains of climate, science, and government as configuring one another is that it casts into relief the contemporary historic moment for each of these domains and the relations among them. In particular, the science versus politics casting of the climate problem starts to make less sense. The issue becomes not simply a matter of politics getting out of the way so that people can "listen to the scientists!" (as climate activist Greta Thunberg, among others, has proclaimed), but of considering how climate, science, and government might continue to be configured into the climate-impacted future.

When we take stock of the configuration of climate, science, and government today and try to trace possible trajectories, some are particularly concerning. Issues of *geoengineering* and *climate security* within the context of climate crisis loom large here, and they are therefore a focus of investigation in this book. Let us take climate crisis first. In developed societies, enduring political challenges to confronting rapid and unprecedented climate warming are the systematic denial of the issue, the relative failure of political will and coordination to mitigate global warming by reducing greenhouse gas emissions, and the hollowing out of social capacity to contest an economic system that runs on fossil fuels. Recognizing this dire situation, social movements and media framings have turned to action featuring discourses of *climate crisis* and *climate emergency*. These discourses give voice to real suffering and despair and are anchored in climate science projections of more severe impacts along with other sources of concern regarding risk and vulnerability. They also wager that proclaiming a crisis, even apocalypse, may make the wheels of government change course, forcing a choice between addressing a climate emergency or facing the threat of extinction.

But what if this wager fails? What then might efforts to govern climate crisis actually look like? Geoengineering and climate security, introduced in turn below, are emerging issues at the intersection of science and government that offer some plausible, if problematic, hints with respect to these questions.

Geoengineering, specifically what is termed "solar radiation management," includes using technologies to alter the climate system through technical intervention to reduce warming on a planetary scale. Scientists and policymakers are increasingly looking to geoengineering as a possible "Plan B" in the event of climate policy failure and increasingly disastrous climate change impacts to society. Climate security includes harnessing innovations in earth system modeling (which "downscale" global climate change projections to specific areas, sectors, or populations) to project who can be secured in the face of climate-driven catastrophes, and who will likely be insecure and hence treatable as threats to the would-be secured. Geoengineering and climate security represent two possibilities that take hold of climate crisis and incorporate scientific developments into new ways of governing climate. There are other possibilities, some more or less desirable, which are explored toward the end of this book. The initial point is that these possibilities do not result from a science *versus* politics battle, but rather the synthesis of the two.

This book takes the perspective that the configuration of climate, science, and government is a fundamentally and deeply historical one. The paths that scientists, policymakers, governments, and others are presently traveling, paving, or imagining (geoengineering, climate security, or otherwise) are built with tools, materials, and ideas developed in the past. As such, they are dynamic and open to sometimes transformational change. To be sure, contemporary global society has no historical analog for how to confront global warming. Even so, tracing how science and government have approached issues of climate over the historical long term provides a means for explaining our current circumstances and for approaching the contemporary moment in terms of trajectories and opportunities (rather than, say, apocalypse).

To pursue this argument, this book turns back not to the 1970s (when, according to Rich and others, we "could have stopped" climate change), but rather to the 1770s. By stepping into the worlds of scientists, their public audiences, and governmental actors of all kinds from around that time until the third decade of the twenty-first century, it is possible to gain a fuller picture of modern societies' wrestling with the problem of how to govern climate.

GOVERNING CLIMATE

Scientists have long sought to measure, know, and understand climate. In doing so, they have tried to make sense out of some of the most capricious elements of the world: the winds, skies, oceans, storms, and seasons.

People have also variously tried to make society legible, especially so that social groups and trends may be made governable by central authorities. The history of climate science and of government, in these basic terms, is often told separately. This is a mistake. Debates and struggles about climate within the scientific enterprise have intersected with practices of government through a process that has resulted in distinct configurations of climate knowledge at different historical periods. Rather than being determined by scientific inquiry or climate patterns alone, these configurations can be understood as distinct logics of government. In this book I refer to these as logics of "meteorological government." Over the course of the research represented in this book, I have come to define meteorological government as the process by which actors deploy meteorological knowledge to know climate and govern social order simultaneously. Defined as a logic, the concept implies that people across time and place have different rationales and strategies for building knowledge and governing society with respect to climate. Meteorological government is therefore necessarily a social process, rather than a formal social institution.

In this book I use the case of the United States from around 1780 to the 2020s to investigate both the formation and the consecutive transformations of logics of meteorological government. To get a bird's-eye view, let us briefly preface each of these logics. The first, which I locate in the late eighteenth to mid-nineteenth centuries, centered on understanding climate as a domain primarily located between human bodies and "airs." Under this logic, climate knowledge, broadly within the purvey of natural philosophy and medicine, could inform practices of moral discipline and social difference unique to the diffuse colonial and protonational political order that characterized the period. Dominant by the post–Civil War period, the novel profession of climatology marks a second logic. Although hardly universalized in transnational climate science, this logic rejected previous understandings of climate—especially the earlier central concern with climate change—and instead treated climate as a geographically and historically "stable" category (that is to say, something that does not change). Climate knowledge thereby could be readily integrated into the governing of industrial capitalism, which thrived in part by making land, spaces, and natural resources legible in numeric, graphic, and monetary terms. A subsequent logic, which I locate with the mid-twentieth-century emergence of climate science and post–World War II government institutions, centered on understanding climate as a global system undergoing radical destabilization because of global warming. Developments in twenty-first-century meteorological government can be interpreted as responses to the challenge of destabilization. On the one hand, it might be possible to restabilize global climate change by engineering the climate. On the other hand, governing climate in the face of climate crisis operates through a logic in which the impacts of global warming may entail catastrophic social dislocations that—to remain governable—must be vigilantly anticipated through practices of climate security.

The aim of this book is to analyze the historical transformations between these logics by demonstrating how competing approaches to climate knowledge intersected with developments in government at critical societal junctures. As alluded to already, this process is ongoing, perhaps at a crossroads. The book therefore concludes by addressing movements of science and government that may involve the unequal distribution of climate risk between people who are secured against climate change impacts and those who are unsecured.

If these movements represent plausible future trajectories, then it stands to reason that climate science is not inevitably tied to progressive social causes or environmental sustainability. Indeed, history suggests otherwise. Informing social difference and hierarchy, as I show, has long been a hallmark of climate knowledge. Such an outcome is today, as in previous periods, not inevitable but rather dependent upon struggles that run outside the boundaries of common discourse linking the hard-won facts of climate science to a progressively sustainable society.

METEOROLOGICAL GOVERNMENT

Starting with Power

"Climate," generally defined as the character and dynamics of atmospheric states and patterns across a range of time- and spatial scales, is a physical environmental force of nature that shapes human society and history. Even so, rather than holding as its compass simply the facts of climate history, power orients this study. This orientation stems from a few major questions that have preoccupied social scientists for quite some time. The first question concerns how environmental forces shape or determine social trajectories, and how culture, knowledge, and power relations characterize environmental and social outcomes. The second question concerns how and why scientific knowledge changes over time. Does nature present itself unmediated to science? If so, are human reason and innovation a sufficient explanation for scientific change? Or rather, are social factors significant, even decisive, in shaping the dynamics of science? A third major question is how science may be caught up in power relationships. Do scientists, ideas, and scientific institutions have authority? If so, how does it relate to other kinds of authority and power in modern society?

Historians of U.S. meteorology and climate science have until recently generally studied science as a relatively set-apart domain that features rational debate, technical innovation, and growth. Social relationships of power were therefore not a central focus. Although applied to social and political issues, science has been more typically seen as fundamentally driven by the quest to make better knowledge. Scholars have especially focused on several periods of change, such as the advent of physics-based theories of atmospheric motion and weather prediction in the latter half of the nineteenth century and, later, mathematical modeling of weather patterns and climate systems in the mid-twentieth century (Feldman 1990; Fleming 1990, 1998a; Nebeker 1995; Monmonier 2000; Harper 2008). Supposing progressive stages of development, scholars have depicted meteorology and related fields as forming into a bounded domain of scientific inquiry. Over time, this narrative establishes, science came to capture climate realities more successfully. Complete with standardized instruments, networks of meteorological observations, agreed upon conventional methods, and theoretical controversies that superseded speculative dogma, the established narratives conclude, meteorology came to produce a more thoroughly *scientific* account of weather and a more circumscribed physical account of climate. Based on such narratives of scientific progress, scholars have generally sought to explain how modern meteorology emerged from an "art" into a "science" (Harper 2008:1–2, 61), or in Monmonier's (2000:x) terms, more dismissive of former practices, how real scientists replaced "wannabe forecasters" of the late nineteenth century, who had "studied maps rather than physics." Still other scholars have developed a similar historical narrative to show how climate-change science developed through a logic of discovery, a process that only later began "breaking into politics" (Weart 2008:138; Le Treut et al. 2007). As Rich (2018) characterized the early 1980s context, "What started as a scientific story was turning into a political story." And it makes sense that the prevailing historical accounts of climate science would draw hard lines between science and non-science. In a world where some vocal critics crudely decry climate change science as "unsettled" and primarily "ideological," climate scientists and intellectual allies have responded with historical accounts emphasizing consensus and objectivity (Oreskes 2021; Cook et al. 2016; Bolin 2007; Le Treut et al. 2007; Weart 2008). At the center of battles between scientific fact and politically mobilized anti-science values lies this point, well put by historian of science Naomi Oreskes: "Facts do not speak for themselves, people need to speak for facts" (cited in Gilles 2017; see also Oreskes 2021). Because climate change is a politically charged issue, historical and sociological accounts have tended to "speak for [the] facts" and thus drawn distinctions between the social world of climate science and the world of politics and ideology. One result is that we know a lot about the social construction of ignorance and the political mobilization of climate denial, as well as about political battles regarding climate mitigation and adaptation policies in local, national, and intergovernmental forums. But another result is that developments, both historical and contemporary, may go unnoticed especially across the science-politics divide.

In an alternative approach taken here, the configuration of knowledge and power stands at the center of an explanatory account of the jagged history of climate science. Narrating history in this manner, to be clear, provides no fodder for rejecting the authoritative consensus among scientists regarding global warming (Cook et al. 2016). Humans have unequivocally caused global warming and fundamentally altered the earth system. When brought to bear on historical patterns, however, analysis runs into problems when it rests upon an implicit separation of the climate, knowledge about it, and the social world with its power relations. It is at times more difficult to separate out actors and institutions of science and government than it is to recognize how they shape one another. The wager taken here is that, upon investigation, problematizing scientific practices in relation to government provides a better explanation for changes in climate knowledge than appealing to boundaries between science and broader social developments.

I have already introduced the concept of meteorological government by defining it as a process by which both social and climatic order are established among those who, in various historical circumstances, possess the authority to speak on behalf of the climate and to organize social activity according to that authority. This concept weaves into existing ideas about the nature of power and government.

Sociologist Max Weber has defined power as control over the means of coercion, the chances of subordinating challengers, and the ability for some people to dominate others regardless of resistance. *Government*, inspired here by the work of Michel Foucault, is a specific case of power. It is marked principally by a quest for a legible social order that can be improved, modified, integrated, or subordinated through deliberate action. Government in these terms is not unique to any specific political formation (a democratic state, for example), although bureaucratic states hold great capacity for government over wide regions and large populations.

How might government, in this more expansive way of thinking, relate to knowledge? Of course, people may exercise power without knowledge, as in rageful physical violence. But not so when it comes to governing. To resist the coercive power of physical violence is to fight back; to resist government is to be out of *order*. Conversely, to govern is to make order where it did not exist before and to exercise power to preserve it. Governing thus almost always requires knowledge about people, places, resources, and territories. Although the will to know and the will to power can be considered separate human motivations, it is not possible to trace a history of government without incorporating a history of knowledge.[1]

If the concept of meteorological government is generally oriented by a concern with how science relates to power, then it draws from a wealth of research in the social sciences. The first theme concerns the *coproduction of science and society*. Scholars of coproduction investigate how social dynamics inform changes in the sorts of ideas, questions, emphases, and institutions that characterize science in a given context. The second, which I label *imperial geographies of climate knowledge*, evaluates how climate science, in historical and contemporary contexts, has participated in or been marked by imperialism, colonialism, and state-making. The third is *historical sociology*, which takes aim at explaining the historical bases of durable social institutions, including capitalism and state formation, which are highly relevant to this study.

The Coproduction of Science and Society

Since the late 1980s, scholars working in the field of science and technology studies (STS) have theorized the "coproduction" of science and society. Sheila Jasanoff (2004:2) has defined coproduction as "the proposition that the ways in which we know and represent the world are inseparable from the ways in which we choose to live in it." Political sociologists Scott Frickel and Kelly Moore (2006) have likewise argued that to explain scientific authority, controversies, and their effects, researchers must pay attention to formations of power expressed in historically variable or novel institutional dynamics. Such dynamics might include the commercialization of science, for example in medicine or through intellectual property, or political tensions around scientific credibility, for example in the case of the COVID-19 pandemic.[2] Historically oriented scholars deploying the concept of coproduction are especially concerned with how science and power coalesce to shape the process of modern state formation (Carroll 2006, 2012; Mukerji 1997, 2009; Jasanoff 2004) and capitalism (Lave 2012; Lave, Mirowski, and Randalls 2010; Moore et al. 2011). The focus afforded by this work demonstrates that practices of governing territories, state agents, subjects/populations, and natural environments (including climates) are not given by nature, nor by brute force or command, but more so by projects that diverse actors carry out to make relations between social and natural worlds "legible." James Scott outlines the concept of "legibility" in the following useful terms: "Legibility is a condition of manipulation. Any substantial state intervention in society–to vaccinate a population, produce goods, mobilize labor, tax people and their property . . . requires the invention of units that are visible" (Scott 1998:186). For his part, Scott was primarily concerned with the contradictions of state-making as a project of legibility. He showed that interventions, say to control an entire watershed or force into place the movements of culturally itinerant groups, have often failed in their attempts to create the "simplified" social and natural order that state officials believed they successfully made visible and hence governable. Attempts to control people and nature, especially in authoritarian contexts, often misrecognize the more complex realities on the ground. More generally, the concept of legibility can usefully be extended to account for how and why social actors go about collectively simplifying phenomena, describing them in relation to various "units" and by doing so coproducing knowledge and government.

Governing climate can be understood with reference to specific projects that make climate legible in relation to social interests or political goals. The legibility of weather and climate is without a doubt a hallmark achievement of climate science. As a material accomplishment, across time climate science has rested upon a "vast machine" of atmospheric and related measurements, which in turn rely upon a web of technological infrastructures that allow people to "see" otherwise invisible atmospheric patterns and capricious effects (Edwards 2010). The concept of legibility is useful because it provides a way of thinking about knowledge and government as involving efforts to establish *order*, principally through moving back and forth between the messiness of the natural and social worlds on the one hand, and the neatness of written inscriptions and representations on the other.[3]

How states and sciences in a given period look for and exploit the order of nature is typically related to how society is organized economically. Quests for legibility are especially important to the material processes that have characterized capitalism as a social and economic system. Scholars working in the vein of Marxist political economy bring such connections to light. Marxist accounts of environmental science conceptualize the "production of nature" as a process of carving up nature into legible component parts that can circulate as exchange value, that is, commensurable with money (Smith 1984). The legibility of nature, whether made by prospecting, mining and processing minerals, digging up and refining fossil fuels, or inventing synthetic chemicals in a laboratory, is often a precursor to making and exchanging commodities. Unlike minerals or chemicals, weather and climate are relatively ephemeral. Even so, as scholarship on the history and present of financialization of weather and climate indicates, the economic legibility of atmospheric phenomena has long been a part of the commodification of nature more generally (Randalls 2010; Randalls and Kneele 2021; Braun and Whatmore 2010).

Tracing the tangible, material legibility of atmospheric and social phenomena, especially through analysis of the construction of meteorological data infrastructure and related technologies for "observing" climate, will be important to any account of meteorological government. Yet "legibility" is not only a material process. Whatever individual people represent on a government census or whatever surface temperatures are recorded at a given location, the very categories involved in legibility projects hold deeper, symbolic value.[4] The category "climate" may take on special meaning, especially when it provides a way to represent human populations, environments, and their entwined trajectories. Consider, for example, the many symbolic values pinned to a major contemporary idea, namely that global warming has ushered in "the Anthropocene." The Anthropocene originally was coined as a technical designation of a new geologic epoch marked by human impacts on the planet now evident in the geologic record. Some scholars date the Anthropocene era as beginning with thermonuclear bomb tests in the 1950s, others with the patenting of the Watt steam engine, and still others with the arrival of African slaves and settler-colonists in North America (see Callison 2024). Others have discarded the term, instead preferring Capitalocene or other monikers to critically highlight the groups and economic processes responsible for human planetary impacts (Moore 2017). The Anthropocene concept is not simply technical in nature. It variably signifies a new scientific paradigm for what is now called earth system science, a historical break marked by the arrival of human planetary domination, a cultural zeitgeist centering on the scourge of modernity, and impending apocalyptic collapse. What the controversial idea of the Anthropocene shows is that representations matter, and they can carry political implications. They can form an aspect of power relationships, what sociologist Pierre Bourdieu terms "symbolic power." Following symbolic power implies paying attention to the variable definitions, meanings, and interpretations of climate, thereby assessing their relevance to practices of government.

Imperial Geographies of Climate Knowledge

A second body of work involves the imperial geographies of climate knowledge. Work in this vein traces the coproduction of imperial and colonial order on the one hand and climate knowledge on the other. This work is especially important for helping to situate efforts at governing climate in the United States in a comparative-historical and international (or transnational) context.

Scholarship among geographers, environmental historians, political ecologists, and historians of science has sought to problematize the basic category of "climate." A key question in this line of research involves how the spatial and temporal dimensions of climate (e.g., "the climate of New England" or "global climate in 2050") relate to the social circumstances of those who create or use scientific knowledge (Mahony and Hulme 2016; Brace and Geoghegan 2010; Baker, Ekstrom, and Bedsworth 2018). In short, how does the *who, where,* and *when* of science and scientists relate to the *what* of climate science? This general question has resulted in a range of recent studies. A core focus has been on how colonial and imperial political structures have historically enabled meteorological and climate sciences by virtue of their economic interest in exploitation of diverse places, as well as their expansive maritime networks, territorialization efforts, and aims to settle territory with foreigners while exploiting allegedly inferior local people (Mercer and Simpson 2023; Mahony and Caglioti 2017; Livingstone 2003; Coen 2018).

The growing body of work on science and colonialism has found that climate knowledge has had real consequences for, and has been integral to, political domination. Scientific practices and technologies have facilitated the unequal appropriation of natural resources, thus creating imperial, uneven geographies (Goldman, Nadasdy, and Turner 2011; Robbins 2012; Forsyth 2002; Li 2007; Vandergeest and Peluso 2011). Although imperial science is chiefly expressed through economic utility and exploitation, colonial forms of power that invest in subjugating or incorporating Indigenous and other cultures have other features. Namely, science can inform narratives of historical change and visions of the future that can serve to interpret or shape social activity. For example, Melissa Leach and Robin Mearns (1996) explain the persistence of a colonial "desiccation" environmental narrative, which held that arid African landscapes had once been green but were degraded, in part because of allegedly backward local cultural practices. In their study, the inherited narrative has roots in historical colonial land use patterns, colonial European ideas about how native land use led to historical "degradation" especially in North Africa, and gross misunderstanding of local livelihood practices (Davis 2016). By the late nineteenth century, as Philipp Lehmann (2022) has found in the case of French and German imperial projects in the Sahara region, entrenched historical views of desiccation could then inspire grand visions of climate engineering that could transform arid lands and dry climates into wetter, lusher, and more productive landscapes, ripe for European settlement or exploitation. Such historical visions of aridity and regional climate change continue to shape environmental management policies in North Africa, including climate policy aimed at preventing desertification.[5]

Visions of past or future climates generally shaped colonization strategies, imperial practices, and the relationship between scientific ideas and other (often subjugated) knowledges. Historians and geographers have found that geographically expansive forms of rule, especially colonialism, informed basic theories of climate as they were developed from the seventeenth to the twentieth centuries (Golinski 2008; Grove 1995; Livingstone 1991, 2002; Coen 2011, 2018; Zilberstein 2016).

First, modern colonial empires covered increasing amounts of the planet, which mattered for how climate was understood in terms of scale. In the case of the mid-nineteenth-century British Empire, Mahony (2016) finds that attempts to create new ways of coordinating and calculating global atmospheric measurements were technically accomplished because of British imperial reach. Climate science was likewise defended as a science supportive of empire-building amid climatic, geographic, and social (including racial/ethnic) diversity. Coen (2018) finds in the case of the Habsburg Empire that climatologists successfully linked local to regional patterns by virtue of the territorial integration and cultural amalgamation of the empire. Cushman (2011) finds that regional climate patterns, like the El Niño Southern Oscillation (ENSO), was made legible to nineteenth-century climatologists by virtue of how "creole" scientists could translate local/Indigenous knowledge in communication with allegedly cosmopolitan colonial scientists like Alexander von Humboldt.

Second, colonial authorities often conceptualized climate as a deeply social concept, thus relating to other common social distinctions in colonial contexts. For example, in a range of European colonial settings, "climate" provided a general discourse for explaining social and civilizational forms, understood in resolutely hierarchical terms. Scholars of British and French colonization, for instance, have shown that an imperial construction of climate knowledge informed evaluations of the "torrid zone" of equatorial, tropical, and arid climates (Arnold 1996; Beattie, O'Gorman, and Henry 2014; Davis 2016; Duncan 2007; Harrison 1999). The intersection of colonialism, racial hierarchy, and environmental science looms large here. Beginning in the eighteenth century, scientists helped to articulate a colonial discourse that understood climate through a lens of environmental determinism, an intellectual position holding that climate strongly determines historical, physiological, and sociohistorical developments (Locher and Fressoz 2012; Peterson and Broad 2009). Intellectuals, colonial officials, and scientists who elaborated the climatic character of social order effectively coproduced a worldview in which European whiteness and civilization were something that must be anxiously protected. For example, the field and practice of tropical medicine could allegedly help support white bodies in an "unfit" climate. As another example, a civilizing mission of "improvement" (Conklin 1997; Drayton 2000) was expressed through efforts to reshape landscapes deemed unsafe and deleterious to white colonists and settlers while also remediating native "backwardness." Racial understandings of climate and the body played a role in justifying racial stratification and domination (Bashford 2004; Livingstone 1991). From French colonial North Africa (Amster 2013, ch. 2; Locher and Fressoz 2012) to British-controlled India and the West Indies (Harrison 1996), debates about the "acclimatization" of the colonizer and the racial character of the colonized helped to established, for colonial officials, the "natural" basis of social hierarchy that informed policies about hygiene, settlement, assimilation, and military strategy. Although theories of climate, like colonial systems, were eclectic rather than uniform, it is clear enough that scientists built local, regional, and general accounts of "climate" across colonial circumstances. This process reinforced a generally hierarchical worldview about nature and society, while also serving to establish the technical and intellectual basis of climate science as a transimperial (if not entirely global) science by the turn of the twentieth century.

The American case is unique but reflective of colonial cases in other regional contexts. Apprehending or seeking to institute social and climatic order is hardly reducible to colonial projects in specifically "torrid" or "tropical" spaces. White (2020) shows that New World studies of climate from the sixteenth to the nineteenth centuries have a common thread: they were used in a comparative and cosmopolitan manner to develop general theories of climate, for example the effects of prevailing winds on temperature and weather patterns or the relationship between latitude and regional climates. Studies of climate discourse in North America have yet to focus on how colonialism intersected American state formation and capitalist political-economic developments (Golinski 2008; Kupperman 1984; Zilberstein 2016). A focus on the state-science relationship can contribute to the broader international history of climate and colonialism by showing the degree to which meteorological government in the United States reflected practices developed in other contexts, or to what extent the United States presents a unique case because of the historically contingent processes of U.S. state formation, settler colonialism, and capitalist development.

A Climate for Historical Sociology

The field of historical sociology centers around building explanatory frameworks for large-scale social patterns and transformations. As a field it can be nicely, if simplistically, differentiated from history, which aims to truthfully narrate events and processes, and from sociology writ large, which may or may not incorporate historical patterns or changes into accounts of social phenomena. As early as 1983, sociologists and climate scientists together recognized that global warming constituted an "intrusive concept" that was "disturbing [to] the social science peace of mind" (Boulding 1983:3). Historical sociologists are just beginning to take up this "intrusive concept" by incorporating climate (and climate science) into accounts of social change.[6]

The power-knowledge nexus has become a major theme in historical sociology. This is reflected in attention to culture, discourses, and especially symbolic power (Gorski 2013; Loveman 2005, 2012). It is also reflected in postcolonial critiques of mainstream sociology and subsequent attention to critical, nonmetropolitan standpoints in historical analyses. The purchase of a historical analysis of the configuration of climate, science, and government is that it can contextualize ongoing issues of climate change, but also that it can speak back to long-standing issues in sociology, namely in this case the historical process of state formation in comparative context and the specificity of American political development. Historical sociologists and political scientists focused on American political institutions have variously puzzled over American state-making as featuring a "weak" bureaucratic administration from the late eighteenth to early twentieth centuries, but with "strong" features, particularly regarding territorial expansion, legibility of and strategic control over land via land/ settlement policies, and the durability of racial formations (Frymer 2017; Rockwell 2010; Go 2011; King 2019). The interplay of these relatively weak and strong features within American political development may help explain the formation of U.S. meteorology and climate sciences, which were relatively well-developed in comparison to other settings across historical time periods (Fleming 1990). What has yet to be investigated is just how these sciences thus configured central aspects of the U.S. state—particularly, as I show, the prevailing issues of land, settlement, territory, race, population, and economic productivity and dominance.

Scientific capacity and infrastructural development over the course of the nineteenth century demonstrate parallel aspects of U.S. state-making that can also be understood through the lens of climate (Kelly 2014; Carroll 2012). For example, political scientist Andrew Kelly (2014) provides an analysis of the development of networks among U.S. scientists and organizations, with an emphasis on the role of private-public partnerships that circumvented the otherwise strict limits on establishing a stable, scientific bureaucracy. There are two ironic lessons based on Kelly's analysis of the important cases of the U.S. Coast Survey and the U.S. Geological Survey in the mid- to late nineteenth century (both of which are relevant to a historical-sociological account of climate science in the U.S. context). The first is that partnerships between Survey administrators and private parties, including universities, science societies, observatories, and commercial associations, led to a sprawling network of people, organizations, instruments, and infrastructure critical to the development of physical sciences. A "weak" center thus resulted in a strong scientific network. The second irony is that the rise of public-private partnerships, which even political critics recognized as strong and apparently durable, led to core constituencies that could mobilize to support federal science. Thus, not only did the Coast and Geological Surveys survive under the conditions of the nineteenth-century patronage state, but they also *thrived* in creating networks that would become the institutional basis for strong national scientific capacity in later periods. Weather and climate present perennial problems of government that are only capable of being solved, as meteorologists came to successfully argue, through organizing an expansive technical and scientific network. The networks necessary for the scientific project of climate science can therefore be approached in relation to the kinds of political-institutional networks that Kelly and others view as sources of state strength, the entrenched political systems of anti-bureaucratic "spoils" and patronage notwithstanding.

Overall, work in the theoretical veins of coproduction, imperial geographies of climate knowledge, and historical sociology help set up a deeper look into meteorological government. Scholarship in these fields suggests it is important to analyze the social context of scientific practices and concepts. They provide a rationale for evaluating the central role of environmental knowledge in colonialism and imperialism. And they don't lose sight of the macrohistorical significance of capitalist and state-making processes.

METHODOLOGY, OR, HOW TO TELL
THE HISTORY OF CLIMATE SCIENCE

Narrating the past; selecting, excluding, and interpreting data; and applying concepts to historical events are hardly straightforward processes. There is neither a simple nor a singular way to tell the history of climate science, especially if the resulting story is intended to inform the present rather than merely recollect or explain past events. Methodology provides some contours for research, while also making certain aims, assumptions, and limitations apparent. "Political genealogy," which organizes this study, is a method of investigating the history of knowledge as it is configured with environmental and social forces and then representing that history in a manner that helps to decode the present.

Consistent with constructivist approaches in social studies of science (Lahsen 2024), I approach climate as a category of practice rather than a category of analysis (Brubaker and Cooper 2000:5–6). In other words, I choose to find, follow, and interpret the meaning of climate and climate science as they present themselves in various textual, and by extension intellectual, practical, and social, contexts. Climate thus carries multiple meanings for different groups at different times. I do not impose much of an exogenous definition of climate, say, as a global system of physical atmospheric patterns that cause regional weather and related weather phenomena. The central category of analysis, developed to decipher and put together history, if not an objective definition of climate, then, is meteorological government.

Taking this approach, diverse efforts across time and space to govern climate can appear like pieces of a kind of jigsaw puzzle. Such efforts need to be located and turned over. Then they must be individually looked at from multiple perspectives in order to make sense, and finally they must be put together into a coherent whole. I approach the puzzle of climate and government by doing what Hall (1999) labels a *configurational history*. The central task for inquiry within configurational history is to construct a narrative that focuses on significant historical patterns and transformations but does not adhere through teleological or stage-oriented accounts of historical causation.[7] The latter would effectively look at the past as if it prefigured or anticipated the present or was somehow driven by a deeper story, say, one of progress. I avoid that because I believe it would impute anachronistic meaning onto historical actors and yield a very partial picture of how history bears upon the present and future. As much as it is tempting to read the past as prefiguring contemporary issues of global warming, I rigidly refrain from a progressive narrative in which facts are only considered for their significance to later processes. My goal is not to validate modern climate knowledge through documenting its historical murmurings, for example what Feldman (1990:171) calls the "first hints of a discipline" in eighteenth-century meteorology. Regarding science, historical studies inspired by Thomas Kuhn have found that the separation of social forces operating "outside" science from truth unfolding "inside" scientific reason fails under empirical scrutiny (Bloor 1974; Shapin and Schaffer 1985).

In this book, the research underlying each chapter pursues roughly chronological sets of developments. In doing so, the analytic scope of the narrative zooms in and out, for example, from the microcontext of individual biographies to transnational trends, in a way that illustrates and accounts for the configuration of climate, science, and power across multiple empirical domains. Each chapter and its sections thus contain subsidiary investigations that I draw together into an analytical narrative thrust forward by an account of meteorological government.

Analysis in this mode is not primarily a historicist, or fact-settling, enterprise. In practice, this means my use of evidence and data is hardly exhaustive and cannot be so ambitious as to yield a comprehensive, fine-grained history of climate knowledge in particular contexts. The upshot is that larger patterns and historical transformations can come into view, which help bring a kind of historical texture to the present.

So, the point of a political genealogy is to narrate the past with a deliberate concern for the contemporary social context of its narration, but not in a way that imposes exogenous categories onto historical actors and events. In other words, although the aim is not a progressive narrative, it is nevertheless strongly oriented toward representing the problems of contemporary society in a new light (Hall 1980). Walter Benjamin's insights about history and his rejection of "empty time" provide a helpful frame of reference: "History is the object of a construction whose place is formed not in homogenous and *empty time*, but in that which is *fulfilled by the here-and-now*. . . . To articulate what is past does not mean to recognize 'how it really was.' It means to take control of a memory, as it flashes in a moment of danger" (Benjamin [1940] 2005; emphasis added). In terms comparable to Benjamin's rejection of historicism, Pierre Bourdieu (2004) similarly holds that sociohistorical analysis can help expose relationships of domination. Doing so requires strict attention to the social relations that enable or constrain concepts and discursive categories (whether that be the category of "climate" or racial or gender categories, for example). The history of categories can then lay bare, or even affect, the contemporary meaning of those categories. Although concepts signify objective phenomena, the task for historical analysis, as Margaret Somers (1995:15) writes, is to trace "the conditions of possibility, or the conceptual networks within which concepts are framed and constrained." Replacing "empty time" with configurational history, then, a political genealogy recognizes that climate is a concept "marked indelibly (although obscurely) with the signature of time, normativity, and institution-building" (Somers 1995:15). So, what is going on with the pressing issue of contemporary global climate change that may be understood genealogically? Two recent formations (mentioned briefly earlier), namely climate security and geoengineering, are important to reflect upon. Because analysis of these issues only appears later in the book, they deserve an introductory discussion at the outset.

Genealogy for What? Situating Climate Security and Climatic Stabilization Today

Themes of climate emergency and collapse are becoming common in cultural, political, and scientific circles (Wallace-Wells 2019; Ripple et al. 2020). Projections of future climate disorder are likewise increasingly coming to bear on how governments construct at-risk and risky populations. This is especially clear in the case of national security policy that incorporates climate science into anticipations of widespread social disorder. Historical analysis can help to situate the kinds of power and knowledge that circulate through efforts to govern climate change in these terms. If the politics of "denying" climate change dangerously wards off governmental actions necessary to mitigate global warming, efforts to govern climate nevertheless carry on, inviting investigation into the possible trajectories of future society amid the severe dislocating effects of climate change (IPCC 2023). Even if anti-regulatory and anti-science values, embodied in the United States by the former Trump administration, ultimately fail, what forms of governing may come next, not despite but rather *because* of an accepted truth that society may have passed a point of no return?

Most important to this question is the rise, in the past decade, of climate governance aimed at climate change adaptation and climate security. Climate change adaptation programs and practices, although often integrated with policies to mitigate global warming by reducing fossil fuel use, remain distinct because of their central aim of governing climate change *impacts*. Comparing adaptation to mitigation, therefore, the locus of policy shifts from mitigating the socioeconomic impacts to a *geophysical system* (e.g., carbon-emitting industrial activity that warms the climate) to adapting the *social body* to environmental disruptions.

The scalar shift in climate adaptation is generally away from the global and toward the national and supranational. This reorientation in governance centers around the adaptive protection of certain segments of society, limited resources, and given territories, all based on projections of climate disruption. Organizations, programs, and discourse of climate security have become anchoring points for relevant policy and expert fields, especially those related to geopolitics, national defense, and international relations. Climate security proceeds apace with, and is apparently dependent upon, the failure of governmental efforts to curb global warming. I take aim at historicizing the present reality of climate governance in these terms.

If the rise of the global warming multilateral mitigation regime in the 1990s inspired genealogies of "global" environmental knowledge (Coen 2011; Mahony 2016; Miller 2004; Edwards 2006), then the rise of adaptation policy and "securitization" of climate-change impacts should inspire genealogies of what Christian Parenti (2015) identifies as the "environment-making state" that renders climate impacts governable (see also Oels 2005; Baker 2017; Mahony 2014).[8]

Climate patterns may influence, but they do not determine, socio-political responses, including how states make legible and segment populations, groups, places, resources, and ecosystems as worthy of preservation, investment, protection, sacrifice, or treatment as threats in the face of climate disruptions. Climate scholars Mike Hulme (2011) and David N. Livingstone (2015) caution against "climate reductionism," which others have shown may obscure the social and political mechanisms causing vulnerability to climate change impacts. Such mechanisms may be economic precariousness in the case of marginalized groups (Taylor 2014) or the violent rupture of resilient cultures in the case of many Indigenous groups (Callison 2024). In the face of reductionist tendencies, the work of historicizing climate, science, and politics is to demonstrate that alternative political, ecological, and epistemic formations regarding climate were possible in the past and may therefore also be possible today. This is a different way of thinking about the current situation when compared to representations of climate crisis that view climate change as the ultimate sign of the end of the world (Hulme 2017; Klein 2014) or an external fact to which capable sectors of society must simply "adapt" (Swyngedouw 2010; Watts 2015).

Geoengineering is the second emergent issue in climate governance that can be approached genealogically. Although I do not explicate the contemporary context for climate governance in detail throughout the historical narratives of each chapter, it is important to introduce the significance of climate stabilization to climate science and policy today. Since the establishment in 1992 of the UNFCCC (ratified in 1994), the dominant international policy goal has been to reduce greenhouse gas (GHG) emissions and thus "stabilize" global warming and avoid "dangerous interference" with the climate system (UN 1992). Within this discourse, the "relative stability" of the Holocene period (the epoch since the last ice age) represents the "scientific reference point for a desirable planetary state" (Rockstrom et al. 2009; Berger and Loutre 2002). In response to policy failure to maintain or achieve such stability, an increasingly important component of climate governance is to propose, research, and develop feasible possibilities for geoengineering. Proposals to geoengineer the climate by chemical and physical means that alter the reflective and heat-retaining processes of Earth's atmosphere are increasingly accepted as necessary given scientific understandings of the "climatic boundary" of human society (Rockstrom et al. 2009; Kolbert 2021).

Existing prognoses around geoengineering, however, must come to terms with the political, rather than natural, basis of what an optimally controlled or stable climate should look like (Hulme 2014; Schubert 2021; Gunderson, Stuart, and Petersen 2018). Like climate security, therefore, climate engineering is not merely a rational response to technical innovation and the rising costs of climate change. Rather, it entails controversial assumptions about control, risk, and the capacity to set the terms of a desirable planetary climate system. The context for governing climate via "climate stabilization" invites historical assessment of how scientists and other actors have variously viewed climate in terms of instability, stability, and stabilization (Lehmann 2022; Boykoff, Frame, and Randalls 2010). As I argue, the rise or prevalence of concern about climatic stability/ instability is not simply keyed to climatic shifts. Whether in contemporary debates about climate engineering or in past debates about climatic stability and stabilization, such concerns are best understood in terms of meteorological government.[9]

The historical approach outlined here boils down to the principle that narrating a sociological history can inform elements of the present predicament. The overall purchase of a genealogical approach is to help consider a configuration of climate, science, and politics that embeds climate knowledge in desirable forms of life, not because "the climate" (as a "desirable planetary state") demands it, but rather because society can recognize that the present settlement of social and climatic orders can indeed change course. To do so involves changing the futures that are envisioned and around which action is oriented, while critically engaging the futures that authorities envision, anticipate, and enact on peoples' behalf. It is my conviction that human society need not inhabit an apocalyptic present. The proposition that "we" have "lost" is a dangerous one for the future. It can suck the life out of proactive climate politics, replacing genuine public debate over how to live in a warmer world and how to curb global warming with reactionary measures that deal only with consequences (and most likely in a socially unequal manner). What is instead required is a politics of climate knowledge that deliberately reconstructs how science and government can support social equity and harm reduction in an era of increasing risks and social inequalities. The historical methodology deployed here is well-suited to inform parts of this monumental task.

OVERVIEW OF THE BOOK

The remainder of the book is organized in three parts, each with two (or in the case of part III, three) chapters. Each part is devoted to a distinct, basic configuration of climate, science, and government. In the present study, I focus on state formation, rational capitalism, and the authority to delineate official science from nonscientific knowledge as primary forces that characterize each logic of meteorological government while also modulating the transformation from one logic to another.

Part I addresses central concerns about climate change and the (changing) nature of society, roughly over the period from the 1780s to the 1850s. A heterogeneous set of medical meteorologists, physicians, natural historians, and natural philosophers debated the issue of North American continental climate change while bringing scientific controversies to bear on efforts to make legible the basis of a "civilized" society in the early United States.

Chapter 1 investigates two foundational controversies of early American meteorology, roughly proceeding over the period from 1780 to 1840: (1) whether American climates were changing because of "civilizing" practices and (2) how climate related to disease, health, and demographic changes. I analyze these issues using primary source data, including meteorologists' papers and manuscripts, letters, and journals. Analysis in this chapter is highly indebted to the growing secondary literature in the history of science and environmental history in U.S. and comparative contexts. Primary data was gathered through systematic review of existing literature; use of digital repositories to compile historical monographs, scientific treatises, and articles; and coding of compiled textual data. Assisted by qualitative open coding, analysis investigated the meanings, measurements, and focus of scientific practices regarding climate. Unlike some historical accounts that provide a working definition of "modern climate science" (see Coen 2020 for a useful review), in my analysis I followed words, specifically "climate" (and related terms), that permitted study of the changing meaning, valence, and contextual use of the concept. Rather than focus on discourse alone, however, primary and secondary source texts provide for analysis of the institutional structure of climate knowledge production. Based on this data, I reconstruct the aforementioned controversies and find that meteorologists' projects to make a rational science of climate developed through their capacity to formalize knowledge about society. Climate knowledge, by making legible climatic order, informed efforts to try and govern a social order by means of (1) arithmetic legibility of population size, health, vitality, and growth; (2) moral discipline that aligned prescribed practices with environmental realities; and (3) rational organization of social conduct with respect to specific climates and a colonial/state-based categorization of people groups.

Chapter 2 draws on primary print sources to evaluate how the formation of climate knowledge addressed in chapter 1 related to U.S. state formation in the period from around 1800 to the 1850s. The chapter is based on analysis of meteorologists' published writings, data on scientific expeditions, maps, diaries, and raw meteorological data. This data permits inquiry into the governmental significance of meteorological statistics, military-medical meteorology, and what I call racial climatology. "Racial climatology" refers to the use of empirical climatology to explain racial hierarchy, a widespread practice among climatologists around the world at this time. Based on historical analysis in the U.S. case, I argue that scientists helped to coproduce climate knowledge along with early state efforts to evaluate, calculate, and monitor (1) the military body in a context of bureaucratization of the U.S. Army, (2) Western territories in a context of territorial acquisition and providential nationalism, and (3) a stratified population "legible" by biological understandings of racial hierarchy.

What became of this configuration of meteorological government, centered on climate change, and why did it transform? To know and govern climate and society became an altogether different kind of project from around the 1860s to the 1920s. Part II shows that scientific authority regarding climate change and social difference shifted with the demise of environmental medicine, among other factors. The rise of industrial capitalism and formal government administration provided a new context in which those in government and science approached issues of climate. In short, rational capitalism and state administration benefited tremendously by mapping climate onto delineated, static zones, and by forecasting future events that could inform new kinds of economic and financial decision-making. The resulting form of meteorological government, quite distinct from the prior configuration, settled around "stabilizing" climate and making weather events commensurable with economic values. Even though this period included the rise of evolutionary theory and of climate change studies in geology and geography, within the context of U.S. climate governance, climate was largely taken for granted as stable, and climate change slipped quietly into the background.

Chapter 3 focuses on trajectories in climatology, roughly from the 1850s to the 1910s, the period during which climatology emerged as an organized branch of knowledge. The historical narrative traces the rise of climatology both as a professional/institutional project and as a component of a larger governmental logic. Analysis in chapters 3 and 4 rely primarily on data collected from a systematic review (using corpora organized by The Internet Archive—archive.org) of textbooks, periodicals, and science journals in meteorology and climatology, and to a lesser extent economic/trade periodicals, agricultural journals and catalogs, newspapers, popular science periodicals, and almanacs. In chapter 3, analysis of government publications (especially those of the U.S. Signal Service and the Weather

Bureau), professional journals, meteorological logbooks, newspapers, and popular literature provides a perspective on the social organization and regular practices of those involved in producing and consuming climate knowledge. I draw on this perspective to provide a sociological explanation for the emergent "stabilization" of climate as a geographic-statistical category relevant to facilitating capitalism. In brief, I argue that "climatic stability," defined as the basic view that climate is unchanging, linked the interests and practices of climatologists with the broader administration of commercial agriculture, trade, and finance. In this political-economic context, scientists alongside economic and state actors could capitalize on the scientific and administrative legibility of productive capacity denoted by climate "zones."

Parallel to chapter 3 and related to the stabilization of climate, chapter 4 covers how emergent instantiations of climate knowledge helped facilitate and govern national industrial society from the 1870s to 1920s. Over this period, the commercialization of land and society entailed a drive to know nature in its component parts, so they could become incorporated into the circuits of capital. As Karl Polanyi ([1944] 2001) has analyzed in wider historical scope, this incorporation process involved making "fictitious" commodities not initially produced for markets. Weather and climate became a problem for political economy in a Polanyian manner: how to make legible the fluidity of atmospheric phenomena to commercial society and bureaucratic administration? Put differently, could weather and climate, like land, labor, and money, be "economized," or made fictitiously commensurable with dollars and cents? The emergence of weather forecasting and the incipient integration of weather knowledge into economic and financial projections provide fertile ground for answering this question. My argument here is that efforts to economize weather help explain why climate came to be widely considered "stable." Drawing from analysis of meteorological reports, forecasts, instruments, and media, I reconstruct how meteorologists advanced a view of weather on an axis of "normal" and "abnormal" events and meteorological parameters. As a result, scientists and those who relied on climate expertise widely understood climate as a "normal," stable background. They did so in practice, despite the concurrent rise of controversies, which involved climate change theory, about ice ages, deforestation, and to a minor degree anthropogenic global warming.

The chapter links the interpretive process of *normalizing* weather to two social processes. First was meteorologists' struggle to centralize meteorological practices within the U.S. Weather Bureau and secure official representations of weather. Second was the diverse strategies intended to embed weather and climate knowledge into economic rationality, especially in the domains of insurance, finance, trade, and public infrastructural investment.

Such a configuration would eventually unravel. Why? The story of part III again identifies struggles in science modulated by state-making and the political and economic context, extending from around the 1930s to 2022. In chapter 5 I account for the rise of climate science in its roughly contemporary instantiation. Climate science, which I initially analyze for the period prior to when global warming organized the field, was fundamentally concerned with the dynamics (that is, changes) of climate, not its stability over time and space. I draw data from scientific texts across geophysical disciplines, biographical information (memoirs, speeches, letters, and papers), formal organizational information, secondary histories of geophysical sciences/technology, congressional hearings, and federal science policy documents and reports. I also draw upon media reporting of climate science, published materials of global warming skeptics, and secondary data regarding the environmental movement. The analysis shows how climate knowledge provided distinct but compatible resources to those invested in making climate science a formal science on the one hand and building the post–World War II U.S. hegemonic state on the other. For reasons I explore, climate science initially promised rational mastery over climate dynamics. Climate prediction, even engineering, was a potent formation critical to how climate science developed along with a technocratic view of government common to the period from the 1940s to the 1960s. The result was not only the development of a new kind of climate science (eventually centered on global climate change through the wedding of geophysics and computing) but also a "climate state," defined by governmental investment in apprehending global climate dynamics so that they could be managed. Yet the settlement between climate science and state control of the atmosphere in the postwar period was sharply undermined by authoritative claims that the atmosphere was in fact radically unstable. Crisis followed.

Latour (2018:3) elaborates that political denial of climate change signifies a dangerous "lack of a common world," evidenced most strongly by U.S. backlash against international climate agreements, for example former president Donald Trump's 2017 decision to withdraw from the Paris Climate Agreement. Instead of focusing on the battles between ideology and science that have characterized the relative failure of national climate governance in the United States since the 1980s, chapter 6 takes up a more squarely governmental undercurrent in climate expertise and policy along which actors appear to be resolving the dilemma that arose in the post–World War II settlement. Lacking a common world, in other words, is not only evident in denial of climate change, but also in techniques to preserve, protect, and secure *some* groups and places over *others*, that is, through climate security.

The emerging relationship between the national security state and climate expertise in the United States is one domain in which to explore how governing climate is taking the form of anticipatory practices of security. Drawing upon climate-scientific texts, think-tank reports, biographical and organizational information, and media, chapters 6 and 7 focus on the rise of *security technologies* and a field of *climate security expertise*, respectively. I define "security technologies" as scientific products; surveillance, data-gathering, and monitoring activities; and decision-making techniques that rationalize future uncertainty and facilitate strategic action based on anticipated future security environments. In chapter 6 I trace how security technologies developed through the configuration of the national security state and climate expertise. This configuration emerged by fits and starts beginning in the 1970s but converged in the first decade of the twenty-first century to launch a field of climate security expertise. In chapter 7 I demonstrate how relevant, "actionable" knowledge about climate security threats depends on experts' ability to facilitate translation between climate science and national security policy. This skillful process has partly succeeded insofar as climate security experts populated an organizational ecology of think tanks nested at the intersection of science, politics, media, and the state.

The book concludes by relating the genealogy of meteorological government to possible future trajectories.. The first trajectory regards contemporary policy discourse of "climate stabilization." Although grafted onto ambitious climate policies at present, the roots of goals to treat climate as a technically stabilizable entity run deep in the history of modern efforts to govern climate. Geoengineering, as in past cases of interventionist climate stabilization, demonstrates the limitations of a planetary, technocratic sort of legibility. The stability of climate in relation to society is best achieved, I argue, by struggling to incorporate critique of the social-climatic inequalities that modern, particularly colonial and capitalist, climates have long entailed. Such a critique calls for a radically different sort of legibility, necessarily supported by a different institutional configuration of climate knowledge and government. The book brings into focus diverse historical and contemporary programs on the peripheries of climate science, which may amount to a *critical climatology* that works to reconfigure climate science around the legibility of social inequality and justice-based modes of governing climate risk.

The second, and related, trajectory concerns climate security. As climate science develops in ways that make legible the likely impacts of climate change on regional and local population-sector-specific scales, it becomes possible to anticipate climate impacts like natural disasters and, in response, prepare for and adapt to such impacts as they unfold. Recognizing that the anticipation of future risk is a vector for governing climate, approaching climate through the lens of equity and justice demands that governing in terms of climate security not entrench existing social inequalities. As the current iteration of the climate movement shows, such a demand entails mobilizing politically around climate policies that transparently address human rights and social inequalities while criticizing those who presume a future world of gross inequality is inevitable or acceptable. But the problem is not only political. It further entails monitoring and redirecting climate science when the field incidentally or intentionally serves some "customers" and groups while others are made invisible within climate science or underserved in ways that prevent climate risk management, resiliency, and policymaking (Otto et al. 2020). I argue the enterprise of science can more seriously engage social anxieties and grievances, especially through intellectual work that bridges climate and social sciences and in communication with the political processes shaping climate science and policy.

PART I Climate Change
and the Coproduction
of Meteorological
and Social Order

1 Governing Climate in Early America, 1770–1840

What may we not expect from this harmony
between the sciences and government?

—Benjamin Rush (1774:73)

In 1776, physician and meteorologist Lionel Chalmers provided the following medical prescription to the citizens of Charleston in the North American colony of South Carolina: "If the weather be either sultry or showery, or cloudy and close, and sometimes calm, intermixt with a gentle southerly wind, as often happens towards the end of summer, low and what are called nervous and putrid severs will appear, more especially among corpulent people and others who are of a weak or lax habit" (Chalmers 1776:150–51). In his manuscript, Chalmers proceeded to organize a complex analysis of diseases according to season, and within each season he specified the symptoms, complaints, and therapeutic techniques relevant to people's "habits," or more generally, their bodies' "constitutions" and their relationships to the Carolinian climate. For an entire decade, to fulfill his parish medical duties, Chalmers kept a detailed weather diary and took daily instrument measurements to record temperature and other weather variables. What came of his efforts?

On the basis of his work, Chalmers came to conclude that the American climate was changing for the better. He determined that further clearing of the land and the resulting climate change would help alleviate citizens from the diseases unique to the Carolinian heat, moisture, and seasonal wind patterns. Far from eccentric, Chalmers's work represents elements of early American meteorology par excellence. He linked medicine to geographically specific territories and populations, defined disease in relation to climatic causes, and tabulated meticulous meteorological observations toward a defense of the increasing health, or "salubrity," of American climates. Meteorological and medical professionals later came to dismiss Chalmers on numerous accounts. They mobilized climatological data to argue that American climates were not becoming more temperate, that climatic causes of disease were misinformed, and that South Carolina was in fact a dismally unhealthy place in Chalmers's day. Meteorologists and physicians accomplished this by the late nineteenth century, and for this reason Chalmers and his colleagues have been all but erased from memory.

Let us instead take Chalmers, along with others involved in projects to understand climate in relation to society, at face value. From the 1770s to around 1840, I argue, meteorology developed through legibility projects that linked rational accounts of climate to governmental evaluations of the (very uncertain) prospects of a healthy, moral, and well-ordered early American society. What was meteorology in early America all about? As Chalmers's assertion exemplifies, if one could ask physicians or scientists in the latter part of the eighteenth century why they took detailed barometer, thermometer, and other instrument readings, then they would be hard pressed to find some justification that resembles what people now think of as "climate" and "meteorology." Although meteorology today is largely synonymous with weather forecasting and the scientific principles that underlie forecasters' techniques, predicting the weather was not the central pursuit within meteorology at the time. Nevertheless, political, scientific, and popular writings in the late eighteenth and early nineteenth centuries regularly featured discussion of climate (Fleming 1990; Meyer 2014). Climate knowledge was a major concern in American science and government during this period, especially for those seeking to evaluate the future of American population growth, settlement, trade, and territorial expansion. The task here is to situate the fervor around climate without focusing exclusively on the impact of this activity on later developments in modern meteorology. Relevant questions about government also remain unanswered. What was the role of early

American meteorology in the development of American political economy and medical expertise? How did meteorologists help make legible the basic categories of government—especially regarding issues of population, social order, and "salubrity" (defined as the health and suitability of an area)?

The present chapter offers an answer to these questions by showing how rational approaches to "climate" emerged precisely through their capacity to help make legible American population and territory, evaluate their healthfulness and productivity, and envision the human-climate relationship as a governable one. In what follows, I first describe how climate emerged as an object of science. Second, I argue that Western meteorology partially emerged through scientific and political evaluation of New World "habitability." Third, I show how, in the early- to mid-nineteenth century, self-consciously American natural philosophers and physicians brought international debates about American climates into their efforts to establish regional and national meteorological standards of practice. The chapter concludes by showing how meteorological investigations into issues of climate change, disease, and health culminated in practices to further civilize American society through "improving" the land, maintaining healthy and productive bodies, and instituting moral discipline. In an era that lacked any recognizable administration of "climate policy," meteorologists nevertheless advanced a scientific basis for governing the relationship between climates and socially stratified human populations.

WHAT IS CLIMATE?

Classical meteorology had held as its domain *meteora*, in Aristotelean perspective encompassing all the events that occur in the realm between the earth's surface and the moon, including storms and precipitation as well as earthquakes, waterspouts, "fire balls," and other, often extraordinary, occurrences. These special *meteoric events* were held to be caused by earthly, especially mineral, "exhalations" of various mixtures. In Aristotelean meteorology, the four elements—earth, water, air, and fire—compose two exhalations, one wet and vaporous, the other hot and smoky. These exhalations then circulate between the surface of the earth and the moon, a space that forms the liminal point between terrestrial and celestial realms (Frisinger 1983). The elements and the exhalations were considered the material causes of meteorological phenomena (Martin 2006). Inquiry on the subject involved qualitative, local narratives that were often generally integrated—especially, as Jankovic (2000) shows in the English case—into parochial and regional natural histories and imbued with theological, moral, and political significance.

The classical form of meteorology persisted alongside developments in meteorological instrumentation, even while qualitative reporting of extraordinary events gave way over the course of the late eighteenth and nineteenth centuries to more rigidly quantitative representations of regular meteorological patterns (Rusnock 2002; Frangsmyr , Heilbron, and Rider 1990). Chemistry and physics, as well as the law-like mathematical system incorporated into astronomy and other physical sciences, were not effectively introduced in meteorological science until the mid-nineteenth century, even though pneumatic experiments on "air" emerged much earlier among experimental philosophers (Anderson 2005; Fleming 1996; Frisinger 1983). As the physician Stephen Williams concluded in his 1836 "Medical Topography of Massachusetts": "The subject of climate is dark and intricate, and all our speculations in regard to it must ever remain vague and uncertain." Many of Williams's contemporaries came to similar conclusions, such that despite the rich meanings of "climate" and widespread usage of the term in scientific and medical discourse, it remained fungible as a category of scientific knowledge. Climate's darkness and intricacy notwithstanding, scientists sought to collect, centralize, standardize, and distribute meteorological observations and, in doing so, rationalize aspects of climate knowledge.

CLIMATE AND WEATHER AS CATEGORIES OF SCIENCE

Efforts among European philosophers to formally develop "rational accounts" of weather extend back at least to the origins of private weather diaries in the seventeenth century and, later, the common use of thermometers, barometers, hygrometers, rain gauges, and other instruments

(Golinski 1999; Jankovic 2000). In the late seventeenth century, Enlightenment philosophers associated with the newly formed British Royal Society, including Robert Hooke, Robert Boyle, Christopher Wren, John Locke, and William Petty, recorded such instrument readings diligently and with utmost discipline. The Society's *Philosophical Transactions* served as a forum for meteorological correspondence among European, and later American, diarists and observers. Scientific instrument-makers also helped to establish the new experimental-philosophical program, even though the actual use of their instruments was highly variable and poorly standardized (Golinski 2010; Shapin and Schaffer 1985). Nevertheless, meteorological instruments circulated first through science societies, especially the Royal Society and the French Académie Royale des Sciences, and later among clergy and nobility, either as tools in constructing regional natural histories (Jankovic 2000) or as what was called "philosophical furniture" in aristocratic circles (Golinski 2010). Efforts to construct a rational view of weather and climate critically involved early innovations in getting weather and climate to *inscribe* themselves on instrument readings, so that they could be written down, recorded, and tabulated. Regular observations could then be compared with or sent to others to interpret.

Over the course of the eighteenth century, instrumental observation, experiment, and the emerging social positions of natural historians and metropolitan natural philosophers partially displaced popular, often spectacular or religious, interpretations of meteoric events. In his *Rational Account of the Weather*, for example, English philosopher and minister John Pointer (1738:i, iv) compared the "philosophical reasons" for meteoric events to "astrological cant and jargon," in which "the eyes and ears of the unthinking vulgar" were dazzled and deceived by their "glittering starry notions" because they possessed "neither opportunities nor abilities of examining or enquiring into the reasons of things." Along with Pointer, other observers had sought to ground meteorology in empirical observation and naturalism rather than religious and philosophical speculation, including traditional "weather-wising" folk practices common among farmers (see Golinski 2010:70–76).

Empirical observation of meteorological phenomena remained a theologically and politically laden project. European and, later, American meteorologists came to the fore carrying, rather than abandoning, their theological and political concerns. First, weather remained an expression of divine providence, which carried a double meaning. On the one hand, in line with classical meteorology and more mystical religious orientations to nature, meteoric providence could mean special signs based on unusual occurrences. On the other hand, a "rational account" could signify the divine, but stable and regular, provision of natural order (Jankovic 2000:128–130). Such interpretations were hardly provided by nature. Rather, they were a matter of how observers brought forth their observations before their audiences in a given cultural context.

In reality, emphasis on natural order and spectacular occurrence persisted side by side for some time. Many eighteenth-century observers and diarists routinely represented extreme weather events as portents, even as they advanced inductive, quantitative, tabular, and comparative studies. What might be called a theological meteorology itself displaced other, primarily occult and astrological, accounts of weather. Eighteenth-century almanacs, as Peter Eisenstadt (1998:148) writes, were "forums for occultism *and* a source of scientific information." Although Eisenstadt finds that almanacs' occult meteorology greatly declined over the course of the century, later nineteenth-century texts, such as Robert Bailey Thomas's *Farmers Almanac*, continued to provide weather knowledge based on the experiences of farmers. Almanac publishers intersected with weather diarists through a process Jaffee (1990) labels "the village enlightenment," a popular spirit combining naturalistic and traditional knowledge and religious beliefs. Valencius et al. (2016) demonstrate that scientific journals, almanacs, newspapers, and other print sources formed a unique print culture in the early U.S. that synthesized scientific studies and controversies before a popular audience. Almanacs and the popular press ultimately helped to productively relate traditional, especially agrarian, folk knowledge to the emergent meteorological naturalism.[1]

Like theological truths or moral signs, political stability was something that became legible through the languages of providence and meteorological order, similarly raising problems for scientific authority. When he had cited the "impolitic" possibilities of "free" interpretations of extreme weather events, Pointer in his *Rational Account* highlighted a long-standing tension over who had authority to interpret special, mete-oric phenomena. Referring to the popular practice of interpreting storms, meteors, and other events as "battles" among human or spiritual armies, Pointer (1738:198) cautioned that to make "these phenomena and unusual appearances in the air, to be miracles and prodigies, is very impolitick, because oftentimes of dangerous consequence to the State." Pointer up-held divine revelation through the weather, but he did so by emphasizing "the known and ordinary will and power of God established in the course of nature" (1738:196). Sound government, he argued, demanded careful control over perception and interpretation, a control afforded by scientific inquiry.

In general, what observers termed "rational accounts" were cast in opposition to common or "vulgar" reportage of extreme events, not be-cause natural philosophy grounded accounts of weather and climate in a more adequate framework, but because discipline characterized scien-tific knowledge while also feeding into political and religious beliefs cen-tered around order and regularity. Meteorological order, in this sense, follows the larger process that Steven Shapin and Simon Schaffer (1985) uncovered through their analysis of the famous controversy in the late seventeenth century between Robert Boyle, the "father" of experimental science, and Thomas Hobbes. Hobbes critiqued experimental science as politically dangerous because it did not provide a successful means for creating agreement on what Boyle claimed were "matters of fact," which were established through "artificial" experimental conditions rather than demonstrated in nature. Tighter control over interpretations of nature were needed, particularly in the wake of religious and political turmoil that rendered the basis of authority unstable and elusive. Meteorologists engaged in similar debates regarding natural and political order. Political order, many meteorologists argued, could be informed by a securely ratio-nal account of weather, that is, as an aspect of a natural order observable in its regularity and accessible only by disciplined study.

In effect, meteorologists' embrace of numerical inquiry, their Ba-conian faith, and their discipline in gathering a "mass of facts" forged their major claim to scientific rigor (Forry 1842a:20). The change in how weather was recorded and depicted was substantial, transitioning from qualitative reportage to increasingly quantified, tabular representations of matters of fact. To demonstrate this comparison, consider the work of William Whiston (1716:48–49), an English minister who compiled many accounts of meteoric events in the early eighteenth century: "Hour past Eleven. About Twelve, a bright Globular Body appeared, as big as, and like the Sun at his Rising, but not quite so clear. Indeed it was the most astonishing Sight I ever yet beheld. During this time, the Light was such, that I my self (tho' almost sixty Years of Age) and another Clergyman did read several Titles of the Books in the Bible, without any Use of Art." Based on compiled reports, he sought to decipher the significance of special events, in this case the spectacular appearance of a "bright globular body" that provided enough light for two men to read without artificial light. Other than time of day, Whiston rooted his account not in rational representation but rather in impressions explicitly meant to evoke wonder.

As a comparison, the register compiled in 1823 by Edwin James, who edited the scientific materials from the Stephen Long expedition to the American West, was organized in a table and ordered by time and standardized instrument measurements. The register, however, contained significant space for "remarks" of various kinds. Analysis of remarks throughout the register find qualitative discussion of meteorological parameters ("fresh gales in the middle of the day"; "hard shower this forenoon") and also attending to "remarkable phenomen[a]" that could not be incorporated in the conventional table, for example, the appearance of comets and meteors (1823:xlvi; see table 1).

Two decades later, Surgeon General Thomas Lawson's 1844 *Directions for Taking Meteorological Observations* more strictly ordered phenomena numerically by time and space, and his detailed instructions attempted to quantify observations, including of winds and cloud formations, even when instruments to measure these phenomena were at that time unavailable. For instance, Lawson directed that "the force of the wind will be expressed by figures from 0 to 10," whereby "0 will signify a calm," "1 will signify a barely perceptible breeze," and so on (Lawson 1844:12). These examples point to a stark shift in meteorological representations from the early eighteenth to the mid-nineteenth centuries, when observers increasingly reined interpretations of meteoric events into a systematic, quantitative, and tabular form.

Table 1 Excerpt from Edwin James's "Meteorological Register," 1823

Day of Month.	MORNING.		MID-DAY.		EVENING.				REMARKS.
	Temperature.	Wind	Temperature.	Wind.	Temperature.	Wind.	Mean Temp.	Mean Temp. at Germantown.	
1	69	Calm	83	W. S. W	83	Calm	78	—	Fresh gales in middle of the day
2	71	Calm	74	W. by S.	80	N. E. by N.	75	—	Fresh gales in middle of the day
3	67	N. N. W.	80	N. W. by N.	78	N. E. by E.	75	70	Fresh and variable brs.—night fair
4	64	N. N. W.	81	E. S. E.	77	E. S. E.	74	70	Fresh breezes
5	65	S.E. by E.	80	E. by N.	79	E. N. E.	74	69	Light breezes
6	76	Calm	81	W. S. W.	84	E. N. E.	80	69	Freq. showers of R. during the day
7	72	E. N. E.	86	N. W. by W.	86	S. E. by E.	81	65	Hard shower this forenoon
8	76	Calm	90	S. W.	88	E. N. E.	84	67	Light brs.—mackerel sky in even.
9	78	E. N. E.	83	S. W. by S.	85	S. W.	82	77	Hard shower this forenoon. L. in

Source: James (1823:xlvi).

CLIMATE, LATITUDE, AND "ZONES" OF HABITABILITY

The shift in meteorological observation depended not only on an inherent power of numbers but also on a specific rationalization of climate, focused around political and intellectual evaluations of "civilization." As Gilbert Chinard (1947) outlines, the primary context for developing accounts of American climate involved Enlightenment era evaluations of "habitability," that is, attempts to arrive at a naturalistic view of how human settlement can proceed in different geographic areas. In diverse colonial contexts, physicians and other observers who devoted their efforts to regular study of local temperature, pressure, precipitation, winds, and tides recognized patterns and variation in not only their local weather, but also climates. It was in colonial contexts that people most resolutely used the tools of weather observation—taking instrument readings, constructing tables, and sharing weather diary collections—to investigate climatic variations across space and time.

Craig Martin (2006) argues that Western observers investigating dynamics of climate and habitability initially upheld an Aristotelian perspective, principally with respect to latitude. This perspective would hold important consequences for the development of climate knowledge in the United States and other non-European climates encountered through the process of overseas colonization. In Aristotelian and Ptolemaic meteorology, knowing climate was a matter of defining latitudinal zones and assessing their associated habitability. In the foundational text for ancient, medieval, and classical meteorology, Aristotle's *Meteorologica* divided the globe into five latitudinal climatic zones, which included two habitable zones, called *oikoumenai*, bordered by the two polar or "frigid zones" and the equatorial or "torrid" zone. These latter three zones were considered unfit for human habitability (see figure 4 in chapter 2; Martin 2006; Frisinger 1983; Hiatt 2007). As Aristotle explained concerning the temperate zones, "These sections alone are habitable. Beyond the tropics no one can live: for there the shade would not fall to the north, whereas the earth is known to be uninhabitable before the sun is in the zenith or the shade is thrown to the south."[2] This view of climate zones remained the dominant view for centuries within mediaeval and Renaissance philosophy and on into Enlightenment natural philosophy.

New World exploration and settlement as early as the sixteenth century shook up Aristotelian views on climate, latitude, and habitability (Martin 2006). As White (2020) shows in the case of Spanish, and later French and English, colonial science in the Americas, seasonal extremes, along with unfamiliar wind and weather patterns, effectively challenged prevailing views defining climate by latitude. Likewise, the "frigid" and "torrid" zones proved clearly capable of sustaining human societies, which opened new questions about climate, habitability, and bodily and social differences.

The work of the French naturalist and American immigrant Constantine Volney illustrates how latitude helped to organize early meteorological accounts of American climate. Comparing American observations to studies of ancient and European natural history, Volney (1804:102–3) marked a general concern over the meaning of latitudinal climates: "By *climate*, strictly taken, we ought to understand nothing more than the degree of latitude." However, he wrote, "the *temperature* of a country" is not only regulated by latitude. "On the contrary, it seems to be modified by, and sometimes wholly to depend upon, various circumstances of the surface." Volney's comparative analysis of latitude and temperature reflected this framework. For example, in one table Volney (1804:102) represented Salem, Massachusetts, along with the cities of Rome, Marseilles, and Padua, which when triangulated make up a similar latitude. He then recorded the highest and lowest temperature measurements and the simple variation between them in each location, with Salem recording a minimum of –12°F and a maximum of 102.75°F. To fellow observers, such depictions rendered American climate anomalous and without a clear causal explanation. As Volney (1804:11) argued, "The atmosphere is, by turns, very dry or very moist, very hot or very cold, very turbulent or very still; so capricious, that the same day will freeze with the colds of Norway, scorch with the ardours of Africa, and present to you, in swift succession, all the four seasons of the year."

Natural philosophers and physicians found American climates, especially compared to European climates, extremely puzzling but of prime importance, given the central significance of climate to the proposed habitability of any place. Therefore, investigators generally followed the kind of reasoning employed by English physician James Johnson ([1813] 1827:3) in his study of tropical climates and European constitutions: "We should study well the climate, and mold our obsequious frames to the nature of the skies under which we sojourn."

The Aristotelian heritage of identifying climates as latitudinal zones of "habitability" (or inhabitability) helped to frame natural-historical investigations of American climate into the early nineteenth century. Observers, chiefly physicians, initially improvised understandings of climate by combining a focus on latitude with other factors, especially surface topography, forest cover, and prevailing winds.[3] Such innovations, some scholars argue, helped advance the international study of climate-as-average-weather by linking geography to meteorology (White 2015).

Along with discussing natural topographic variations, including mountain ranges, altitude, and proximity to oceans, observers focused on explaining the American deviations from latitudinal climates by turning to anthropogenic factors alleged to change climates. These factors included those related to agriculture, reclamation from forests and wetlands, and other land use practices. Overall, natural philosophers packaged their evaluations of climate zones along with concern for climate variability over time and across diverse geographic spaces. Discourse about human-caused climate change, by "improvement" or otherwise, provided a way for actors to reconcile views of "zones" with understandings of climate change. Observers in America stopped short of understanding "zones" as delineated, stable entities. They were not concerned with delimiting climates geographically for the purposes of rational administration, a project that climatologists would later advance. Rather, they brought the new tools of quantitative measurement, experiment, and comparative natural history together with deep-seated ideas regarding climate, and as explored later, complementary ideas regarding climate change and its role in the progress of civilization.

BUILDING A CLIMATE FOR REASON

As Bruce Mazlish (2004) argues, the category "civilization" formed a keystone in the arches of Enlightenment natural and political philosophy. For their part, meteorologists drew upon and constructed the fungible concept of civilization as an evaluative discourse of historical change that, if understood rationally, could make way for efforts to govern the social relationship to climate. To clarify, I do not treat civilization as an ideal type (as typical in histories of world civilizations) or as a category limited to explicitly social conduct (Elias ([1939] 2000); Foucault ([1965] 1988). Elias's study roots the "civilizing process" in constraints on and modifications to human behavior. Environmental historians have emphasized a broader "civilizing" process that involves two major components. The first is artificial modification, especially improvement, to land and natural resources (Drayton 2000). The second is moral restraint as the basis of individuals being properly governed, productive, and healthy in accordance with socially stratified bodily "constitutions." Natural philosophies of climate stitched these two components together in ways that appeared to make sense, particularly in settler-colonial situations where the fate of settlers and the character of Indigenous populations were glaring concerns.

Although Enlightenment thinkers postulated an allegedly universal capacity for reason within their discourse on human nature, the uneven geography of human nature remained an unsettled philosophical problem. Climate was one way to consider the possibilities of enlightenment. In *The Social Contract*, Jean-Jacques Rousseau ([1762] 1913:64) argued: "Liberty, not being a fruit of all climates, is not within the reach of all peoples. The more this principle, laid down by Montesquieu, is considered, the more its truth is felt." Rousseau concluded, "We find, in every climate, natural causes according to which the form of government which it requires can be assigned, and we can even say what sort of inhabitants it should have" ([1762] 1913:65). For Montesquieu and Rousseau, climate, understood as zones of temperature that dictated alternative forms of human existence, preceded government, liberty, and enlightenment.[4] Climatic accounts of civilization broadly held that temperate climates were the best suited for the development of civilization because, as Scottish historian William Robertson put it, they provoked "industry" and "improvement" through which human "powers are called forth" (Robertson [1777] 1817:30). Regarding North American climates, Robertson argued that Native Americans in Robertson's comparison to Europeans, held "an aversion to labor" and hence "waste their life in listless indolence" ([1777] 1817:97), giving way to a generalized natural and moral situation that "form[s] their minds to servitude" ([1777] 1817:144).

After the American Revolution, the climatic prospects of "civilization" in an independent United States became a significant concern. The state of land held direct implications for the basic habitability of American climates by white Europeans. Lands that remained uncultivated were held to languish, producing poor air that directly impacted people's capacities and social development. William Robertson ([1777] 1817:16–17) thus argued, "When any region lies neglected and destitute of cultivation, the air stagnates." Accordingly, "America, when first discovered, [was] found to be remarkably unhealthy. . . . It appeared to them *waste*."

Climates, in this view, produced fundamentally different kinds of people, articulating with predominant efforts at social classifications in the early United States.[5] Robertson ([1777] 1817:61), for example, proclaimed that American climates had made Natives to North America "resemble beasts of prey, rather than animals formed for labor." This angle on American climate reached its most vicious form in the "degeneration" narrative launched most prominently by French naturalists, especially Montesquieu, Comte de Buffon, and Abbe Reynal, against whom early American philosophers and statesman framed their naturalistic defenses of American climate and civilization (Jefferson [1785] 1794:63; Chinard 1947; Cassedy 1969; Golinski 2008). Buffon, like others, argued that the New World climate "opposes the amplification of animated Nature," evident to him in the inferior character of plants, animals, and especially Native peoples. By the same principles of American climate, he claimed, even European immigrants "shrink and diminish under a niggardly sky and an unprolific land, thinly peopled with wandering savages, who, instead of using this territory as a master, had no property or empire" (translated in Chinard 1947:31). By its very nature, the backward American climate would spell the degeneration of Europeans.

Buffon's scathing narrative of American degeneration rested on the same comparative view of European civilization that can be found in the writings of Dunbar, Falconer, Robertson, and Rousseau, among others. Not only did European nations possess temperate climates as gifts of nature, they also clearly marked the land, reclaimed it, and mastered it in ways held to formally civilize the climate and make it more temperate over time. Because this civilizing process took centuries to complete, such narratives concluded, American civilization was unlikely to succeed as American philosophers, revolutionaries, and statesmen claimed. Social order, much less an independent political and economic organization, was not possible. Debates about degeneration resonated outside the U.S. context. In French colonial natural history, many worried that degeneration was a central barrier to the civilizing process: non-Europeans were not fit for civilization, and non-Europeans in foreign settings might in turn degenerate. Yet others, for example the anthropologist and voyager François Peron, sought to experimentally disprove the view that "the physical degeneration of man is in proportion to his state of civilization!" (Peron [1809] 2012: ch. 12; emphasis in original). In his case, Peron ranked ethnic and racial groups (including Europeans and aboriginal groups across Australia and Southeast Asia) by measurements of physical strength, finding that Europeans were not losing vitality and strength.

CIVILIZING AMERICAN CLIMATES?

American scientists generally rejected critical views about degeneration, instead reinforcing the view that civilization could bring about climate change. To do so, they engaged Montesquieu, Buffon, and others through formal correspondence and debate. The terms of debate helped to enliven the emergent national consciousness of early American statesmen-scientists, who actively contributed to scientific debate in the early national period (Valencius et al. 2016).

Pursuing a broad natural history of "civilization" imputed a central role to powers of reason, industry, and enterprise. Defenders of an American republic specified these powers by revising the received European colonial narrative. Thus, Robertson's ([1777] 1817) text *History of America* claimed that human efforts, "when continued through a succession of ages, change the appearance and improve the qualities of the earth." On the same grounds, Philadelphia physician William Currie (1792:88–89) claimed that "[there] is no unreasonable prospect" that "when [America] comest [*sic*] to be diversified by vast tracts of cleared land," the resulting temperate and "salutary" climate may lead to a "glorious" and "enviable" era. When it came to the development of American science, James Dunbar (1781) even modified his otherwise climatic-determinist stance: "The

New World, from its connection with the Old, opens to the arts and sciences an *opposite career,*" which he argued might contradict the "first arrangements" and the "apparent order of physical laws" by which climate determines civilization. Improvement of uncultivated wilderness, others argued, would temper climates, making them healthier and suitable to the development of civilized society. Advocates who claimed America was both civilized and quickly civilizing sought to demonstrate the degree of improvement as a reflection of American liberal government and salubrious climate, while projecting a vision of American science not bogged down by speculation and unfounded continental criticism.

Fortunately for postindependence scientists, a long history of writers, journalists, and scientists had established that American climates had been changing because of European settlement. Meyer (2014) and Anya Zilberstein (2016) show this view was common among New World colonists and settlers. Since at least the mid-seventeenth century, observers had argued that American climates were improving, even as new climates remained a source of general anxiety concerning health and disease (Kupperman 1984).[6] In 1634 New England settler William Wood ([1634] 1764) claimed that the climate, understood in terms of temperature and seasonal weather patterns, was changing dramatically even within the relatively short time since Europeans had appeared.

Several factors beyond direct controversies over civilization and climate converged to make climate change a central issue. First, drastic changes to local ecologies were a central fact of early settlement and colonization, and any such changes came to bear heavily on the fate of settlers and Native people alike (Cronon 2003). Second, colonial reportage that sought to evaluate and promote exploration and settlement was a primary venue for communicating information about climate relevant to settlement, navigation, and resource extraction. Regarding commercial interests, alongside the interests of missionary and ethnic societies, promoters publicized the health of specific regions, perpetuating views that climates were improving as settlers gradually introduced civilization.

Postindependence meteorological observers, especially those connected to the newly formed American Philosophical Society (APS), took up the narrative about climate change and civilization with a distinctly patriotic emphasis, characteristic of what Kaufmann (1998) has labeled "naturalistic nationalism." They initially argued that no one, especially European skeptics, had effectively provided rational accounts of the causes and consequences of American climate change.[7] In this context, Thomas Jefferson set out to record wind patterns and map the surface topography of Virginia. He concluded that "as the land becomes more cleared, it is probable [prevailing East winds] will extend still further westward," tempering extremes of hot and cold (Jefferson [1785] 1794:79). Currie (1792:85–86), who traveled the United States and conducted experiments to evaluate the effects of forest clearing on soil temperature and moisture, likewise claimed, "It may be rationally concluded that when in the course of time, this continent becomes populated, cleared, cultivated, [and] improved ... the bleak winds will become more mild, and the Winters less cold in the middle states." Others constructed historical accounts. For instance, the New England historian Samuel Williams ([1794] 1809:73) used historical texts to argue that, from 1630 to 1788, the rise in the average temperature of Boston "must have been from ten to twelve degrees."

Collectively, these natural philosophers helped to construct American environments as relatively contained spaces in which their expertise could be socially validated as a sound, empirical approach to solving the politically laden issue of American climate change. By claiming to speak for (especially regional) American climates, APS-affiliated scientists and physicians helped to advance a protonational climate expertise, embodied chiefly in the APS. They established mutual credibility through incipient standards of systematic observation, instrumentation, measurement, and quantitative representation. And they built a network of investigators through formal correspondence, publication, and citation.

These scientists could then claim that American climates were increasingly coming to reflect conditions generally supposed to be the best for cultivating a civilized society. Referring to early English settlement in North America, Chalmers (1776:176) argued that the disease gout had become less frequent because "the climate must have been rather more sultry and damp at that [prior] time than it is at present." He clarified the mechanism of change, which others also generally emphasized, to be forest clearing and cultivation: because the land was deforested, the air "consequently has a freer passage through the inhabited parts of the province." Williams ([1794] 1809:74), in his *History of Vermont*, formulated a similar argument: "The earth and the air, in the cultivated parts of the country, are heated in consequence of their cultivation, ten or eleven degrees more, than they were in their uncultivated state." In effect, Williams argued, in a matter of only several years, the process of clearing that created arable land out of wetlands transformed local climates.

Considering the international context in which interpretations of climate change (especially because of forest clearance) exercised scientific authority in the United States helps to show the political contingencies surrounding climate change knowledge. On the European continent, anxieties about forest clearance and the human impact on climate were becoming more pronounced by the turn of the nineteenth century. Fabien Locher and Jean-Baptiste Fressoz (2012) show that the drying effect of deforestation was a particularly salient problem in France, given intermittent bad harvests from the 1790s to the 1820s and political turmoil. In the French case, deforestation signaled a dire problem for—not a solution to—social order. In France, land practices were much more regulated than in North America at the time, opening a space for governing the negative environmental consequences of climate change through forest conservation policies. In the United States, a similar debate about forest clearing commenced much later, when deforestation presented an ensemble of problems. The work of George Perkins Marsh in the 1850s and 1860s gave voice to a concern that forest clearing was facing natural limits, and climate deterioration was beginning to set in (Marsh 1864; Bramwell 2012).[8] Although such a shift in climate knowledge and forestry occurred decades previously in France, in the late eighteenth century, modification of climates in the American context was predominantly interpreted as a social gain, consistent with the larger expression of postcolonial nationalistic naturalism.

"SALUBRIOUS" CLIMATES: MAKING AMERICANS HEALTHY

Although meteorology had begun to take on institutional form in the APS and through scientific correspondence, it was in the emerging medical profession that meteorology more decisively took hold as a science and, in the absence of state or other bureaucratic institutions, as a component of government. If for natural philosophers reclaiming and improving land meant civilizing climates by means of thermometric changes, climate change held profound medical implications for a healthy, productive population.

For context, physicians focused their studies broadly on relationships between the "constitutions" of bodies and their environments. This orientation to medical practice stemmed from Hippocratic thinking, specifically as it was reinvented by the mid-seventeenth-century English physician Thomas Sydenham. Sydenham ([1686] 1742) applied Hippocratic views of the humoral body and its relationship to "airs, waters, and places" to the study of epidemics and what he termed the "constitution of the atmosphere." As Sydenham ([1686] 1742:495–96) explained, "The change of a constitution depends principally on some secret and hidden alteration in the bowels of the earth, communicated to the whole atmosphere." As an example, Sydenham used this approach to explain the "departure" of a widespread fever that was caused by "a severe frost" in 1664 but was "soon succeeded by a pestilential fever, and, in a short time afterwards, by the plague itself" ([1686] 1742:496). Sydenham's work was avidly promoted by John Locke and then widely popularized by the late eighteenth century (Anstey 2011). American physicians developed medical practices and conventional accounts of diseases roughly consistent with this tradition by combining neo-Hippocratic theories of disease formation, or etiology, with increasingly quantitative approaches to measuring climate.

Physicians' journals and societies settled around a particular convention for conducting studies and publishing results, namely "medical topographies." These studies emphasized the geography and environmental and meteorological context of specific places, with an emphasis on either explaining a given disease or generally characterizing the healthfulness of the place. This convention for medical expertise fed into local and regional "medical geographies" through the 1850s (Drake 1825, 1850; Mitman and Numbers 2003; Valencius 2002).[9] Physicians understood their intellectual foundation to hinge on evaluations of climate. Daniel Drake (1850:447), the influential Ohio physician, later summarized climate as a category for medical topographic practice: "Climate occasions Diseases. —No fact in etiology is more universally admitted, than the influence of climate in the production of disease. He who would understand the origin and modifications of the diseases of a country, *must study its meteorology.*" "Meteorology," as Drake laid out, rather than being a narrowly defined topic, signified the complex characteristics of a place, and as such one that must be integrated into the broader understanding of each place's medical topography.

Concern for climate and health formed the central professional and interpretive framework for studying climate change. New England physician Job Wilson's framework for explaining an epidemic of spotted fever helps to show how scientists integrated measurement of climate change with accounts of salubrity, disease, and civilization: "The very rapid increase of population, the strength, agility, stature and complexion of the inhabitants of these north-eastern States, are sufficient evidence of the salubrity of our climate." However, he tempered this enthusiasm, arguing that "the [recent] changes of our climate have had an important effect in producing the present epidemic, if not the principal cause" (Wilson 1815:125–26).[10] For physicians, the task to understand health required that they also consider climate change.

Before Wilson's analysis or Drake's (1850, pt. 2) formal approach to what he termed "climatic etiology," physician Hugh Williamson had believed that climate change demanded systematic medical attention. A member of the APS who later served as a delegate to the U.S. Constitutional Convention and as the surgeon-general of North Carolina during the American Revolution, Williamson (1771:280) argued that "it is certainly the duty of every Physician, to be careful to trace the history of every disease, and observe the several changes they undergo," his logic maintaining that "the cultivation of the colonies, and the consequent change of climate, has such effects on the diseases of the human body." Such a history of diseases, Williamson argued, allowed physicians the capacity to provide climate-specific therapies for inhabitants within each physician's local area or region. The Harvard-educated physician Edward Holyoke (1793), who later served as president of the American Academy of Arts and Sciences, had also evaluated climate change and its implications for health. He defended meteorological observation as a way to construct comparative "experiments" on the changing nature of the American air: "it must appear highly probably, that America is furnished with sources of depholgisticated air,[11] which are now exhausted in Europe," a comparison that "may be determined most satisfactorily from meteorological observations" (Holyoke 1793:75). He conjectured, regarding America, that "most probably, its atmosphere is really more pure."

Holyoke's argument for pure American air rested on its physical properties. Others were more concerned with the ill effects of heat and moisture in the context of American settlement practices. After complaining that America was nearly "one continued forest," Chalmers (1776:8) argued that the sultry climate produced "excessive exhalations" that arose from the vast "quantities of funk, fenny and marshy lands."[12] It was "almost needless to mention," he concluded, "that these exhalations do not consist of simply aqueous particles, for they must partake of the qualities of the several bodies that emit them." This perspective, widespread at the time, connected heat, moisture, and uncultivated land into a general paradigm of climate-based disease etiology centered on "exhalations," often labeled *miasma* or *miasmata*. Miasma was held to arise from stagnant water, wet soil, and rotting organic matter.[13] By extension, it was "a fact well known," as Forster ([1817] 1829:98) later claimed, that "cutting down large forests, draining off stagnant pools, and cleansing cities of filthy sewers, will add to the wholesomeness of the places so freed from sources of illness." Improving the health of a place, Forster and his predecessors argued, would require coordinated efforts to govern social relationships to miasma—an effort that required both meteorological and medical expertise.

GOVERNING CLIMATE CHANGE, MIASMA, AND THE BODY

The conclusion that collective efforts to civilize the land would produce healthier populations was widely shared, but physicians themselves did not have the capacity to implement the policies they prescribed, which ranged from promoting forest clearing, swamp drainage, and agricultural productivity to building city streets and planning residential settlements in a way to deal with local miasmas, wind patterns, and typical seasonal changes. The road from the documented meteorological order to the prescribed social organization was thus unclear. Given the credibility of meteorology within American intellectual, medical, and scientific circles, however, the process of meteorological government (however limited it might be) was well underway, even before the turn of the nineteenth century.

Their lack of capacity notwithstanding, meteorologically minded physicians did intervene in government through planning efforts and military strategies, principally regarding how to contain the effects of miasma on human bodily constitutions. To inform military organization, for example, American physicians in the southern states (often held to have tropical-like climates) drew from the experiences of other physicians, especially those stationed in the British West Indies, where malaria and yellow fever were also common. The influential Scottish physician Robert Jackson ([1791] 1795:279), for instance, studied the situation among British troops in Jamaica, many of whom were dying and diseased, and he recommended ways in which encampments could protect against "the progress of those noxious vapors." He prescribed the following: "interpose woods or rising ground," "burn a chain of fires," or raise "a parapet wall over top the barracks." Jackson's work was republished in the United States and Germany, and he continued to influence military medicine, including the "discipline and economy of armies," which rested on physicians' strict observation of climate (Jackson 1804).

If containing miasmic exhalations arising from "insalubrious" climates was part of the physicians' battle, this issue of a body's particular constitution complicated their goals of knowing, blocking, and treating diseases.[14] Physicians typically treated the body as systems of fluids, related to the ancient Hippocratic theory of humors, the balance of which made up a body's "constitution." Therefore, changes in the environment influenced constitutional equilibrium. A constitution could be "hardened" or "seasoned" to a given climate on the one hand or weakened by drastic changes and prolonged interaction with unaccustomed climates on the other (Seth 2018; Valencius 2002). Achieving equilibrium depended on both environmental and constitutional factors, a situation that called for climate-specific medical expertise and vigilance on the part of individual subjects. For example, in 1793 an outbreak of yellow fever plagued Philadelphia, killing at least five thousand people. Benjamin Rush (1799:6–10), then the vice president of the APS, explained the outbreak as a function of three causes: "putrid exhalations" (i.e., miasma); "an inflammatory constitution of the atmosphere"; and "an exciting cause" of the human constitution, especially constitutions weakened by being hot, cold, morally intemperate, or fatigued from activity. The problem, Rush put forth, was American air—the "spark of fire" that lit the "gunpowder with which our citizens are charged" (Rush 1799:10). Rush and his contemporaries viewed human constitutions, like climates, as fragile and dynamic.

Understanding the "constitution" of American bodies and (changing) American climates helped to organize early American medicine by establishing a domain of facts and methods that bound inquiry and secured a measure of professional jurisdiction both within and across regions, states, and locales. In 1789, in an opening address to the APS, Nicholas Collin (1789:iii), framed "Medical Inquiries" as the foremost concern of the society, stating, "All countries have some peculiar diseases, arising from the climate, manner of living, occupations, [and] predominant passions." Therefore, Collin (1789:v) explained, diseases were "national evils" that demanded nationally organized scientific attention. APS members, and medical meteorologists broadly, followed up on this charge.

MORAL CLIMATES: FORMING
AND DISCIPLINING POPULATION

Individuals concerned with governing the body-climate relationship through medicine faced a basic question: How does one, in fact, *know* a climate to be healthy, and an area (let alone an entire territory) to be civilized? Apart from making civilization legible on the land and temperature legible in the mercury of a thermometer, measurements of *population* became critical metrics of climatic and civilizational progress. Population dynamics occupied a central place more generally in liberal political economy (Foucault 2004). John Graunt (1662) had famously pioneered the construction of life tables based on the London Bills of Mortality, and William Petty ([1687] 1890) and Gregory King ([1696] 1804) later developed population statistics as a means of conducting "political arithmetic" and measuring the wealth of the English state. Benjamin Franklin (1755) adopted the same principles in his theory of American population dynamics and their implications for colonial wealth.

William Barton ([1791] 1793:25) presented the general issue in the context of the early United States: "There is not, perhaps, any political axiom better established, than this—that a high degree of population contributes greatly to the riches and strength of a state." Within Barton's political economy, a prosperous nation and growing population required more than a steady flow of immigrants and child-producing marriages. It also required a climate suited to those goals. The degeneration narrative, its rebuttal vis-à-vis studies of America's changing climates, and concerns over the effects of climate on European constitutions all suggested that a productive and healthy population in the United States could hardly be assumed. Drawing on innovations in population statistics, Barton compared population growth and fertility across countries, representing "Probabilities of the Duration of Human Life." He concluded that the productivity of the population "arises from the salubrity of the climate," alongside the "fruitfulness of the country," "the benign influence of our government," and the "virtuous and simple manners" of American citizens (Barton [1791] 1793:26). Unlike English political economists and physicians who had developed medical meteorology by using the London Bills of Mortality, U.S. individuals and groups at the time collected neither national vital statistics nor meteorological statistics. Notably, Barton's and others' demographic figures were published prior to those from the decennial U.S. Census (first conducted in 1790 to produce population counts, as mandated by the U.S. Constitution). Vital statistics were only irregularly collected at state and urban levels even up to the 1840s and were not mandated at the national level at the time (Cassedy 1984; Gehlbach 2016). Thus, Barton's tabular representation of the salubrity of American climate and political system took part in making "population" a specific "branch of science" (to use Barton's terms) that could be the basis for economic government.[15]

Political and economic goals of establishing a civilized, productive, and healthy population were fundamentally wrapped up in the problem of American climate dynamics. The geographies of disease, formed as they were at the nexus of bodily and atmospheric constitutions, could not only be managed through land and sanitary practices. They were also amenable to *moral* intervention insofar as disciplined activity helped to preserve a state of body-environment equilibrium. In order to enhance the salubrity of American populations, physicians adopted socially reformist agendas as a component of their meteorological expertise. Physicians studying climate argued that "moral" and "artificial" causes intercepted what others mistakenly held were purely "physical" laws by which climate determined an area's salubrity. Benjamin Rush (1786:164) thus rejected a view of the inherent "unhealthiness" of American climate: "Perhaps no climate or country is unhealthy, when men acquire from experience or tradition, the art of accommodating themselves to it." For Rush, effectively connecting climate knowledge to health required moral interventions that could ensure proper "accommodation."

Physicians involved in working out the climate-disease relationship gave special attention to positive and negative implications of the civilizing process. If one medically significant dimension of civilization involved the natural (and climatic) production of socially distinct human "constitutions," then another dimension involved monitoring the deleterious effects of "civilized" living. The latter entailed evaluating and disciplining conduct on the basis of social difference. For example, Rush (1774:38) claimed that "civilization rises in its demands upon the health of women. Their fashions, their dress and diet—their eager pursuits and ardent enjoyment of pleasure—their indolence . . . are all so many inlets to diseases." In the shadows of civilization, therefore, *intemperance* emerged as particularly dangerous to the body-climate relationship.[16] Many diseases, Rush (1774:63) lamented, "have been produced by our having deserted the simple diet and manners of our ancestors." The medical meteorology championed by Rush, and later by his student Daniel Drake, provided a way to govern the intimate moral relationship between socially categorized bodies and (changing) climates (Rush 1786; Dorn 2001). Temperature and "temperance," then, had a related meaning, which Rush helped to make explicit, for example, through his "Moral and Physical Thermometer" (Rush 1790:12). Rush's "Thermometer," published as part of a temperance article, was chiefly a political metaphor for how moral conduct mediated the body-climate relationship. Just as a Fahrenheit scale could measure a rise in atmospheric temperature, so could his thermometer measure moral virtues arising inversely to alcohol consumption, as measured by scales of "Vices," "Diseases," and "Punishments."

Beyond political and ethical commentary, Rush consistently worked to establish the general perspective that moral discipline and attentiveness to climatic influence could protect citizens from disease. Rush and his followers extended medical-geographical expertise to provide both detailed prescriptive therapeutic regimens for individuals as well as public moral reforms (Drake 1825, 1850). The civilized subject, as it surfaced in what historian Michael Dorn (2001) calls "medico-moral geographies," had to employ self-discipline, supplemented by physicians' medical intervention based on what experts determined to be subjects' constitution and climatic situation. Knowing climate in a systematic way helped physicians and public reformers configure a differentiated social and moral order and prescribe how different categories of people should act in specific environments.

Medical meteorologists would later help to enact sanitary and public health reforms, resting on a discourse of hygiene as meteorologically-informed interventions into the social body. As James Pickford (1858:vii) outlined, hygiene became a novel way of regulating the relationship of bodies and their environments: "Hygiene embraces, therefore, all those matters which, when properly used, contribute to make up, to perfect, and to sustain health, but which, on the contrary, when abused, retained, taken in excess, in insufficient quantity, or in a vitiated or corrupt state, lead to disorder or disease." To establish meteorological order, even in the context of deep concern for climate change, was simultaneously to work toward establishing and improving social order. Drawing from developments in population statistics unavailable to the generation of medical meteorologists that included Benjamin Rush, Drake (1850:649) extended the work of political economists and physicians to study "the modifying influence of moral causes on our national physiology." In his medical geography of the Mississippi Valley, Drake combined demographic, meteorological, and medical statistics to explain, among other outcomes, why some bodies "fail to reach the standard size" in various climates and among certain ethnic constitutions (see Drake 1850:650). He labeled this approach to describing human difference "statistical physiology." Drake reasoned that he and other medical geographers could figure out how national constitutions interact with changes of climate, then use these discovered variations to promote reforms related to diet, clothing, bathing, exercise, and other behaviors.

Moral conduct was particularly important to those concerned with the high mortality costs of migration, a process often framed by physicians as

a "change of climate" (Barton 1837:5). Physicians in sending regions compiled medical guides for travelers (e.g., Darby 1818), while those in receiving regions developed means of "acclimating" and "seasoning" the physical
and moral constitution of new inhabitants. Like Drake, physicians made
moral prescriptions for travelers in a way that emphasized ethnic difference. Louisiana physician and meteorologist Edward Barton (1837:9), for
example, addressed the "national physiognomy" of the English, French,
and natives of New Orleans: "The English cockney, in contempt of what
he calls native effeminacy," Barton argued, had failed to properly adapt to
hot climates compared to the "more accommodating Frenchman." Barton warned that "so with our northern brethren, this [Louisiana] climate
has to stand answerable for all the sins of juleps and champagne—beef
and bacon!"[17] Thus Barton prescribed modes of "acclimation" that corresponded with alleged national moral orders and cast, and in this case, in
gendered terms.

TOWARD THE PROFESSIONAL CAPTURE OF CLIMATE

Physicians formalized the study of moral and climatic dynamics of American populations through the American Medical Association (AMA), established in 1848. Like the APS's framing of "Medical Inquiries" pertaining to
"national evils" decades earlier, efforts via the AMA to organize accounts of
American bodily and atmospheric "constitutions" provided grounds for a
more unified profession that had hitherto been strongly fragmented along
state and regional lines. "The American constitution must be studied by
itself," an 1848 committee of AMA physicians (including Daniel Drake)
argued. As they reasoned, "The American climate remolds the European,
and casts a new die of humanity—will it not generate causes of disease different from those of the Old World?" (Holmes et al. 1848:186).[18] The AMA
worked to organize a diffuse professional field through the principles of
populations' constitutions. As Barton argued, "It is, then, by becoming ac-
quainted with and analyzing the various circumstances affecting the condition of every people that we can ascend to the *great law of causation*"
regarding disease (Barton 1852:581). He envisioned his work as "the key to
interpret the laws of morbid action" establishing how "climate or mode of

life," as well as the "laws" of human constitutions, produce "the geography of diseases." Such was a central project of the new, national organization of the medical profession in the United States.

Barton, Drake, and other physicians devoted their careers to developing a national system of medical meteorology, arguing that it could unify medical inquiry and promote climatic- and population-specific hygiene on regional and national scales. In effect, those who treated population as a "branch of science" through medical-climatic investigation linked the paramount "political axiom" concerning longevity and productivity, the potentially unstable "national physiology," and the physical and moral bases of health to the problem of American climate.

The process of meteorological government had taken root in the early U.S. context. By the mid-nineteenth century, a wide array of social actors recognized developing climate knowledge as a simultaneously governmental and scientific imperative. Concern with climate changes, their effects on productivity and health, and the appropriate social and moral response served to centrally organize meteorological science. This is clearly reflected in the accepted meaning of the category of climate. Writing in the 1840s, meteorologist Samuel Forry helps demonstrate this point: "Climate, in a word, constitutes the aggregate of all the external physical circumstances appertaining to each locality in its relation to organic nature" (Forry 1842a:127). It followed for Forry that to "deduce from this knowledge the influence which they exercise on the physical and moral state of man, such is the wide field which climates present to our investigation." Climate was a fundamentally relational concept, occupying a domain between meteorological phenomena and "organic nature," including especially the physical and moral situation of human bodies. Such approaches made climates legible as "patterns" of "influence," to use Forry's terms, more than simply patterned states of physical and chemical properties of the atmosphere.

COPRODUCTION OF CLIMATE, SCIENCE, AND SOCIETY?

Two major conclusions follow from the arc that I have traced about the emergence of scientific and governmental apprehensions of climate change, disease, and civilization in the early United States. First, meteorological rationality emerged by virtue of its forming a governmental rationality. Meteorologists helped create rational accounts of environments and populations that made the vagaries of American climate, land, and bodies legible to science in terms relevant to governmental efforts to establish a new U.S. society. Questions of political economy were intertwined with the climate problem. Leading figures in natural philosophy, medicine, and politics grappled seriously with the relationship of climate to a "civilized" order. Those who won the capacity to speak for American climates influenced and even consolidated the earliest accounts of U.S. population growth and longevity. They established local, regional, and national medical expertise. They provided detailed means of protecting and disciplining medical subjects. In doing so, the first meteorological experts in the United States organized moral evaluations of Native people; of national, racial, and gender differences; and of the broad trajectories of "civilization" in the country. Bodily constitutions, they held, were variously susceptible to disease, and these bodies were mobile across unaccustomed, unknown geographies and allegedly changing climates. How could these complex relations be governed? Insofar as actors oriented to this question in practice, a relatively coherent logic of meteorological government emerged.

It is important to recall that in the early U.S., many spheres of social life lacked formal government administration. Science or "the" state had yet to make climate legible from any central vantage point. Physicians' ability to articulate a vision of a morally disciplined society is thus remarkable for having addressed prevalent issues in political economy. Meteorologists, specifically as they became organized in scientific organizations, from local medical organizations to the APS and the AMA, provided a way to bring together the obvious flux of people and the atmospheres they inhabited. Meteorology could take this flux, order it in neat tables, and shape practices considered essential to civilizing society, improving population productivity and health, and securing salubrious territory. By doing so, rational meteorology helped to develop a rationality of government despite the absence of a bureaucratic administration.

For a second conclusion, tracing meteorological controversies shows the inadequacy of studying history in relation to the boundaries of meteorology and climate science today, especially the basic categories of weather, climate, air, and atmosphere—all dynamic concepts that underwent profound changes over the course of the eighteenth and early nineteenth centuries. To provide a genealogy thus far required uncovering the dynamics of the scientific enterprise, in this case primarily of natural history/philosophy and the embryonic medical profession. But such an account also had to include investigation into coproduction of knowledge and power, in this case the governmental logic of developing a civilized social order in the context of an independent republic.

Analysis of the governmental orientation of climate knowledge in the early United States also demonstrates the usefulness of investigating the basic category of climate, rather than assuming that climate only exists "out there" and thus does not need to be problematized in analysis of the climate-society relationship. If one were to neglect the significance of meteorological government and only evaluate early American meteorology in today's terms, we might conclude, as historian Theodore Feldman (1990:175) claims, that "the program of late Enlightenment climatology had failed," because "meteorology had not brought the perfection of the sciences of agriculture and medicine," and because "no useful correlations had been discovered between the weather and agriculture and disease." Such a conclusion would suggest that studying the period offers little empirical or analytic leverage toward a genealogy of anything except the curious failures of an allegedly premodern meteorology. Yet Feldman's perspective stems from a tenuous historical premise that implies science contains an internal logic of its own separable from problems of government. Within the development of American meteorological practices there was a confluence of knowledge from diffuse streams, forming a context in which American climates—as a bundle of relationships between bodily, social, moral, and atmospheric constitutions—emerged as objects of government.

LOOKING FORWARD: WHITHER METEOROLOGICAL GOVERNMENT?

Existing accounts of early U.S. meteorology (Fleming 1990, 1998a; Golinski 2008; Feldman 1990; Dalezios and Nastos 2016) recognize climate change and disease as central features of the eighteenth and early nineteenth centuries. However, these accounts suggest that controversies around climate change and civilization were peculiar to the period and were finally overcome through meteorologists' empirical findings (Fleming 1998a:45–54).[19] In this perspective, meteorologists emerged professionally around the 1840s and 1850s to study storms, build theories of weather patterns, and try to predict the weather. These self-conscious experts thus displaced the previous generation of philosophers and physicians who had been primarily concerned with climate changes, and finally made meteorology into a science. In parallel, over the course of the nineteenth century physicians organized their profession less around medical meteorology and more around theories of contagion (and later, microorganisms), transforming the basic categories of disease, constitution, and climate (Cassedy 1986). Modern medicine, in other words, displaced medical meteorologists at a time when predicting storms was already proving to be a more fruitful direction for meteorologists. But were established facts about storm patterns or disease, on their own merit, the primary cause for redirection in climate knowledge?

As I address in chapters 3 and 4, developments in scientific agriculture, transportation, and trade converged with interests among meteorologists to develop the capacity to map climates and predict the weather. By studying climate as "fixed quantities" within geographic regions (what I later label "climatic stability"), climatologists over time either rejected theories of climate change or, through a focus on stable historical patterns, rendered them superfluous. Moreover, Manifest Destiny and its associated nationalism, along with the rise of an increasingly robust American capitalism after the Civil War (1861–1865), displaced some of the previous ecological anxieties about the future of an American civilization. Mid-nineteenth-century studies among professional meteorologists and the changing orientations of climatologists and physicians may help to explain why we would expect subsequent U.S. meteorologists to focus less on climate change, civilization, and disease. However, a narrative centering around the inherent superiority of professional meteorologists comes up short in several important ways.

First, the rise of storm theories and a narrower weather-focused meteorology developed alongside a persistent concern with climate change theories. Indeed, in the case of German and French colonialism in Africa, Lehmann (2022) argues that the late nineteenth century marks the rise, rather than decline, of concern with climate change, colonial acclimatization, and wide-ranging climate anxieties. Lehmann (2022:ch 1–3) analyzes colonial engineering projects to block climate change or intentionally alter regional climates, like the entire Sahara-Mediterranean area, arguing that climate change thought was coming of age in the late nineteenth century, albeit outside the limited professional jurisdiction that meteorologists had carved out for themselves.

Second, if concerns about weather patterns and physical climates did, in fact, displace or reorient aspects of climate knowledge regarding broad topics of civilization and "salubrity," this reorientation is not neatly captured by the terms of meteorological science alone. Climate change as an issue did not rise and fall as a neatly bound scientific controversy. Instead, meteorological rationality can be explained vis-à-vis the governmental significance of climate change.

2 Meteorological Frontiers

CLIMATE KNOWLEDGE, TERRITORY,
AND STATE FORMATION, 1800–1850

We mounted the barometer in the snow of the summit,
and, fixing a ramrod in a crevice, unfurled the national flag
to wave in the breeze where never flag waved before.

—John C. Frémont (1845:69)

To govern the relationship between society and climate, people need to know climate, make it legible, interpret it as meaningful, and render it relevant to social conduct. We have already seen diffuse movements in this direction in the late colonial period and during the emergence of an independent United States. In this chapter, I trace how such efforts at climate legibility and governance related to early U.S. state formation. The goal is to see how the incipient creation of a bureaucratic state incorporated the previous concerns about climate, and also to consider how climate science may have supported, challenged, or otherwise influenced elements of American governmental institutions and projects, roughly over the period from 1800 to 1850. The analysis primarily draws upon meteorological scientific texts (books, articles, meteorological tables and maps, and letters), published accounts of government-sponsored expeditions, and records of the U.S. Army Medical Department. I use this data to trace the governmental significance of meteorological statistics, military-medical meteorology, and racial climatology. These three central components of climate knowledge coproduced state efforts to evaluate, calculate, and monitor, respectively, (1) the military body in a context of bureaucratization of the U.S. Army, (2) western North American lands in a context of territorial acquisition and providential nationalism, and (3) a stratified population "legible" with reference to racial hierarchy.

Let's begin with a single moment that can bring into focus some of the basic minutiae of climate knowledge and the intricate relationship between meteorology and government. In the summer of 1820, while a group of men were trailblazing an overland return eastward from the Rocky Mountains and got lost on the vast Arkansas plains, they deserted the detachment of their commanding officer and the Stephen Long expedition party. These soldiers fled on horseback, their saddlebags filled with stolen goods, including incidentally the various scientific journals of Thomas Say, a naturalist dispatched by the APS on what was labeled a "military and scientific expedition." The journals likely included some of the meteorological tables ordered to be kept by the expedition party.[1] These important materials, "utterly useless to the wretches who now possessed them," as Major Long wrote in his journal, "were probably thrown away upon the ocean of the prairie." He went on to lament: "The labor of months was consigned to oblivion by these uneducated vandals," these "worthless, indolent, and pusillanimous" deserters.[2]

One can imagine the pages of meticulously recorded meteorological observations baking on the grassy plain, carefully inked numbers running together in the first rains to strike them. The Long Expedition—with its perils of route finding and desertion on what members famously termed the "Great American Desert" (what is now called the Great Plains)—exposes some of the challenges through which new spaces came to enter American government, territory, and science (Goetzmann 1966). However much meteorological knowledge was successfully recorded, much was lost. Meteorologists faced tremendous difficulty in aligning a vast network of objects and interests: the vicarious weather, often inspiring poetic description more than disciplined measurement and numerical comparisons; the delicate instruments, inclined to break or lose calibration; variable forms of accounts by Natives, settlers, and traders; and the duties of underpaid troops prone to disease, drunkenness, and in some cases, desertion, theft, and mutiny.

Indeed, within scientific surveys and the journals of military missions during the early nineteenth century, both western lands and the weather came into view as anything but conquered or remotely governable. Long's journal account presents a picture of precarious wandering, not mastery. More than an isolated event, however, the Long Expedition desertions evoke the broader issue that concerns us here: how knowing climate and efforts to craft an expansive territorial state developed alongside one another within the U.S. context. Beyond providing insight into the development of meteorology from 1800 to 1850, tracing the coproduction of science and the state can show how the dynamics of climate knowledge—centered as they were around climate change, medicine, and social categorization—actively incorporated efforts to govern climate in the process of state-making.

WAR, TERRITORY, AND RACE

The previous chapter gave scant attention to the state, even if it traced the formation of dispersed practices aimed at securing civilized land, climates, and bodies as objects of government. Analysis of these practices did not assume that practices of government begin and end with the state as a formal social institution. The problems of a productive and healthy population, of the cultivation of a salubrious atmosphere, and of civilization as a climate-dependent process informed diffuse efforts to govern human relationships with climate, but not as a largely centralized plan or technique. The state was not the center for governing climate.[3]

In this chapter I argue that over the course of the first half of the nineteenth century, climate knowledge came to be articulated with U.S. national state formation, producing a centralized network of meteorological knowledge, an expansion of control over territory, and an increasingly national legibility of American populations. My point here is not to identify how the state used or promoted an independent scientific field or body of facts, as historians have done by detailing the origins of state weather services and observation networks (Fleming 1998a:33–44; 1990). Rather, I want to show how central aspects of state formation linked up with the logic and structure of climate knowledge and thus further organized the process of meteorological government that had previously been much more diffuse.

Inquiry into the state-science relationship is supported by scholarship that has investigated the work of experts, scientific knowledge, and technical infrastructure in shaping, expanding, constraining, and some-times resisting the power of states (Ash 2010; Rueschemeyer and Skocpol 1996; Mukerji 2003, 2011). Scott (1998), Anderson (1991), and Carroll (2012) argue that states form a governable world only insofar as state actors make legible complex geographies, environments, and social practices. Thus, state formation has pivoted on classificatory and representational prac-tices pertaining to people and identities as well as geographies and nature. If the state governs by "seeing" the world, for example through territorial mapping or census categories, "it" does not do so from a central vantage point. Rather, the state is marked by competing actors with shifting alli-ances and priorities, comprising what Pierre Bourdieu (1994, 2015) calls a "bureaucratic" field. Bourdieu argues that states have two basic features. First, the state organizes a "division of labor of domination" both between formal government organizations and within society writ large (Bourdieu 1996b:165).[4] Second, the state serves as an "instrument of unification" that shapes society through a monopoly of coercive and symbolic power (Bour-dieu 2015:226; Loveman 2005; Dezalay and Garth 2002). As Bourdieu (2015) explores, symbolic power is manifest in the taken-for-granted na-ture of social reality, especially significant when it is "misrecognized" by virtue of what many take for granted as "official."

Although the state is primarily made through formal political struggle, at times scientists are important, especially insofar as they engage the bu-reaucratic field as a terrain of their professional struggles. Existing studies of state formation and expertise have often found that competition in and over the state impacts how "official" knowledge is produced and how ex-perts are able to intervene in society (Steinmetz 2008; Dezalay and Garth 2002; Fourcade 2006). Boundaries between science and nonscience, ex-pert and lay knowledge, and "savage" and "civilized" ways of knowing are all instances in which official designations may often be misrecognized not only by participants but also in historical analysis because the out-come tends to privilege positions that were, in fact, won through struggle.

Regarding the antebellum United States, recognizing the role of scien-tists in making way for administrative government challenges claims by Stephen Skowronek (1982) and Theda Skocpol (1995), among others, that an American bureaucratic state is a product only of post–Civil War poli-tics and late-nineteenth-century industrialism. I contend that the United

States in the period from 1800 to 1850 presents an important case of state-science coproduction. In particular, discourse of "civilization" transected political and scientific efforts regarding the legibility of social progress in relation to natural order (Mazlish 2004). Of course, U.S. state formation is reducible neither to a civilizing process nor to a scientific program. Rather, as Michael Mann (1993) and Charles Tilly (1990) emphasize more generally, coercive power and war are critical in shaping state territoriality, bureaucratic expansion, and the capacity for state action to organize society. Indeed, the geopolitics of the U.S. frontiers, complete with continental expansion and imperial wars, arguably marks the initial consolidation of U.S. state power. By the 1810s, those seeking to protect trade, settle new land, subjugate Indigenous people, and control settlers and government agents called for a formal military, expanded taxation, and related demands placed upon citizens, as well as novel bureaucratic institutions to organize administration of coercive powers. These powers, at the same time, served to integrate western land, peoples, and nature as governmental domains and therefore to encompass the more diffuse, "civilizing" governmentality already activated by the physicians, statesmen, reformers, and others addressed in the previous chapter. Although it makes sense that scientific discourse would be significant to such statecraft, including both organizing various means of coercion and integrating a wider governmental apparatus, little is known about how climate knowledge related to U.S. state formation in these terms.

The rise of a U.S. territorial state and concurrent developments in science is complicated by racial stratification in the antebellum period. In the ideology of Manifest Destiny, territorial expansion represented a state-facilitated providential nationalism, upheld by a cultural frame of white superiority with officially sanctioned rights for whites to claim and exploit land and to legitimately and brutally subjugate Blacks, Indigenous people, and racialized immigrants (Horsman 1981). Contested issues of citizenship and political representation, in conjunction with a volatile politics of slavery, consolidated around a newly elaborated racial category of Anglo-Saxon whiteness. The American state emerged in this way as a *racial state*. The American racial state was anchored not only in racist violence but also in the symbolic power to naturalize hierarchy, through appeals not only to political philosophies of citizenship rights but also to scientific defenses of racial order (Goldberg 2002). As Paul Frymer (2017) has shown, territorial rule expanded across the continent through government officials' capacity to facilitate white settler-colonial expropriation through land and settlement policy, while also negotiating the territorial limits of a white racial empire that many had envisioned. Even so, the racial frames that shaped political, administrative and military strategies amid territorial expansion in North America in turn informed practices of U.S. overseas imperialism by the turn of the twentieth century (Go 2011; King 2019).

Many scholars have investigated such elements of state formation without reference to meteorology or climate knowledge, while most historical accounts of meteorology and related sciences in the United States have yet to coherently conceptualize state-making practices. Recent focus on climate science as an "imperial science" in the nineteenth and twentieth centuries helps situate the American case in transnational context. Philipp Lehmann (2022), in *Desert Edens*, shows how late-nineteenth- and twentieth-century anxieties around both desertification and colonial control in North Africa led to large-scale climate engineering proposals and efforts that could allegedly block climate change and ensure stability for colonial regimes. Racial difference in relation to climate was perhaps particularly salient to climate knowledge in geographic contexts marked by settler colonialism and by incomplete territorialization and social control (e.g., an expanding, continental empire). Here, the legibility of domination is unclear, and discourse of climate and natural hierarchy can inform colonial narratives and policies and settler practices. In stably controlled territories or those marked by imperial extraction rather than settlement, climate knowledge may orient more toward the legibility of natural resource development and economic productivity (e.g., in agriculture or navigation), even as the racialization of labor in imperial contexts still invoked climate determinism when justifying a racist division of labor (Asaka 2017; see also Mercer and Simpson 2023).

U.S. climate knowledge in relation to government was hardly a purely American affair. Although their ideas and practices were unique, U.S. meteorologists developed them through a more general pattern of transnational correspondence between climate scientists working across different regions and within distinct imperial structures. Take for example the nineteenth-century concept of acclimatization, what Michael Osborne

(2000) calls a "paradigmatic colonial science." As Osborne (2000) and Richard Drayton (2000) show in the cases of zoology and botany in the French and British empires, acclimatization linked metropolitan and colonial knowledge, raising questions about how life works—or may be improved—when it is taken to different climates. The concept was likewise taken up by those invested in understanding social hierarchy, race, and the body, giving way to intellectual exchange across human and life sciences and between empires, including the United States. Politically inflected debates emerged. For example, Hans Pols (2012) shows that Dutch scientists in the Netherlands and settler-colonists in the Dutch East Indies ultimately countered the prevailing idea in the early nineteenth century that suggested whites had best avoid tropical climates and instead exploit Indigenous groups. This supported settler interests and advanced racialized debates about acclimatization (Seth 2018), which continued to surface for decades in debates featuring climate, civilization, and race in the United States (Huntington 1915, 1919).

ORDERING OBSERVATIONS: MILITARY STRATEGY AND THE METEOROLOGICAL REGISTER

At the beginning of the nineteenth century the prospects of a national U.S. state were unclear. Since the 1780s, the United States remained highly fragmented along state lines. The Continental Army, formed during the American Revolution, had been almost entirely disbanded. Executive power was severely limited in its capacity to establish rule within the original thirteen states and their borderlands. Only through a series of wars with independent and foreign-backed Indian confederacies and, finally, following the War of 1812 with Britain, did the United States establish a more permanent military-administrative bureaucracy through which to regulate territories and wage national conflicts.

The administrative and military bureaucracy of the United States expanded during the War of 1812, against the preference of many, including then president James Madison. In hastily mobilizing military and economic resources for waging the war, the Madison administration effectively reversed Democratic-Republican political commitments to a weak central state. It instead sought to build a national military, reestablish a Second National Bank, and expand taxation (Hickey and Clark 2016). After the war, military and administrative forces were further developed under President Monroe's secretary of war, John C. Calhoun. The war resulted in U.S. military victories against Britain and allied Indian tribes, yet the number of casualties of war, primarily from disease, were unanticipated and totaled around twenty thousand U.S. soldiers. The security of frontier territories in the face of Native tribes, imperial powers, and trading companies remained precarious, especially because U.S. officials taking advantage of the power vacuum left in the wake of British retreat hastened land grabs and military advances against tribes on the southeast and northwest frontiers. These geopolitical and territorial efforts, by routinizing a level of military-administrative bureaucracy, extended the realm of federal government concerns.

Among these developments, the Military Reorganization Act of 1818 expanded the Army Medical Department, supervised by Army Surgeon-General James Tilton. Tilton, an army surgeon since the American Revolution, had ordered in 1814 that standardized meteorological observations be recorded at military posts. The network of observers was expanded through Secretary Calhoun's 1818 Reorganization and the appointment of Tilton's successor, Joseph Lovell, in the same year. Protecting the health of soldiers, especially those posted in unfamiliar and insalubrious (or unhealthy) climates, was an eminent concern at a time when diseases appeared a worse enemy than Native tribes or competing powers.

The health of soldiers was an explicitly meteorological affair because health and disease were primarily understood to be a matter of behavior within particular climates. For example, army medical observations used "systems of climate," numerically represented, to understand variation in disease types (see table 2). Army Surgeon Joseph Lovell's ([1817] 1873) report, "Remarks on the Sick Report of the Northern Division for the Year Ending June 30, 1817," amounted to a treatise on military-medical reorganization in light of the perils of hasty, undisciplined, and disease-ridden warfare. In it, developing a national military, especially in the broader nationalist context of post-1814 American politics, aligned directly with physicians' concerns over climate-based medicine and over how to organize their profession within the War Department.

Table 2 Excerpt from Samuel Forry's *Statistical Report on the Sickness and Mortality of the US Army, 1840*

Systems of climate.	Deaths per centum per medical returns.	Deaths per centum per Adjutant General's returns.	Ratio per 1,000 of mean strength under treatment annually.	Ratio of cases per 1,000 of mean strength.						Respiratory organs.			Totals.
				Intermittent fever.	Remittent fever.	Synochal fever.	Typhus fever.	Diarrhœa and dysentery.	Catarrh and influenza.	Pneumonia.	Pleuritis.	Phthisis pulmonalis.	
North'n lakes Atlantic coast	9–10	1 3–10	2,185	193	33	16	4	253	300	19	30	9	358
Stations remote from ocean and inland seas	1 5–10	2	1,912	36	26	43	5	170	233	22	26	9	290
*†Average	8–10	1 4–10	3,103	151	24	45	5 9–10	300	552	17	28	5	602
*†Average	9–10	1 5–10	2,660	143	26	37	2 4–10	269	439	18	28	7	490
†Average	1 1–10	1 6–10	2,400	217	28	35	3 3–10	243	362	19	28	8	412

Source: Forry (1840:172).

Note: Cross-tabulates disease type by "systems of climate."

*† The former average expresses the result of a comparison between the aggregate strength of the three divisions and the aggregate of cases and deaths. The latter shows the mean of the results obtained in each class. The first exhibits the actual ratios given by the statistics of the post, and the second supposes an equal number of troops in each division.

Lovell organized the first central collection of U.S. meteorological statistics. Initially, he decried the underutilized expertise of army surgeons: "It is from a knowledge of *minutiae* . . . that the experienced officer and surgeon becomes so much superior to the undisciplined recruit. It is almost entirely in order to acquire this kind of knowledge, that a military establishment is kept up in time of peace, and it is an undoubted fact that in no department of the army is it so slowly acquired and therefore so deficient as the medical" (Lovell [1817] 1873:105). Noting the failure of the military to keep soldiers healthy, alive, and disciplined during the War of 1812—a failure "too well and too publickly known to need comment"— Lovell called for regulatory reforms and an expansion of military medicine. "It is therefore suggested," he claimed, "whether such alterations be not required in the regulations, as are calculated to produce a system of medical police" ([1817] 1873:104–5). This was precisely the system that he later instituted to govern the training, appointment, and monitoring of military physicians; their military hospitals and posts; and the bodies and behavior of soldiers.

A policy of "medical police," present in European military-medical departments influenced by German physician Johann Frank's ([1799] 1976) *A System of Complete Medical Police*, became increasingly popular among physicians' societies in the United States around the time of Lovell's appointment as surgeon-general. Medical policing was an appeal by physicians for government intervention into the regulation of unruly subjects and their environments, especially through hygiene laws and regulations of the medical profession. As Andrew Abbott (2005) discusses for a later period using the case of medical licensing, a discourse of medical police served as a hinge, connecting the logic of medical professional reform with an emerging, if limited, national military bureaucracy.

As Lovell ([1817] 1873:104) outlined, medical policing should be "best calculated to remedy the evil" of disease, considered chiefly as "the effect of climate." Referring again to mortality during the War of 1812, he emphasized: "It must have been the climate—the weather—that produced the mischief" ([1817] 1873:103). Lovell espoused widely held claims linking climate and disease, yet he was in a novel position to carry this concern into the realm of U.S. federal government policy. He thereby worked to incorporate knowledge of mischievous climates into how the central government treated military bodies and U.S. territories. On this basis, Lovell ([1817] 1873:106) instructed, "It should be made the duty of every surgeon, together with his quarterly report to transmit an account of the local situation of his station, of the climate, the diseases most prevalent in the vicinity, and their probable causes, the state of the weather during the time reported with respect to temperature, winds, rain, etc." Through such reports, later compiled within *Meteorological Registers*, army surgeons would "be enabled to give such an account of the diseases that had occurred, their causes and his treatment, as would be the best possible criterion not only of his medical abilities, but also of his industry and attention to duty" (Lovell [1817] 1873:106).

The governmental logic of mandating the meteorological register was to evaluate and, to use Lovell's terms, "police" soldiers and the "industry" of officers with respect to weather and climate variation across space and time. In effect, those invested in meteorology coproduced a new observational network while also upholding administrative efforts to regulate relations between soldiers' bodies, officers' duties, and the potentially disruptive forces of climate, especially at frontier military posts. Lovell's scientific developments combined a meteorological and disciplinary gaze toward soldiers in their relations with allegedly unhealthy climates and dangerously unregulated behaviors.

This approach was novel in its formality in the United States, but borrowed directly from military meteorology in other imperial settings. U.S. officials concerned with military discipline and strategy in unknown or allegedly unfit climates borrowed from the lessons previously learned through British military medicine, especially the work of Sir John Pringle (1753) and colonial medicine in the tropics. For example, the famous Philadelphia physician Benjamin Rush annotated and republished Pringle's *Observations on the Diseases of the Army* in 1808, and it was widely used thereafter among American physicians (Cassedy 1986:15). The work of British Army surgeon Robert Jackson ([1791] 1795) helps demonstrate how military medicine combined the sort of meteorological and disciplinary gaze that Lovell and others intended to develop in the U.S. Jackson's ([1791] 1795) solution to the problem of dying armies was to monitor soldiers' behavior and institute medical discipline. He rejected the widespread belief that "European constitutions" were incapable of safely pursuing hard labor in "hot climates," and instead argued that soldiers were susceptible to the climatic causes of diseases because of their poor *moral* situation. "It is known to every medical person," Jackson ([1791] 1795:260) explained, "that the fevers of hot climates are most dangerous in full and plethoric habits." However, because "soldiers have little self-command, and seldom resist the gratification of their appetites," Jackson ([1791] 1795:260) directed officers toward practices of "great vigilance and attention," including the daily inspection of meals and of sleeping, drinking, clothing, sexual habits, and exercise regimens. Jackson ([1791] 1795:280–81) thereby rejected a policy of making "people of color do the drudgery of soldiers," reasoning that indolence would only further degrade the moral constitutions of European soldiers. In other words, proper surveillance and discipline would help acclimatize the constitutions of European soldiers and secure a successful colonial policy in the region, the inherent unhealthiness of the climate notwithstanding. In the early nineteenth century, Lovell helped to formalize this approach to military discipline, reflecting the work of his contemporaries, particularly those working for the British Empire (Seth 2018; Arnold 1996; Bashford 2004; Livingstone 2002).

In the United States, on the eve of executive efforts by officials of the War Department to institute military bureaucracy and imperial-continental ambitions, climate knowledge moved into the military body and onto the frontiers. In September 1818, just months after publishing "Remarks on the Sick Report of the Northern Division" ([1817] 1873) and subsequently being appointed surgeon-general, Lovell issued "Regulations of the Medical Department of the United States Army" ([1818] 1873). These regulations outlined bureaucratic roles and procedures aimed at systematically examining "the diary of the weather, medical topography of the station or hospital, account of the climate, complaints prevalent in the vicinity, etc., [and] suitable inquiries concerning the clothing, subsistence, quarters, etc., of the soldiers" (Lovell [1818] 1873:111, 114). Although Lovell's orders were not always followed, they were more successful than calls in the decades prior for coordinated systems of meteorological observation, including parallel efforts by the General Land Office. By the standards of meteorologists at the time, Lovell's network provided a remarkably stable basis for evaluating the nature of American climate-disease relations and their legibility to government.

After Tilton's and Lovell's initial struggles to order regular collection of meteorological data, army physicians continued to succeed at building the first national network of meteorological observations. Beyond pointing back to the War of 1812, or in Tilton's ([1781] 1813) case, to the debacle of military life during the American Revolution, medical and military meteorologists expanded their range of concerns. First, meteorological knowledge could solve outstanding debates about allegedly ongoing climate change and its relationship to the civilizing practices of white Europeans. Second, meteorological networks could provide the basis of a "medical geography" of western lands, which could be used to evaluate the health of military posts and the fitness of certain areas for white settlement (Dorn 2001; Valencius 2002). By integrating these concerns over climate change and disease into governmental practices, Lovell and his colleagues effectively developed their professional visions for meteorology within the Army Medical Department. Issues of climate change and disease clearly frame the first Army Medical *Meteorological Register* (see figure 1). These registers were widely circulated and cited among U.S. and European meteorologists and physicians and were updated periodically even as other meteorological centers were also established (Lawson 1844).[5]

The opening lines of Lovell's (1826:3) first *Meteorological Register* link the problem of medical meteorology directly to the active concern that American climates might have been changing—a process widely held to impact the health and future habitability of certain areas: "On the question whether in a series of years there be any material change in the climate of a given district of country and if so, how far it depends upon cultivation of the soil, density of population, etc., the most contradictory opinions have been advanced." U.S. expansion into the uncivilized West, he continued, presented a unique "opportunity" to evaluate possible climate changes wrought by land clearing and related practices in the eastern United States. This opportunity, for Lovell, was a relatively urgent matter. Because of increasing white settlement in the West during the 1820s, he claimed that the chance to establish the degree of climate change in civilized regions, "like that for recording the [customs] of the aborigines of the country, is fast passing away" (1826:3). He reasoned that "both these sons of the forest and the interminable wilderness they inhabited will, for all useful purposes, be as though they had never been." Considering his

No. II.   **FEBRUARY.**

PLACES OF OBSERVATION	THERMOMETER							WINDS									WEATHER				
	A.M. VII.	P.M. II.	IX.	Aggregate Mean Temperature	Highest degree	Lowest degree	Range	N. Days	N.W. Days	N.E. Days	E. Days	S.E. Days	S. Days	S.W. Days	W. Days	Prevailing	Fair Days	Cl'dy Days	Rain Days	Snow Days	Prevailing
Hancock Barracks	9.92	20.71	14.82	15.15	40	−7	47	3	12	8	.	3	.	2	.	N.W.	16	.	.	12	Fair
Fort Snelling	8.07	11.95	2.50	7.51	33	−30	63	2	5	.	.	.	3	3	15	W.	22	2	.	4	Cl'dy
Fort Howard	−3.17	16.74	7.07	6.88	41	−32	73	6	2	4	1	1	4	6	4	S.W.	12	16	.	.	Fair
Fort Preble	12.07	29.21	20.64	20.64	34	−2	36	.	3	3	4	2	4	5	7	W.	17	8	1	2	"
Fort Niagara	19.10	25.89	19.96	21.65	52	0	52	1	3	3	.	.	4	8	8	S.W.	15	11	.	2	"
Fort Constitution	16.89	29.07	22.42	22.79	40	5	35	.	4	4	1	.	1	3	6	N.W.	19	9	3	.	"
Fort Wolcott	20.03	30.82	24.07	24.97	42	9	31	1	13	1	.	.	.	3	11	W.	17	2	.	6	"
Fort Armstrong	7.67	18.17	11.38	12.41	50	−16	66	8	8	2	4	7	7	.	8	N.	24	2	.	2	"
West Point	17.10	29.39	23.00	22.83	40	2	38	4	2	1	1	.	.	3	.	N.W.	18	9	2	.	Cl'dy
Fort Columbus	21.57	30.07	24.67	25.44	41	11	30	.	16	4	.	2	1	4	6	N.W.	19	10	3	6	Fair
Washington city	22.07	35.42	28.38	28.62	46	8	38	1	13	9	1	8	1	1	.	N.W.	12	9	3	2	Cl'dy
Jefferson Barracks	14.28	27.03	20.25	30.52	70	1	69	1	16	3	1	.	.	3	2	N.W.	19	8	6	3	Fair
Fortress Monroe	32.21	39.86	36.14	36.07	54	22	32	7	9	5	2	11	1	2	3	N.E.	11	8	1	3	Cl'dy
Fort Gibson	22.85	37.51	25.58	28.65	66	0	66	3	6	8	.	2	.	.	1	S.E.	19	11	10	.	Fair
Augusta Arsl.	37.00	49.89	43.92	40.27	64	25	39	.	3	7	1	.	1	6	.	N.W.	10	7	1	.	Cl'dy
Canton. Jesup	36.85	53.85	45.10	45.27	72	18	54	6	15	8	2	.	.	.	1	N.W.	16	11	9	1	Fair
Canton. Clinch	47.42	55.82	53.57	52.27	70	30	40	13	10	4	.	3	2	4	.	N.	11	7	1	.	Cl'dy
Petite Coquille	48.25	56.10	49.17	51.17	70	28	42	4	2	4	6	.	10	.	5	E.	15	11	9	2	Fair

Figure 1. Excerpt from Joseph Lovell's 1829 *Meteorological Register.*
Source: Reprinted in Lawson (1840:32).

results uncertain, however, Lovell argued that only a longer series of observations would help "ascertain what changes, if any, have taken place,
either in the mean temperature, the range of the thermometer, the course
of the winds, or the weather, in the Atlantic States" (1826:4).

Lovell remained inconclusive regarding civilization and its effects on
climate. His reorganization of the medical-meteorological bureaucracy
within the War Department was a more enduring legacy. Despite his reluctance to proclaim closure to controversies over climate change, and despite the fact that his programs were implemented inconsistently and his
office even nearly abolished in 1830 (Gillett 1987:27–52), those who followed him continued to pursue military meteorology in terms of climate
dynamics, disease causation, and medical policing.

TERRITORIAL EXPLORATION:
MAPPING CLIMATE FUTURES

Military medical meteorology represents one aspect of governing climate
in the early nineteenth century. Territorial exploration and expansion
in relation to climate forms yet another aspect. An entry point for understanding efforts to know and govern frontier climates in these terms
can be found around the time Lovell was just beginning his education
at Harvard College and U.S. diplomats were negotiating the terms of the
Louisiana Purchase in France. On June 20, 1803, President Thomas Jefferson sent detailed instructions as Captain Meriwether Lewis and the
Army Corps of Discovery prepared for their surveying campaign in the
soon-to-be-purchased territory. Jefferson's instructions to Lewis make
clear that projects to extend governmental order and what Jefferson
(1780) had earlier called "the Empire of liberty" were to include not only
topographical information but also "civilizing" interventions: "Considering the interest which every nation has in extending and strengthening
the authority of reason and justice among the people around them, it will
be useful to acquire what knowledge you can of the state of morality, religion, and information among [Indians], as it may better enable those
who may endeavor to civilize and instruct them, to adapt their measures
to the existing notions and practices of those on whom they are to operate"

(Jefferson 1803). Civilizing in these terms meant to "instruct" and "operate" through "acquiring knowledge."[6] However, for Jefferson the prospects of civilization, like the health and discipline of soldiers, were partly climatological matters. Jefferson and his contemporaries widely agreed that civilizational trajectories were climate dependent. In *Notes on the State of Virginia*, Jefferson ([1785] 1794) argued that climates in eastern lands were changing because of cultivation. Jefferson pursued related concerns within the APS and, during the time of Lewis and Clark's expedition, with the visiting naturalist Alexander von Humboldt.[7]

Insofar as it was tasked with mapping the historical and possible future relations between climates and civilization, the expedition represents an early U.S. effort to institute a meteorological gaze over vast unknown areas of the continent. Jefferson (1803) instructed Lewis to record all encounters with "climate, as characterized by the thermometer, by the proportion of rainy, cloudy, & clear days, by lightening, hail, snow, ice, by the access & recess of frost, by the winds prevailing at different seasons" and other factors.

Joel Kovarsky (2014) has explained the meticulously planned scientific mission and Jefferson's concern over American climate change by emphasizing Jefferson's position as a philosopher-statesman with a background in geographic surveying. However, expanding the frontiers through military, geographic, and meteorological intelligence was less about individual ambition and more about the broader science-state coproduction of frontier legibility. Jefferson's instructions and his scientific interests were consonant with justifications before Congress that the expedition should serve to protect American economic interests against British and French traders encroaching on western territories. Measuring western lands, through boundary surveying, military documentation of possible threats, and meteorological and ethnological intelligence, provided a metric for the prospective expansion.

The production and distribution of frontier cartographies formed a major component of frontier representation and strategic expansion (Schulten 2012; Frymer 2017). Given prevailing concerns regarding climate and the general significance of cartography and surveying to statemaking (Carroll 2006; Short 2001; Biggs 1999; Anderson 1991), it makes sense that representing climates would help to explain and evaluate the distribution and prospects of settlement and territorial expansion. Efforts to map prospective futures—onto western lands in particular—brought together cartographic innovations within state surveying with parallel developments in meteorological mapping (see Anderson 2005:171–233; Schulten 2012).[8]

The federal government and professional societies (chiefly the APS) often jointly organized the scientific missions of expeditions to the U.S. West. Knowing nature articulated with political and military strategies to survey, evaluate, settle, protect, or usurp territory otherwise used or claimed by foreign governments, trading companies, and Native tribes (Frymer 2017; Rockwell 2010). Of course, securing frontier legibility did not always proceed as planned. Yet practical failures, as much as successes, demonstrate the coproduction of climate knowledge and state territorial expansion.

An instructive case of such mutual failures of meteorology and state legibility concerns the fate of Zebulon Pike's meteorological tables during his expeditions from 1805 to 1807. Departing from Saint Louis and commissioned by General James Wilkinson to survey boundaries of the Louisiana Territory, the expedition party lost its way and ended up in Mexican territory, where they were captured by Spanish authorities and taken to Santa Fe. The expedition was embroiled in political controversy, including possibly imminent war with Spain and federal investigation of Aaron Burr's conspiracy (implicating Pike's commanding officer, General Wilkinson) to secede and annex Spanish lands in the Southwest.[9] This situation left Pike and his meteorological records in a precarious position.

Pike worried that his charts and tables (such as figure 2), stored in a single trunk when he was captured, would be found by the Spanish and expose his party as spies. So, as Pike narrated in his letters (reproduced in Maguire 1889:390), he strategically "caused [my] men to secrete my papers about their bodies, conceiving this to be safer than leaving them in the baggage." However, "in the evening, finding the ladies of Santa Fe were treating them to wine, &c., I was apprehensive their intemperance might discover the secret."[10] Pike retrieved the papers, but they were promptly discovered and confiscated by Spanish officials. After Pike's release, in an 1807 letter to Spanish governor Salcedo (in Maguire 1889:386), Pike appealed to him: "Your Excellency may be induced to conceive that the

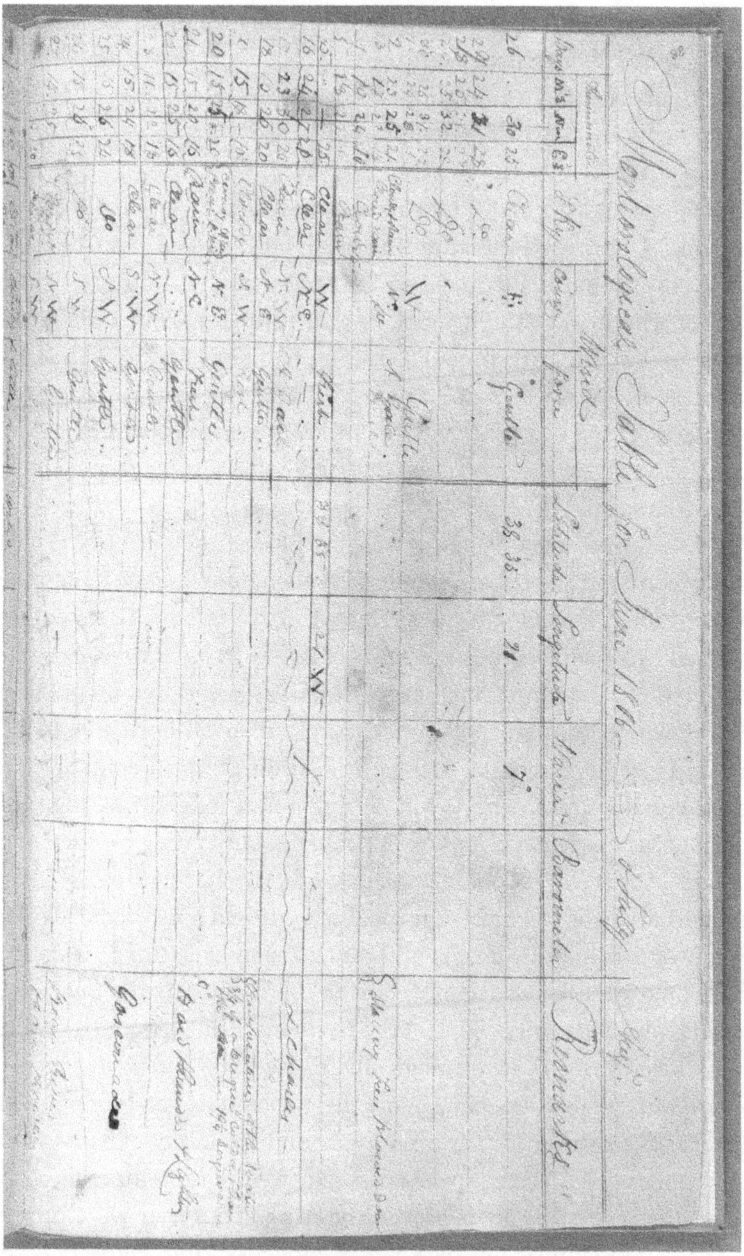

Figure 2. Zebulon Pike's "Meteorological Table for June 1806," among those seized by Spanish officials. *Source:* Pike (1805–1807).

measure of seizing my notes, plans, meteorological and astronomical observations . . . may not be justifiable." He argued to Salcedo that his papers "would enable the executive of the United States to take some steps to ameliorate the barbarous state of the various savage tribes whom I visited" and "would have added in some small degree to the acquirements of science, which are for the general benefit of mankind" (Maguire 1889:386). Salcedo, claiming that Pike's papers violated international agreements, promptly refused. Neither Pike's appeal to science nor efforts to have superior officials intercede on his behalf succeeded in making it possible to incorporate meteorological observations into incipient efforts to construct a national view of climate. Moreover, his instruments, "most of them . . . ruined in the mountains by the falling of the horses from precipices, &c," were sold during his captivity to avoid having to pay for their carriage back to Louisiana (Maguire 1889:393).

The materiality of meteorological observations was intimately tied to successes of state action. The preceding example vividly shows just how difficult and partial the construction of climate through scientific intelligence was in the early nineteenth century. With this difficulty in mind, Secretary of War John C. Calhoun instructed Major Stephen Long, when he was preparing to set off on a later expedition up the Missouri River in 1819: "The object of the Expedition, is to acquire as thorough and accurate knowledge as may be practicable, of a portion of our country, which his daily becoming more interesting, but which is as yet imperfectly known. With this view, you will permit *nothing* worthy of notice, to escape your view" (James 1823:37). The Long Expedition, introduced at the beginning of this chapter, collected detailed meteorological instrument readings throughout the West. Among hosts of "interesting" facts, meteorological records might be a curious appendage to territorial expansion. Over time, however, efforts to stitch climate knowledge into larger tapestries of western settlement and progress helped to construct western territories as sites of agricultural productivity, "salubrious" landscapes, and possible settlement. Explorations and the factual observations that they circulated to state officials and hopeful settlers helped construct western spaces as knowable, habitable, and governable.

The influential maps produced by the Army Corps of Topographical Engineers of the Frémont Expeditions (1842 to 1845) show how meteorological

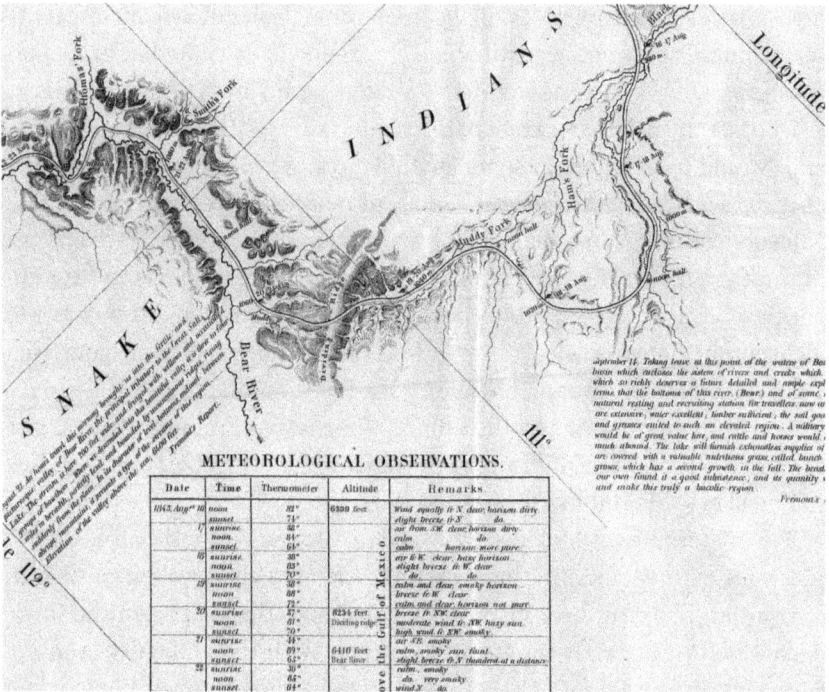

Figure 3. Excerpt from John Charles Frémont and Charles Preuss's "Topographical Map of the Road from Missouri to Oregon, Section V." *Source:* Frémont and Preuss (1846).

facts became integrated into prospective accounts of western territories as habitable and governable spaces. The expedition was explicitly oriented to facilitating westward migration. The published maps and associated reports contained not only topographical data but also novel cartographic representations of future settlement. John C. Frémont's cartographer, Charles Preuss, recorded minute notes, often reproduced from Frémont's journals, on the exact number of Indian "warriors" in various villages; observations on water, fuel, soil, grazing land, and "Indian" problems; and meteorological data collected during the expedition (see figure 3).

Those who crafted a system of meteorological recordkeeping worked alongside surveyors, military strategists, and settlers to make new cartographic constructions of the nation that could in turn inform narratives pairing geographic expansion with future progress. Frontier cartographies, like accounts of climate, often involve appropriation of local knowledge. U.S. topographical surveyors were not simply corps of men, instruments, and logbooks. Rather, their work comprised assembling chains of what John R. Short (2009) analyzes as "cartographic encounters." Maps and tables often began as a patchwork of Native, settler, trader, and explorer testimony and measurements. State actors translated and consolidated diverse information and then distributed it widely to government agents and, through print media, to settlers, often accompanied by adventure narratives, western boosterism, and medical advice. As exploration proceeded, popular "emigrants' guides," as they were often called, synthesized expedition reports and other information on western climates.[11] Emigrant guides relied on federal expeditions for data and credibility, and they translated climate knowledge into accessible geographical and medical terms that projected settlement futures in western territories. Timothy Flint (1826:188–89), for example, combined his own travel narrative with official reports to portray the "exhaustless fertility" of the Missouri plains, on which he envisioned that "the climate will grow salubrious with its population and improvement" and hence "will arise the actual '*Ne plus ultra*.'" Formalizing the provision of government data, the Fremont Expedition's reports and maps were uniquely printed by legislative order for direct distribution to the public.

RACIAL CLIMATOLOGY

By the mid-nineteenth century, meteorology had developed into what its leading participants and, later, historians have called an empirical and modernized science. Given the governmental significance of racial stratification in the decades leading to the American Civil War, was race a component of scientists' efforts to establish empirical meteorology? To consider this question, I here trace the work of a leading figure in the emergent scientific view of climate, Samuel Forry, and contextualize his work and climate knowledge with respect to racial ideologies in the pre–Civil War decades. As a "true science," to use Forry's terms, climate knowledge advanced elements of American racial state formation.

Samuel Forry, Race, and a "True Science" of Climate

Forry, a meteorologist and army surgeon, worked under the direction of Thomas Lawson, who was appointed surgeon-general after Joseph Lovell's death in 1836. Forry's work on American climate was internationally well regarded because it effectively used new data from across the North American continent (see Forry 1843c:116). It was hailed a century later as the "first general work on the climate of the United States, and thereby also our most important climatological incunabula" (Leighly 1954:335). Recent historians have mostly viewed Forry's work as a culmination of developments in the Army Medical Department. Fleming (1990:68–70) situates Forry's statistical work within the context of storm-related studies, especially those by the meteorologist James Espy that were commissioned by Thomas Lawson in 1842, after Forry had resigned from the Army (Espy 1843; Forry 1843a). Such accounts miss some of the immediate context and significance for Forry's work.

In his famous 1842 text entitled *The Climate of the United States and Its Endemic Influences*, Samuel Forry summarized the status of meteorology: "Numerical analysis applied to governmental objects soon bestowed the character of a science upon political economy [and] the doctrine of averages has been not unaptly styled the mathematics of medical science. . . . So [in] meteorology, nature has found faithful interpreters content to observe facts and to trace their relations and sequences, thus bestowing upon it the characters of a *true science*" (Forry 1842a:26; emphasis added). In other words, "numerical analysis" of atmospheric patterns, disciplined by "facts," "relations," and "sequences," meant meteorology was achieving scientific status. On this basis, Forry rendered the United States within one statistical view of climatological space (see figure 4).

To what end did meteorology construct such a view? Forry echoed Lovell's framework for meteorological observation: "A *mass of facts* thus accumulated will prove of immediate practical use to the philosopher, the physician, and the agriculturist; and to future generations, it will serve to determine what changes, if any, time may effect upon the climate of a particular region" (Forry 1842a:20). Forry argued that his predecessors and contemporaries who had promulgated a theory that civilization was actively changing American climates had in fact made "premature

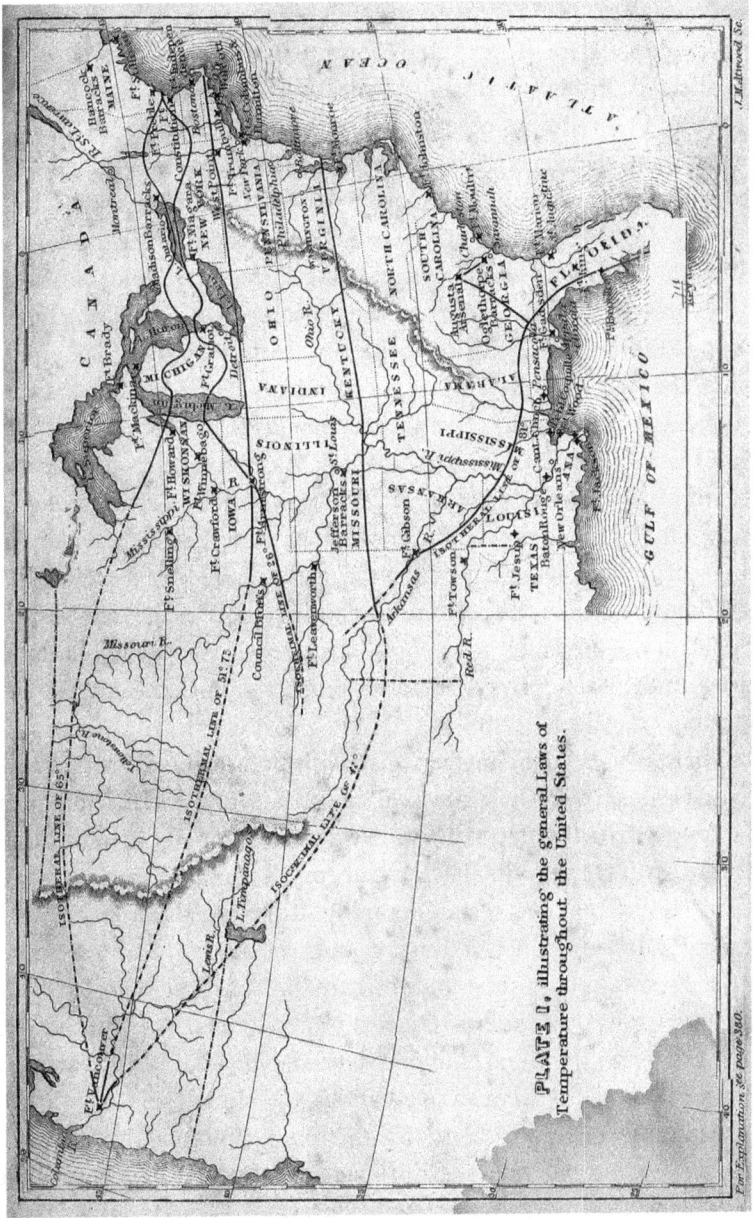

Figure 4. Frontispiece to Samuel Forry's *The Climate of the United States,* showing isothermal lines, intended to represent "general laws of temperature." *Source:* Forry (1842a).

deductions" that relied on the "testimonies of travelers" as opposed to the "thermometrical data" he provided (Forry 1842a:103). Based on decades of data collected through the Army Medical *Meteorological Registers*, Forry (1842a:108) granted that "climates are susceptible of melioration" by "the labors of man." However, "these effects are extremely subordinate, compared with the modification induced by the striking features of physical geography." In other words, Forry concluded that the activities broadly constitutive of civilization—especially clearing, reclaiming, and "improving" land—expressed minimal effects on the trajectories of American climates.

Despite following Lovell's inconclusive results with ones that appeared to refute the view that climate had progressively changed, Forry nevertheless upheld a strong civilization-climate perspective especially adapted to issues of racial stratification. He reasoned: "As climate not only affects the health but modifies the whole physical organization of man, and consequently influences the progress of civilization"; it followed that comparing "systems of climate" would "reveal to the medical philosopher much that is now unknown, and to the political economist many of the circumstances that control the destinies of a people" (1842a:95). Forry, like his contemporaries, understood "physical organization," "civilization," and "people" in increasingly racialized terms.

As analysis of military-medical discipline and territorial expansion already indicates, the "whole physical organization" of bodies, lands, and civilization was entwined with categories endogenous to government. By the 1840s, race was just such a category. This category shaped state officials' task of policing social hierarchy; it informed partisan interests vehemently for or against national expansion, slavery, and the rights of free Blacks and immigrants; and it figured in scientific theories, including among medical and meteorological professionals. As I will show, race altered the terms by which climate knowledge proceeded to evaluate the dynamic relationship between populations and climates.

In addition to helping to solidify a national climatological field of vision, like his contemporaries in European colonial settings, Forry's foundational texts in American climatology presented comparative statistics that he and others could draw upon to evaluate the natural-historical and probable futures of racial order. For example, Forry drew upon European contemporaries to make a common point that "the superiority of the warlike nations of Southern Europe over the effeminate inhabitants of Asia" had resulted from "this all-pervading agency of atmospheric constitution" (1842a:22). Climate, in this view, endowed "nations" with properties that explained their character and trajectory. Turning to Indigenous peoples and "the political horizon of North America," Forry (1842a:22) further concluded: "If we look upon history as philosophy teaching by example, it requires not the gift of divination to foresee the destiny of Mexico and the States south of it, whose inhabitants, enervated by climate, conjointly with other causes, will yield, by that necessity which controls all moral laws, to the energetic arm of the Anglo-Saxon race." Elements of Forry's conclusions were hardly new. African slaves had been consistently portrayed by slaveholders and scientists as having bodily "constitutions" suited for laboring in southern heat without succumbing to the diseases threatening the white constitution in such climates (Johnson [1813] 1827). Arguments about differentiated constitutions rationalized a longstanding fear that whites were unfit for "intemperate" climates while also building a scientific basis for enslavement of Africans in the U.S. South (Kupperman 1984; McCandless 2011; Puckrein 1973). Yet Forry's view of the "energetic" Anglo-Saxons was not one of an *unfit* race, but of one endowed with relatively stable racial features that exhibited exceptional biological qualities that could transcend climatic influence.

Written several years before the imperial Mexican–American War (1846–1848) and amid rising political tensions regarding slavery and territorial expansion, Forry's arguments concerning western climate and racial fitness illustrate the broader conceptual import of what we can call "racial climatology." Advancing a complex climatic theory of racial domination, Forry helped to build the contours of racial theory, and his authority as a meteorologist and his control over climate statistics were important to such developments. Forry's view of climate enacted a kind of "racialization," what Omi and Winant (2015:13) have defined as "the extension of racial meaning to a previously racially unclassified relationship." In this case, the alleged relationship between a historicized racial type (the "energetic" Anglo-Saxon) and climates informed the construction of racial hierarchy.

In a series of articles concerning "racial polygenesis," or the theory that races lack common ancestry, Forry (1842b, 1843b, 1856) explicitly developed the political (and theological) significance of racial climatology. Although he rejected the polygenesis theory by affording a degree of flexibility and "adaptation" to racial types, his work nevertheless reified a superior Anglo-Saxon race. Having traced the Anglo-Saxon racial type to premodern population and climate dynamics in northern Europe, Forry posited that degrees of intelligence, civilization, and morality were products of three major factors: climate, social organization, and, drawing positively from phrenologists, what he termed "cranial organization." Forry (1842b:130) concluded, "Can it be supposed that [the] noble developments of the present races contrasted with the low forehead, diminutive stature, and deformed figure, of some of the northern hordes who overran Southern Europe, are not owing mainly to the influence of civilization and a more genial clime?" Answering affirmatively, Forry nevertheless emphasized that prior accounts of civilization and climate were mistaken insofar as they proceeded "without any knowledge of the functions of different parts of the brain" (1842b:131). Regarding "Caucasians," he concluded, "it is among these nations that the progress of civilization and the development of the anterior portion of the brain, each exercising on the other a mutual influence, have gone hand in hand" (1843b:38). Unlike prior climatic theories of human difference dating back to Montesquieu's *De L'esprit des Lois* and Buffon's *Histoire Naturelle*, Forry synthesized the influential phrenological and anatomical-racist theories of his day, many of which challenged the entire paradigm linking climatic causes to racial formation.[12]

To summarize, Forry provided an account of racial types as outcomes of interaction between, on the one hand, the physical and social effects of climate on degree of civilization, and on the other hand, the moral and "cranial" effects of the civilizing process. Recall that an initial governmental logic for pursuing climate knowledge concerned the degrading or unhealthy effects of unknown or changing climates on white European constitutions. By linking intelligence and morality to relatively stable categories of "cranial organization," however, Forry's analysis did not problematize white "constitutions" as frail or at risk. On the contrary, Forry's racial climatology was oriented to evaluating what he saw as inevitable racial domination. In other words, Forry was working out the political implications of how historical climates had inscribed *brains*—not only bodies, civilized minds, and moral behaviors—with qualities that were more stable than the bodily "constitutions" as they were defined by prior climatic theories of race. If Anglo-Saxon whites were racially stable by biological fact, then regardless of climate, they were by nature fit for continental expansion. Such implications articulated with American racial state formation in relation to territorial expansion and settler colonialism in the U.S. West (Horseman 1981; Frymer 2017; Goldberg 2002).

Understanding the racial aspects of climatology does not suggest that science strongly drove racial politics, rather that racial state formation and empirical meteorology were not entirely separate enterprises. Racial climatology is but one dimension of meteorological government, although it reveals configurations of climatological categories that had real consequences for the governmental logic of the racial state. Racial climatology, in effect, helped undergird and naturalize broader discourses concerning Native, immigrant, and black races that transcended political debates over slavery, expulsion of Native people, and other forms of state-sanctioned racial subjugation. Asaka (2017) shows how the discourse of tropicality created political geographies separating temperate climates from tropical climates, with the former being the province of white settlers and workers and the latter being reserved for Black labor. By focusing on the cases of British-Canadian and U.S. free Blacks, including Black emigration and African colonization policies, Asaka demonstrates how climatic determinist thought formed a pillar supporting racial formations in the nineteenth century. The delineation between climate zones and the alleged impacts on racialized bodies was clearly political. Thus, states of the U.S. Northeast and West could be considered a "temperate" space unsuited for free Black migration and better suited for white citizens or settlers. If the tropics formed a natural region for Black people, then, "Black tropicality gave advocates of [African] colonization a biological ground for eliminating free blacks from the free-soil western frontiers" (Asaka 2017:140). Alternatively, a climate determinist stance could be refuted by those interested in preserving a nonwhite labor force or by abolitionists (including Frederick Douglass) who argued that nonwhites had become acclimatized to temperate regions.

Alongside coercive racism and policies that incentivized white nation-building (Madley 2018; Frymer 2017), civilizational discourses as developed through racial climatology contributed to a diffuse biopolitics aimed at an allegedly "productive" engagement with climate and racial difference (cf. Foucault 1980:119). Increasingly popular among urban reformers and western physicians, "medical geography," and what Forry (1840; 1842a:28; 1848) and others termed "hygeiology" and "state medicine," later helped to work out a governmental and medical synthesis of climatological and racial difference.[13] Studies in medical geography were highly racialized. Daniel Drake, a prominent medical geography and Ohio physician, is an important case. In his magnum opus, *Principal Diseases of the Interior of North America*, Drake (1850) provided a medical program based on climate statistics, racially stratified biomedical indices, and therapeutic techniques. He intricately mapped geographical and climatological areas, framing his analyses according to the "Caucasian," "African," "Indian," and "Esquimaux" races (1850:vi). Drake primarily surveyed the medical geographies of the Caucasian race, stratified by European ethnicities. Thus, he studied geographic influences on the "physiological etiology" of Caucasian diseases. He likewise evaluated how westward migration modified racial constitutions and their medical consequences. Drake suggested that racial destinies would culminate in "amalgamation": "The homogeneous millions, with which time will people the great region between the Appalachian and Rocky Mountains [are] thus destined to present the last and greatest development of society" (1850:647). Yet he too reified racial difference, particularly by focusing only on "the history of the diseases of the Caucasian races" and putting aside other races (1850:638).

Medical geographers diverged politically regarding slavery and reform movements. The politics of racial climatology was thus not reducible to the overt partisan divides fomenting in the antebellum decades. Valencius (2002:246) finds that Drake supported slavery and reserved his vision of racial "amalgamation" only for European nationalities. Polygenists who supported slavery, for example Josiah Nott (1851), rejected arguments of monogenists like Forry by denying the relevance of climate knowledge to scientific racism altogether. For his part, Forry rejected slavery as immoral, reasoning that "Negroes in the lowest stage of civilization are the ugliest." Yet he concluded that these "most ferocious savages—stupid, indolent, and sensual," could be "elevated in the scale of social condition," leading to possible "improvement in their physical feature" (Forry 1842b:118). How might such improvement proceed? Forry wagered that "political institutions and social organization often struggle successfully against climatic agency," the Anglo-Saxon race forming his primary historical case in point. As he argued, "The superior endowments of a more fortunate race should be exercised in extending the blessings of civilization" (1842b:132).

Asserting racial superiority as a basis for allegedly necessary, beneficial, or inevitable social domination had long-standing consequences. Perhaps the most important in the U.S. context were Democratic political movements to justify slavery and naturalize racial exclusion through such landmark political developments as the 1857 *Dred Scott* Supreme Court decision. Yet a diffuse, "civilizing" racism emboldened Anglo-Saxon superiority as political and climatic destiny. Racial climatology in the United States, moreover, advanced a providential zeal, distinct from other colonial contexts that featured racial anxieties around protecting whiteness (see Anderson 2006a; Parsons 2014). The role of science in constituting "whiteness" and its racial others as sociobiological categories has been widely documented (Kendi 2016; Anderson 2006a, 2006b). The present analysis shows that some meteorologists provided a climatological basis for racial hierarchy in an expanding United States. Although perhaps less vehement when compared to biological justifications for slavery or state-sanctioned violence toward those deemed nonwhite, racial climatology nonetheless advanced an ideology of providential nationalism that placed racialized territorial expansion on a scientific footing. Racial climatology thus formed a component of the broader coproduction of climate knowledge and government that emerged in the United States prior to the Civil War.

The Enduring Legacy of Racial Climatology

Racial climatology did not end in the mid-nineteenth century, although my analysis of later periods does not focus on its trajectories into the twentieth century. The logic of racial climatology was especially evident in late-nineteenth-century colonial "tropical medicine" in the British and French empires. Geographer David N. Livingstone (1991, 2002) argues not for a declining racialization of climate but rather for historical continuity in "moral climatologies" that imbue climate and its changes with power-laden evaluations of social and racial difference. In the case of U.S.

imperial science in the early twentieth century, human geographer and climatic determinist Ellsworth Huntington (1913) is especially significant. Fleming (1998a:95–106) argues that by the late nineteenth century, such figures became increasingly peripheral to meteorology and climatology as professional fields. Yet racial climatology remained a feature of U.S. science. For example, Robert DeCourcy Ward—a Harvard professor of climatology and once a president of both the American Meteorological Society and the Association of American Geographers—was also a founder of the Immigration Restriction League of Boston. As a eugenicist and climatologist, Ward drew from themes including German geographer Friedrich Ratzel's racial essentialism and environmentally determinist "anthropogeography" to propose what he called "*anthropo*climatology." He developed this approach in his 1908 text *Climate: Considered Especially in Relation to Man*, and, in 1923, at the end of his life, "The Acclimatization of the White Race in the Tropics" (Ward 1929; see also Lavery 2016; Rohli and Bierly 2011; Brooks 1932). Furthermore, during his tenure as director of the U.S. Weather Bureau (from 1895 to 1913), Willis L. Moore (1904:16) wrote an encyclopedia on climate, rejecting the idea that climate had changed over the course of "authentic history" ("notwithstanding the popular notion to the contrary"). Yet he still held that "climate is the most potent of any factor in the environment of races." He supported this position by stating that "Northmen," situated between climatic extremes, expressed the most complete achievement of human strength and society. Notably, advancing a racial climatology for the likes of Robert DeCourcy Ward and Willis Moore, and compared sharply to meteorologists like Samuel Forry half a century prior, meant holding to an understanding of climate as unchanging (or "stable") while also accepting a static biological racism. Among climatologists internationally, races and climates were typically treated as relatively rigid, at least over shorter historical timeframes. The work of the influential German climatologist Wladimir Köppen demonstrates how climate knowledge in these terms reflected the synergy between statistical climatology and imperialism at an international scale. Köppen is credited with the first global classification of climates by temperature and precipitation (Köppen [1884] 2011). He utilized geographic-statistical data on climate (for the United States, reliant especially on Schott 1876; Köppen [1884] 2011:356) and compiled other records to construct seven "climatic zones of the Earth," mapped as "belts." Although he clearly embedded his work in professional climatology, Köppen also constructed a historical narrative in which climate zones produced racial characteristics. Thus, he used the logic of climate classification to narrate "the shift of the center of human civilization towards the cooler zones" in European history, with an "analogue movement of culture from the warm towards the cooler" zones in America ([1884] 2011:357). The hierarchy of Köppen's climate classification, in the end, supported an imperial worldview: with the historical shift of "the entrepreneurial spirit" to temperate zones, the "hot countries" presented "favorable conditions where abundance in natural products allow nourishing large human masses and where the people's inertness makes it possible to master them" ([1884] 2011:357).

Although borne from previous concerns with climate change, then, racial climatology adapted quite well to biological racism on the one hand, and on the other to the increasingly typical scientific notion that climates were geographically delineated and stable over time, a view explored in the following chapters. Köppen's racial climatology thus made little mention of either climate change or racial fluidity.[14] The concurrent rise of Aryanism in German geography, especially the work of human geographer Friedrich Ratzel, further helped to place racial character on a stable geographic basis less beholden to climate changes (Ratzel [1901] 2018). Ratzel argued that climate produces regions of similar conditions, which in turn produce "regions of civilization which are disposed like a belt round the globe. These may be called civilized zones" (Ratzel 1896:28). The imperial optic of global climate classification corresponded to climate determinism: "As the tastes of civilization grew, the belt comprising it shrank into the regions where the great capacity for achievement co-existed with the temperate climates" (1896:29). Ashutosh (2018) has shown that American geographers, including Ellsworth Huntington and Ellen Semple, adopted aspects of Ratzel's and colleagues' racism and climatic thought as tenets of the discipline of human geography. Such thought likewise influenced eugenics, with Ward and Huntington being central to eugenics organizations in New England. Efforts, like Köppen's, to map and classify climates geographically thus proceeded apace of climate reductionist accounts of race and civilizations prominent internationally among human geographers.[15]

CONCLUSION

During the period from 1800 to 1850, the emerging science of meteorology was coproduced with three domains of U.S. state formation: military medicine, discipline, and administration; territorial exploration and frontier expansion; and foundational ideologies of a racial state. Creating a governable social order was simultaneously a matter of making climate a category of scientific knowledge and of making legible the relations between mobile, often racially coded, bodies and their environs. This historical narrative of science-state coproduction has two major implications. The first concerns how to interpret matters of science and government in the period from 1800 to 1850 in the United States. The second takes stock of how analysis thus far provides a general way of considering projects to govern climate, which are taken up in subsequent chapters.

First, regarding historical narratives of meteorology, it is true, as Fleming (1990) and others have shown, that meteorological networks in the nineteenth century were foundational to subsequent developments in meteorology in the United States and internationally. However, the narrative I have sketched of the formation of meteorological networks and climate knowledge in the period from 1800 to 1850 did not involve tracing the prehistory of weather forecasting and later modeling techniques. Meteorologists did not have these future developments in mind, even if they later contributed to them. They were not primarily developing accounts of storms patterns or establishing stable regional characteristics. Rather, climate knowledge succeeded because of actors' efforts to practically link meteorological networks to territoriality and to develop, through a kind of symbolic power, representations of lands and stratified populations. Clearly, what meteorologists established throughout this period is something their predecessors had not yet achieved: a centralized meteorological observation network with widespread applications. However, this network, its precipitating causes for institution, and its meaningful significance developed as they did because those participating in the scientific enterprise had become aligned with the process of state-making. Those who succeeded in becoming spokespersons for regional and national climates, including the nature of climate change and its relationality with stratified human groups, did so by articulating a vision of meteorological expertise as a required component of a governable social order. From the management of the unruly and diseased military body to evaluations and imaginaries of western territorial expansion, to the racial politics of diminishing or "civilizing" inferior groups, climate knowledge tied in with power relations.

Second, regarding science-state coproduction, the present chapter shows that statecraft involved neither an instrumental harnessing of scientific authority for an autonomous political logic nor any inevitable capacity to make bodies, diseases, habitable territories, and state agents governable by a central bureaucratic administration. Rather, state actors and meteorologists learned to govern climate through pulling together diverse ideas, resources, and practices. Governing the social relationship to climate entailed aligning physical networks, including the diverse components assembled in meteorological statistics and the tenuous peripheries of state institutions. In addition, meteorological government involved projecting and evaluating climate-population dynamics, making climatic and social orders legible together and hence categories of governmental concern. Meteorologists' efforts to establish meteorology as a science were also critical to this process, insofar as scientists continued to invest in empirical evaluations of climate change, disease, and human difference. The emerging logic of meteorological science in these terms fit quite well with a logic of state-making that centered around settlement, territorial expansion and security, and racial stratification, despite a relatively decentralized government administration.

PART II

Stabilizing Climate,
Economizing Weather

3 Climate Does Not Change

AGRICULTURAL CAPITALISM, CLIMATOLOGY,
AND THE STABILIZATION OF CLIMATE, 1850–1920

Attached ideas of change to the whole subject
[of climate are] difficult to remove.

—Lorin Blodget (1857:481)

Human civilization began with the stabilization . . .
of the Earth's climate.

— Peter Schwartz and Doug Randall (2003:14)

From the 1850s to 1920, climatology emerged as an organized branch of meteorology and government administration in the United States. Climatology, I argue, emerged as both a coherent professional project and a component of a governmental logic specific to capitalist society and a bureaucratic state. As a result, climatology operated through a discourse of *climatic stability*—if not always intellectually, then functionally. The basic view that climate is unchanging within a delineated geographic space emerged in a way that linked the social interests and practices of climatologists with the broader administration of commercial agriculture, trade, and finance. Through a reorganization of science and the social transformations associated with the rise of industrial capitalism, the process of governing climate became a very different enterprise when compared to prior and later periods.

To explain the transition toward treating climate as stable in the U.S. context, I center my analysis on climatologists' books, papers, maps, data collection, information dissemination efforts, and social organization,

especially within the emerging federal bureaucracy. As in prior chapters, analysis is here informed by a "symmetrical" approach (Bloor 1974)
that treats the institutional structure and technical content of scientific
knowledge—in this case, the rise in treating climate as a stable entity—
without positing in advance the validity or truth of such claims. Transformation of concerns about the stability/instability of climate are not neatly
explained by climatic shifts or the logic of scientific discovery alone, but
also by changes among the domains of science and government. Investigation in this chapter starts by establishing the fact of a major definitional
change regarding climatic stability in the specific context of U.S. science.
I then draw upon scholarship tracing the rise of industrial capitalism to
frame an analysis of how meteorologists and allied actors constructed
"stable" climates and put them to work, on the one hand, for the formation of climatology, and on the other hand, as basic categories of modern
rational capitalist society. As later chapters uncover, climatology in this
period held contradictions. On the one hand, despite a paradigm that centered the timeless stability of climate, climatology internationally nonetheless pursued new climate change studies in conversation with geology
and with colonial anxieties about the sustainability of expansive, imperial
regimes (Lehmann 2022; Mercer and Simpson 2023). On the other hand,
the practical and professional orientation toward climatic stability meant
that later concern with global warming would fly in the face of climatological common sense, particularly in the U.S.

THE PUZZLE OF CLIMATIC STABILITY

A convenient way to locate changes in knowledge is to uncover novel basic
definitions within a field in order to recognize them not simply as given
by nature but as effects of definitional struggle among people situated in
time and place. Scientists' understandings of climate exhibited just such a
definitional struggle and transformation in the latter nineteenth century.
By 1903 climatologist Robert DeCourcy Ward translated the Austrian meteorologist Julius von Hann's influential *Handbook of Climatology* and
revised it for an American audience. Ward (introduced briefly in the last
chapter) was the first professor of climatology in the United States, at

Harvard University. He defined climate as "the sum total of the meteorological phenomena that characterize the average condition of the atmosphere at any one place on the earth's surface" (in Hann [1883] 1903:1). As meteorologist Willis Milham likewise made clear in his influential textbook *Meteorology*, published in 1912: "Weather changes from moment to moment, but climate remains the same" (Milham [1912] 1918:426). The "constancy of climate," as Milham ([1912] 1918:437) put it, meant that climate had not changed in either recent or historical times (he cited seven thousand years).

Definitions of climate as "normal weather," rather unsurprising today, compare sharply with earlier relational conceptions of climate. Recall that just decades prior to the publication of the above-cited definitions, people did not typically treat climate as a geographically stable set of averaged physical parameters, but rather as a set of dynamics relating human populations to their environments. Consider again American meteorologist Samuel Forry's (1842a:127–28) definition of climate as "the aggregate of all the external physical circumstances appertaining to each locality in its *relation* to organic nature," a domain in which Forry included the "dynamics of influence" that climate exercises on "the physical and moral state of man." To know climate was central to understanding how and why it might be *changing*, and with what consequences for equally dynamic human and social developments. In 1857, climatologist Lorin Blodget (1857:481) frustratedly characterized the situation when working to establish a new vision of climatology in the U.S.: "Attached ideas of change to the whole subject [of climate are] difficult to remove."

So a central question concerns how to explain a definitional change in climate and evaluate its consequences. Other scholars have analyzed the eighteenth to mid-nineteenth centuries to emphasize that as a category of science and a set of social anxieties, "climate change" today is hardly new, but rather "the modern revival of the debate in a new version" (Thompson 1981:238), for which analysis must find one or another "valuable historical analog" (Stehr and Storch 2000:13; see also Fleming 1998b; Hulme 2008; Zilberstein 2016). This position faces major limitations when it comes to interpreting climatology in this period. First, historical scholarship on climate change risks presuming a false parallel between the contemporary situation of global warming and climatic theory in prior periods.

For example, Fabien Locher and Jean-Baptiste Fressoz (2012) draw upon French and comparative cases to argue that ideas of anthropogenic climate change can be genealogically traced as part of modern political formations, which have since the eighteenth century been acutely aware and anxious regarding climate change. When a change-centered framework is applied to the late nineteenth and early twentieth centuries, active debates about the climatic effects of deforestation, the geologic discovery of ice ages, and early aims to modify climates through engineering come to the fore (Lehmann 2022; Davis 2016). Although I address these major scientific issues, I believe focusing on them presents a partial narrative that neglects a major, stability-centered, and change-rejecting current, at least in particular political, economic, and intellectual situations.

It is worthwhile to consider whether those who enacted a definitional change regarding a stable climate were simply correct. Recent reconstructions of climate history hamper the otherwise plausible claim that climatologists' data speaks for itself and therefore, straightforward discoveries had closed the "debate" about climate change in the mid-nineteenth century by empirical falsification. In a review of nineteenth-century climate theory, for example, climatologist Kenneth Thompson (1981:238) writes, "Ironically, the period of extended debate on the climate-change issue, when opinion was essentially polarized on either climatic amelioration or climatic stability was actually a period of distinct climatic deterioration for the western world." By "deterioration," Thompson means the cooling period, which climatologists later labeled the "Little Ice Age." Climate historians and climate reconstruction modelers (Bradley and Jonest 1993; Mann et al. 2009) have consistently identified this cooling trend, which was global in scope up to the late nineteenth century (although it expressed temporal regional variation). Recognizing that climatologists' views on climatic stability do not neatly correspond with reconstructed trends suggests that climate data itself provides an insufficient explanation for the transformation, within the logic of climatology, to a "stabilized" climate. This resonates with findings by Anya Zilberstein (2016) and Sam White (2020) on colonial North America, which show how many people believed climate was changing in ways irreconcilable with historical-climatological records.[1] Rather, an unstable and changing climate strongly resonated with the social experiences of those living in an earlier era guided less by standard representations of climate than by the tumult of life marked by immigration, mobility, and social change. Regarding climatology, likewise, the social context can help explain how climate knowledge may have related to the broader rationalization of U.S. society.

CAPITALIST CLIMATES

Internationally, the coproduction of climate knowledge and state- and empire-building in the late nineteenth century benefited from refined techniques for representing and mapping climates geographically, what Deborah Coen (2018) has analyzed as the process of "scaling." Although classical Aristotelean meteorology had understood climates to be latitudinally defined climatic "zones," such zones were more heuristic compared to later cartographic representations that effectively and objectively pinned climate down (Martin 2006; Humboldt 1817; Hann [1883] 1903; Köppen [1884] 2011). By elaborating the spatiality of climate, climatologists working in diverse geographic contexts were able to inform political and economic interests, ranging from land use policy to trade routes to continental and global imperial ambitions. The spatial rationalization of society and its environments, which transformed many elements of natural and social reality into discrete units for the purposes of bureaucratic administration and commercial exchange, was hardly limited to the U.S. context. Coen (2011, 2018) studied Habsburg climatographers and found that they bridged local climate scales and thus represented a politically coherent but culturally amalgamated imperial dominion. Imperial climatologies thus informed the diverse but hierarchical nature of the empire. Katherine Anderson (2005:ch. 6) and Martin Mahony (2016) have argued that the British Empire, toward the height of its imperial reach, supported and drew from a geographically expansive and diverse meteorology that matched imperial ambitions of operating in "all types of climates." Regions, in these intellectual and political circumstances, could be reconceptualized as zones fit for specific, designed purposes, for example, a given commercial crop or planned infrastructural projects. If "seeing like a state" could mean mapping climate zones for administrative and commercial purposes, then other areas could be imagined as sites of climatic transformation in support of profit and state control, as Lehmann (2022) shows in the case of German-controlled North Africa and Russia's East.

Given the configuration of science, capitalism, and state-making, the U.S. from the 1850s to 1920 provides a case of how those working in or connected to the domain of climatology helped to build what can be labeled "industrial capitalist climates." Climates under this conceptualization are "scaled" in a manner that provides a basis upon which markets and capitalist social relations can be progressively and predictably built by economically exploiting land, water, air, soil, and labor productivity. Analysis of industrial capitalist climates can draw from a Marxian understanding of nature, the state, and knowledge as comprising what Allan Schnaiberg (1980) labels "production science," which facilitates the capitalist "treadmill of production." The natural sciences in the latter nineteenth century formed one "hand" of capitalist state formation by continuing to facilitate the legibility and accessibility of territory, land, and natural resources (Morgan and Orloff 2017; Scott 1998). Physical scientists during this time oriented toward the innovative, if destructive, extraction of raw materials and labor-power, among what Karl Polanyi later labelled the "fictitious commodities" upon which capitalist production depends (Polanyi [1944] 2001:187–200). To use political economist James Dunbar's (1781:308) prescient late-eighteenth-century terms, "economic government" must "recover . . . our patrimony from Chaos." The production of capitalist nature must confront, and hence exploit or overcome, nature's complexity. A capitalist order thus entailed making nature and society legible in ways previously unimaginable. This process became even more rationalized within a nineteenth-century international market system in which profits were only secured through competitive production, financial innovation, and efficient trade on relatively open markets (Davis 2004; Cronon 1991).

The period from the 1850s to 1920 marks the industrialization of the U.S. economy. Barrington Moore Jr. (1966) has argued that the U.S. Civil War constituted a bourgeois revolution that consolidated a national capitalist class in manufacturing and finance, paralleled by the "freeing" up of western land to property development and the "freeing" up of labor through the abolition of chattel slavery in the American South. Historian Christopher Clark (2012) has shown that agrarian development complemented the rise of industrial manufacturing through more intensive capitalization of land, property, natural resources, and agriculture, compared to the antebellum period. New state institutions facilitated this process (Skowronek 1982; Frymer 2017) while financing infrastructural and land development, reclamation, navigation, and public works (Carroll 2012). Capitalist state formation therefore sets up a particular character of science, oriented to problems of incorporating nature into market society. An empirical account of this historical process begins with the rise of climatologists who devoted themselves to agricultural development.

WORKING LANDS: AGRICULTURAL DEVELOPMENT AND CLIMATIC STABILITY

Climatologists working in the U.S. began to resettle an otherwise fragmented science of climate primarily by leveraging a capacity to speak for the agricultural productivity of geographically delineated areas. Within science, climatology emerged at a time of deep challenges to prevailing concerns among meteorologists. Beginning around the 1850s, the logic of "medical geography" as the basis for meteorology was already breaking down as developments within medicine had begun to displace miasmic-atmospheric theories of disease (Ackerknecht 1948; Mitman and Numbers 2003). Although variable across contexts, internationally, the contagionist-bacteriological paradigm for disease generally shattered the prospects of a medically centered climatology (Rupke and Wonders 2000). Furthermore, natural-historical accounts of life, earth history, and social organization underwent deep challenges within the emerging social sciences and Darwinian/evolutionary theory. These developments in science provided distinct understandings of long-term developments that were not, as the previous generation largely believed, climatically determined. Active controversies about climate change among geologists by and large supported rather than challenged the view that climate was stable over historical timeframes. Geologists came to understand climate change on the order of millennia, not centuries—a radical departure from previous generations of meteorologists. The earth was not changing quickly. Of greatest significance here, geological theory of climate change did not tie in with human social developments. U.S. climatologists could largely take the *temporality* of climate for granted and focus on the unchanging *spaces* of climate.[2]

Seeds of Climatic Stability within Scientific Agriculture

In the mid-nineteenth century, meteorologists in the United States advanced their profession by explicitly linking it to the larger movement in science and government toward "scientific agriculture." In 1858, meteorologist and founding secretary of the Smithsonian Institution Joseph Henry wrote a treatise, titled *Meteorology in Its Connection with Agriculture.* Henry sought to bridge the development of experimental science with the recent institutionalization of meteorological data collection on a national scale, at the time organized by a Smithsonian-based network of weather observers. The observation network that facilitated collection of meteorological data was connected via correspondence and telegraphy to other meteorological data infrastructure organized by the U.S. Army Medical Department, the U.S. Navy, and the APS (Fleming 1990).

To outline a national meteorology, Joseph Henry (1858:456–457) provided a political philosophy of climate and agriculture: "To our political organization, under Providence our prosperity has mainly been promoted by the ample room afforded us for expansion over the most favored regions of this continent. It becomes, therefore, important for us to ascertain the natural limits, if there be any, to the arable portion of our still untenanted possessions." Henry was beginning to reconsider problems of climate with reference to possible "natural limits" of the nation, a political problem that science could help to address. On this basis, the U.S. Patent Office and Smithsonian Institution began to develop a climatological view of the continent, understanding that "a knowledge of the peculiarities of the climate of a country is an essential requisite for the adoption of a system of scientific [agri-]culture" (Smithsonian Institution 1864:31).

Around that time, in 1858, the U.S. Agricultural Society (1858:37) reported that the Patent Office also partnered with other organizations to circulate forms to U.S. diplomats and merchants abroad with the aim of formalizing intelligence on "economical plants, growing in the countries you may visit." Consistent with the work of comparative botanists and of

"acclimatization" societies focused on the distribution (and transplanta-
tion) of plants (Osborne 2000), reporters were told to describe, among
other features, crops' "periods of sowing and harvesting, the character of
the soil and its elevation above the sea, the mean, maximum, and minimum of the thermometer, and the amount of rain, in inches, each month
of the year, together with the periods of the latest spring and earliest autumnal frosts." By obtaining this data, agricultural reformers aimed to
compare, evaluate, and establish the attributes of climate zones. The effort
of comparative agricultural climatology, haphazard at first, envisioned
that climate knowledge would ultimately facilitate the proper economy of
inputs, agricultural productivity, and trade.

The relationship between the Smithsonian and the federal government
propelled Henry's vision of a national, agriculture-centered meteorology.
On the eve of the Civil War, on January 25, 1860, Joseph Henry wrote to
the U.S. commissioner of patents, W. D. Bishop, in the preface to the joint
Smithsonian-Patent Office *Meteorological Observations*: "The results are
considered as furnishing very interesting and valuable statistics, not only
in regard to the science of meteorology, but to that of agriculture, which
will be of increasing importance in determining the climatology of different portions of the country" (U.S. Patent Office 1861:iii). The budding
expertise of the climatologist was central to developing this vision into a
formal program.

The Emergence of "Positive Climatology"

An important initial effort to formalize the project of climatology can be
found in the work of statistician Lorin Blodget, who worked for a range
of federal government organizations. In *Climatology of the United States*,
published in 1857 and compiled using Army Medical and Smithsonian
meteorological records, Blodget demonstrated a novel approach to science
that he labeled "positive climatology." The positivist approach would reject
climate-change theory and instead advance statistical-geographic representations of the "permanence of special features," meaning spatially demarcated, stable climates. Blodget (1857:vi, ix) organized statistical tables
and maps with a goal to provide "a general discussion of the records from
all sources in the sense of a CLIMATOLOGY" and to verify the statistical

record for "the purpose of using it as a valuable approximation to the various fixed quantities of climate." Blodget contrasted "the illustration of fixed or average conditions" as standing "against the general opinion" and other "historical absurdities and extravagances," alleged to result from "blendings of sagacity and charlatanism which have been always busy in prediction [of] the weather" and provoking anxiety about climate change (1857:vii). If the two-thousand-year time series of climatological records were as available in America as in Europe, Blodget stated, "it would be found to dissipate the apprehensions so frequently entertained that it is becoming more variable or extreme" (1857:24–25). Climate change had no place in a positive climatology.

Blodget's program instead denoted the "permanence" of climate with reference to monthly averages of atmospheric measurements, which upon representation could serve as the basis for regional economic development. "Climatology," in its instantiation by the 1860s, signified claims to expertise about the atmosphere as relevant primarily to the emerging concern to establish scientific agriculture. Climate, thus configured, could provide solutions to regional disparities in "rural economy" (Smithsonian Institution 1859:34), establish the limits to profitable development, and replace meteorologists' work of monitoring climate change with geographic description and comparison.

From the time that Blodget wrote until the mid-twentieth century, U.S. climatology continuously positioned inquiry against concerns about climate change. In 1889 meteorologist (and Weather Bureau official) Cleveland Abbe argued in "Is Our Climate Changing?": "It will be seen that a *rational climatology* gives no basis for talked-of influence upon the climate of a country." He asserted, "The true problem for the climatologist to settle during the present century is not whether the climate has lately changed, but what our present climate *is*, what its well-defined features are, and how they can be most clearly expressed in numbers" (Abbe 1889).

On the one hand, climatologists rejected climate change because they could represent large-scale visions of time and space that had not been available to previous generations of meteorologists. For example, statistician Charles Schott compiled a wide range of sources on temperatures across the United States and published an empirically exhaustive climatological time series in 1876 (see figure 5). Schott's tables were essentially

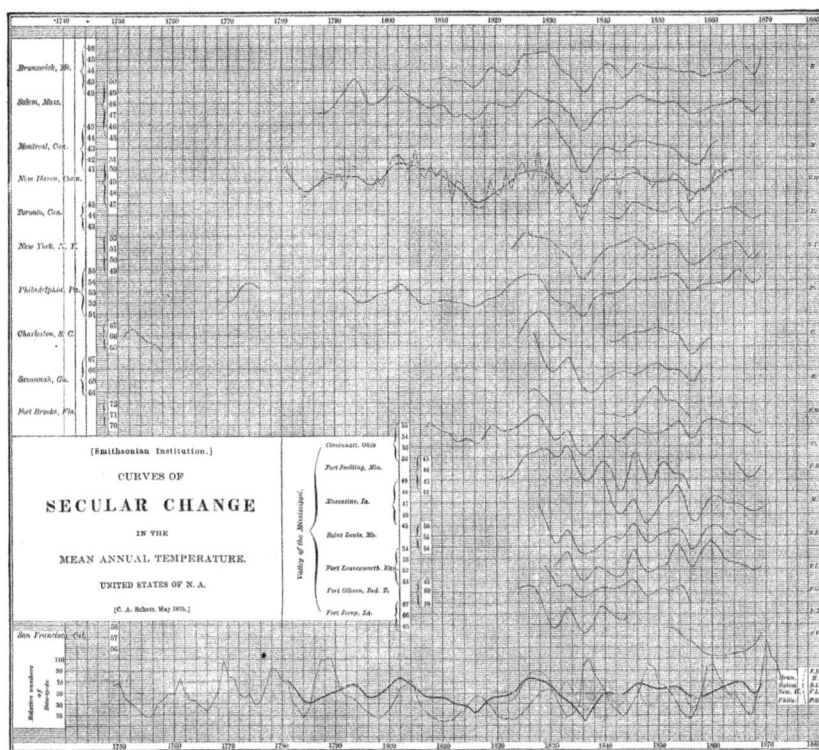

Figure 5. Statistician Charles Schott's "Curves of Secular Change in the Mean Annual Temperature of the United States" from 1740 to 1870. *Source:* Schott (1876:311).

descriptive—in line with a positive climatology—and they did not register climate change patterns. On the other hand, the vision among climatologists to "fix" in a statistical sense the parameters of climate across space and time cannot be reduced to a logic of empirical discovery. Rather, it articulated broader political and economic interests.

Establishing climatology across the United States was beset by significant material and administrative challenges that affected climatologists' capacity to secure the standardized meteorological data they required. The Civil War destroyed critical elements of the Smithsonian and Army Medical Department networks of stations, instruments, and observers, and the social, telegraphic, and mail systems that meteorologists relied upon to formulate climate knowledge. Moreover, the Civil War facilitated profound changes in the regional political economy of the United States, featuring industrial development in the North, continental expansion (especially via the railroads), and political integration in the West, and Reconstruction in the South. It was under conditions of rapid societal change that meteorologists and their allies successfully institutionalized positive climatology within government.

SECTIONAL DEVELOPMENTS: DEFINING REGIONAL CLIMATES AS COMPETITIVE ECONOMIC ZONES

How to restructure the post–Civil War economy became a central concern for those already aiming to designate climatic areas with reference to agricultural production and trade. To take one example, the oceanographer Matthew Fontaine Maury, famous for identifying Atlantic shipping lanes and a system of maritime meteorology, sought to rebuild his native Virginia after the Civil War. In his antebellum work, Maury asserted that meteorology provided a way to economically manage the atmosphere as "a grand machine—perfect in all its parts, wonderful in its offices, sublime in its operations" (Maury 1856, in Corbin 1888:76). The atmospheric "economy," he argued, showed that "supply and demand are in as rigid proportions here as elsewhere" (Maury 1846). The task of meteorology, in this view, involved harnessing the "economy" of nature to the economy of human affairs. Maury sought to reimagine climatically tailored means of economic reconstruction: Thus, "considering the circumstances under which recent events have placed the people of Virginia," he performed "an economic study of the geographical position of the State," with the goal "to develop the physical resources of the State and to point out the great commercial advantages which naturally arise from its situation . . . to the end that industry may be stimulated" (Maury 1869:3, 6).

With a decimated, slave-based plantation economy, labor shortage, and competition from western agrarian marketization, agricultural market development was especially significant for regional growth. Maury worried that railroad development would leave Virginia in economic ruin while other areas prospered from increased immigration and export-oriented trade relations established on more favorable terms. He advanced a detailed account of possible canal- and rail-building schemes to connect the Mississippi Valley, through Virginia, to the Atlantic Ocean, which he labeled the "great highway of nations." He argued that existing transportation routes constituted "a violation of the laws of political economy," marshalling evidence from weather and insurance records to insist: "Two and a half per cent upon the value of all the commerce that has sought a passage [east] from New Orleans and Mobile since the purchase of Louisiana, surely amounts to more than $100 million, and the use of these Virginia routes would have saved much if not all of it" (Maury 1869:34). Maury continued by arguing that insurance and shipping industries were overly susceptible to weather and climate risk. In a context of economic competition in agricultural markets and the development of trade infrastructure, Maury constructed one of the first "climatologies" of Virginia in hopes that it could spur the Reconstruction economy.

The process of performing a "climatology" for a given area of administrative-economic interest was replicated widely by other U.S. states and in the following decades. Knowing climate became especially relevant to establishing productive and profitable regions and to evaluating the risk of financial investment in infrastructure to produce goods and transport them to distant markets.

Within this commercial context, on February 9, 1870, President Ulysses S. Grant signed the joint congressional resolution that established the Division of Telegrams and Reports for the Benefit of Commerce within the War Department's Army Signal Service. The division required the secretary of war "to provide for taking meteorological observations at the military stations in the interior of the continent, and at other points in the States and Territories . . . and for giving notice on the northern lakes and on [Atlantic] seaboard, by magnetic telegraph and marine signals, of the approach and force of storms" (U.S. Signal Service 1873:368–69). This resolution is widely considered the origin of national weather services in the United States.

State investment in facilitating commercial trade through providing "notice" of the "approach and force of storms" articulated with other ways of governing social and commercial engagement with climate. For commercial specialization, climate knowledge provided a metric of tailored

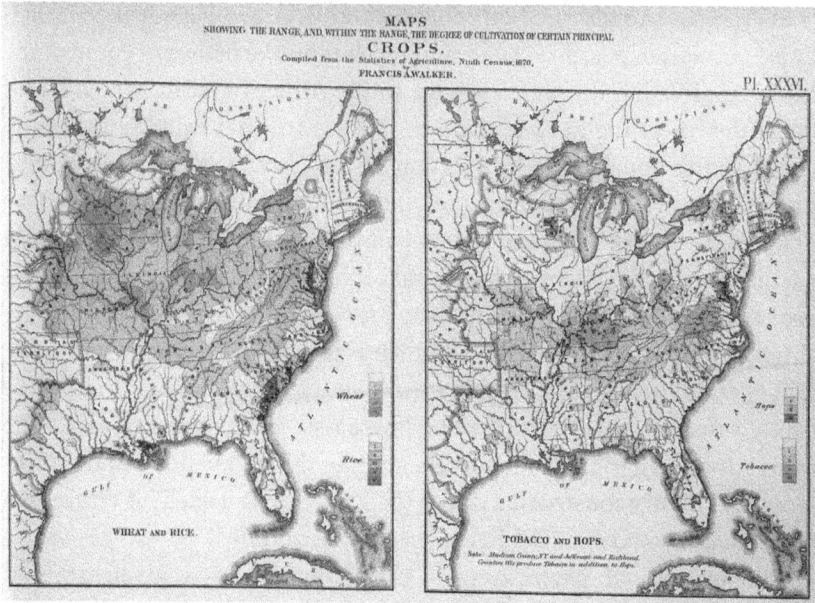

Figure 6. Plate 36 of 1874 *U.S. Statistical Atlas*, showing ranges of crop extent and cultivated yields. *Source:* Walker (1874).

economic development that could be integrated into other statistical analyses of crop yields, labor conditions, and market access. The development of the U.S. land censuses and statistical atlases, in particular, created government data designed to inform agrarian economic and land development (see figure 6).

Efforts to make legible climatic zones for the purpose of commercial agricultural expansion relied on simultaneous efforts within scientific and state institutions. President Abraham Lincoln had established the U.S. Department of Agriculture (USDA) in 1862 alongside the Homestead Act. The USDA formed the Office of Experiment Stations in 1888, following a decades-long struggle over whether and how the federal government would provide services for advancing scientific agriculture. The USDA *Farmer's Bulletins*, regularly published by state experiment stations beginning in 1889, routinely included government climate data in pronouncements concerning the climatic zones suitable for profitably growing crops (Dalrymple 2009; Rosenberg 1997). Around the same time, U.S. states established their own weather bureaus to supplement national data collection and dissemination. For example, an annual report of the Oregon State Weather Bureau (in U.S. Signal Service 1889:5) justified its establishment by claiming that "the climatology in the State has never been observed and recorded in three-quarters of the State, as is necessary for the best promotion of the various industries." Facilitating climate-specific commerce became a clear governmental priority in order to mitigate economic risk: "Now [the farmer] employs implements and machinery which can be made only with large capital and the highest mechanical skill, and by men who make this manufacturing a business." Therefore, "research—the finding out of nature's secrets," is "costly," meaning "the more useful it is to be, the greater must be the outlay of money, labor, and scientific skill. Here, if anywhere, wise economy calls for the best" (1889:6; see also Harrington 1895). "Wise economy" within science and government linked the logic of climatology directly to its commercial application.

Following the logic of commercial agriculture, in 1890 President William Henry Harrison signed into law the transfer of government weather services from the War Department to the USDA, establishing the U.S. Weather Bureau and linking climatological expertise to the task of commercial development. Climatological expertise, once established in government, helped shore up a central principle that "the crude method of tilling the soil common in these days will certainly give way to an exact economical procedure, based largely upon the result of meteorological research, increasing in precision" (Bigelow 1900:85).

In reality, expertise within government weather services often had a complex, sometimes outright antagonistic, relationship to alternative formations of climate knowledge. One example comes from optimistic boosters of settlement on the drought-prone Great Plains. Concerning railroads—as a fixed capital investment—boosters accentuated local climatic niches accessible by rail lines. For example, one Northern Pacific Railroad (1893:13–14) guide advertised: "This great Northern Pacific system of railroads has opened to settlement, during the past few years, one of the fairest sections of the country—a region exceeded by no other . . . in its wealth of natural resources, and not surpassed in any of the conditions of climate or of soil which are best adapted to the well-being of the human race." Although such pronouncements characterized climates of newly settled areas as well-suited for development, railroad companies, settlers, and the U.S. government poorly understood western arid regions.

In the space of uncertainty and boosterism, some considered climate less important to agricultural productivity, for example those adhering to dry-farming techniques (Libecap and Hansen 2002). Hardy Webster Campbell's *Soil Culture Manual*, for example, renders climate subservient to techniques of managing soil moisture (Campbell 1902). Such an approach to climate is also evident in his 1916 publication, *Progressive Agriculture: Tillage, Not Weather, Controls Yield*. Yet those who oversimplified or denied the reality of the arid climate had by that point already been partly responsible for severe crop failures and exodus from the Great Plains during unanticipated drought (Sweezy 2016).

Whether emanating from the Weather Bureau, commercial enterprises, or those who would challenge official constructions of climates, the meanings of climatic designations were primarily oriented to the economic productivity of specific regions. Climate change mostly did not register as significant. A stable climate did. If climate itself could not be "improved" (meaning, changed), as upheld in other contexts, then stable climate zones could still support auxiliary efforts to improve soil, land, water, and settlement patterns. Improvement, in practice, entailed federal policy that could protect access to land and resources understood to be climatologically useful for given purposes. Policies of land improvement included forcible relocation of Native people to reservations in the 1870s and 1880s, exclusion from designated wilderness reserves and settler areas, and destruction of indigenous land tenure through the institution of private property. Especially significant is the Dawes Act of 1887, which reallocated many tribal landholdings to individual Natives and settlers.[3] These processes expropriated indigenous land use patterns (Warren 2002), and as Whyte (2013, 2017) has shown, ruptured indigenous relationships to climate. Climatology thus formed one dimension of the larger governmental logic, at once rational and violent, of making legible and fixing in place the socioeconomic relationship to territorialized climate zones.

MAKING NATIONAL ADMINISTRATIVE CLIMATES

Climatologists learned to relate their accounts of climate "zones" to zones appropriate to bureaucratic administration. Making climates into administrative categories could facilitate government provision of climate information, which users regularly praised as a "vital necessity for the protection and advancement of commercial and agricultural interests" (U.S. Signal Service 1878: 36). Willis Moore, chief of the Weather Bureau, thus prefaces Alfred J. Henry's (1906:5) major publication, *Climatology of the United States*, by stating that the text provided important "comparative climatic statistics for the different portions of the United States." As Henry explains, the Bureau of Plant Industry (like the Weather Bureau, within the USDA) was introducing newly researched seed and plant varietals, meaning that the department must map climates so that "the new plant or seed be placed in a climate closely resembling that of its original habitat." Plans to promote agriculture across the department led Henry to conclude that "the ideal census of climatology, so to speak, is one that shall give the essential features for every county in each political division," in this case to facilitate the USDA's task of distributing seeds and managing programs based on administrative and climatological units.

The task of representing climates as "essential" characteristics of administrative units proceeded beyond the issue of seeding crops. At the state level, the USDA experimental stations, state weather services, and boards of trade regularly communicated with the federal-level Weather Bureau. State-level installations undertook the task of linking administrative governance to suitable observations or representations of local conditions, with the explicit aim of defining state-level climate. How was climate represented in such a context? Ward (1915) discusses how Weather Bureau officials developed twenty-one "climatic subdivisions" representing "all groupings of districts and stations for convenience of administration, of forecasting, of the collection of data, or of reference." Over the first decades of the twentieth century, the division of climatological areas changed because of "practicality rather than on homogenous climate considerations" (Guttman and Quayle 1996:294; Ward 1915; U.S. Weather Bureau 1912).

Practical reasoning and administrative convenience notwithstanding, representations of climatological zones firmly shaped common

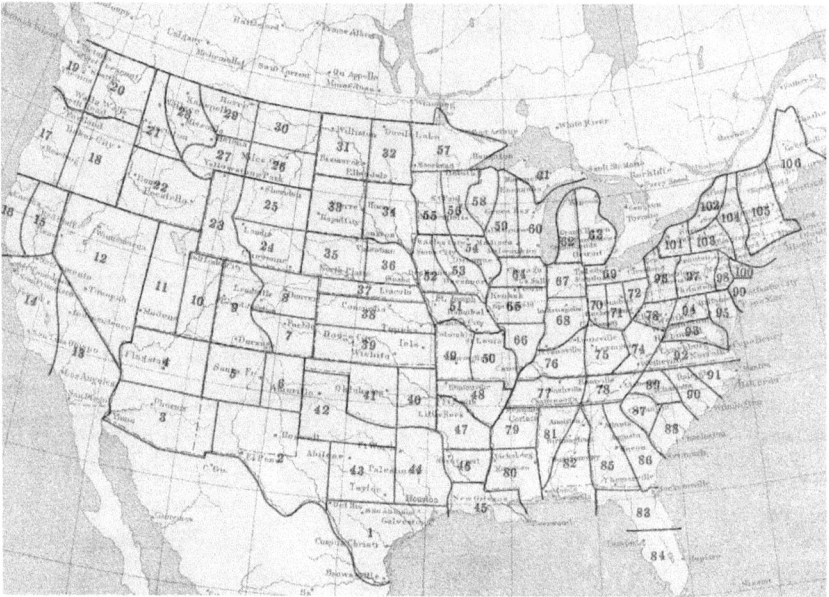

Figure 7. "106 Climatological Sections in the United States." *Source:* U.S. Weather Bureau (1911:Supplemental Charts 2:670).

understandings of climate. They formed the basis for Weather Bureau studies, reports, and narratives that could structure popular discourse and regional economic strategies. Synopses printed in the Bureau's *Crop Bulletins* and in the *Annual Yearbook of the USDA* (e.g., USDA 1906:473–91), for example, draw upon "normal" statistical averages to narrate "departures from the normal by districts." Seeing climate like a state meant that the administrative form supplanted prior understandings of the more open times and spaces that climate had represented.[4]

The invention of climate zones naturalized categorical differences (e.g., see figure 7; see also U.S. Weather Bureau 1912). Pierre Bourdieu (2015) aptly labels this process "state magic," that is, recording reality while consecrating its existence. Similar to how climatologists in other contexts had to confront the "real" versus "invented" nature of their purportedly empirical representations of climates (see Coen 2018), Ward reflects on the arbitrary nature of "official" U.S. climatic designations: "There is no limit to the number of possible classifications, for these depend on any author's special interest or view-point, which may be climatic, or botanical, or physiographic, or one of administrative convenience. Even from the single view-point of climate alone, an almost infinite number of classifications might be proposed" (Ward 1915:675). If such divisions of climate and their relatively stable features appear to make sense, either to Weather Bureau officials or to later generations, it is through the internalization of an administrative logic of territorial government—represented by a superimposed map of the United States—not through the stability or geography of climate itself. The positive climatology proposed by Lorin Blodget in 1857 had come to bear on representations of the climates of the United States. As stable units became linked to administrative units, climate became legible to government actors. The state, because of the legibility of climates, could then *act* upon them: improve them, valuate them, and otherwise facilitate their integration into capitalist society.

PLAUSIBLE ALTERNATIVES? MEDICAL CLIMATOLOGY AS A PATH LESS TAKEN

Could actors in this historical context plausibly have defined climate differently? The case of medical climatology provides some nuance, showing how U.S. climatology emerged in part through professional struggles. Around the turn of the twentieth century, some of those who identified with "climatology" were invested in alternative professional possibilities. In 1891, a diverse group of scientists founded the American Climatological Association. The stated focus of its journal, *The Climatologist*, was "all matters relating to Climatology, Mineral Springs, Diet, Preventive Medicine, Race, Occupation, Life Insurance, and Sanitary Science." The group's mission was to "become eventually an authority upon all subjects which are included in its title" (meaning *climatology*), especially by attempting to synthesize medical geography with bacteriology (American Climatological Association 1892:1). The association's formulation of climatology did not subscribe to a view of stable climates, but rather built upon the previously dominant medical-topographic tradition. While rejecting a narrow view of climate, the association also sought to retain an otherwise weakening paradigm of medicine. As one contributor, Dr. Isaac Platt, put forth in an article on acclimatization: "Nearly all the physical influences surrounding a human being . . . are those which go to make up what we know as climate. The air he breathes . . . the emanations from the soil; the water he drinks; the food he eats; . . . In fact, the climate to which he is subjected is practically co-extensive with his environment" (Platt 1886:104). The holistic "environmental" approach to climatology that would maintain a connection to medical, human, and environmental sciences failed to define the field.

To be sure, policies regarding "sanitary meteorology" and "climatic physiology" persisted into the twentieth century (Meisinger 1921; Edson 1921), just as tropical medicine synthesized climate and medical sciences in equatorial colonial contexts. Adolphus Greely (1888:4), a former Weather Bureau chief, had understood that industrial expansion and urban public health relied on meteorology to evaluate "the fitness of local climates as a means either of extending the scope and extent of national industries or of alleviating human suffering and saving human life." A Weather Bureau circular in 1895 encouraged submission of vital statistics by bureau service members and their volunteer networks to evaluate how "local climatic peculiarities" impacted diseases and to apprehend where invalids and other "health-seekers" might best benefit from "visitation of health resorts and change of climate" (see Michigan Weather Service 1895:4). As Valencius (2002) shows, although medicine was becoming less central to the work of professional climatology, the business of health-seeking resorts remained popular in the latter decades of the nineteenth century. Yet by 1895, as one physician, F. R. Campbell (cited in Mitman and Numbers 2003:398) disparagingly wrote: "Etiologists have at present almost given up the investigation of atmospheric causes of disease. . . . They insinuate that the study of medical meteorology is a subject redolent with the ignorance of the Dark Ages." Medical education and physicians' practice from that point forward decentered medical climatology from accounts of most diseases.

The situation of medical climatology exposes a social contradiction in the designation of "stable" climates within U.S. capitalist society. Although climatology generally served to make legible "stable" climates, profit-seeking created novel, class-based experiences of climate. Government with respect to "good" and "bad" climates reflected social inequalities. On the one hand, mobility of the rich entailed a newfound capacity to obtain "pure" air at resorts or by residence (Mitman 2003). Taylorist attention to labor productivity, on the other hand, meant that the novel climates of industrial factories elevated the significance of regulating what science writer Harry Mount (1921) calls "indoor meteorology" that "makes all the difference between industrial success and failure." When Mount (1921:188) outlines new air conditioning technologies that reduced the sweltering heat of textile factories, he reports, "It is possible to speed up machines and workers alike and at the same time to lessen the hardships imposed on both, simply by the use of manufactured atmosphere." Governing the deteriorating air quality experienced by workers centered on productivity at the time when the consumption of pristine environments became a prerogative of the leisure class (Cronon 1995; Simpson 1992). The changing, socially stratified relationship to atmospheres notwithstanding, climatologists primarily centered around climatic stability, which matched dominant economic and administrative concerns.

THE ENDURING LEGACY OF CLIMATIC "STABILIZATION"

One of the major consequences of positive climatology was that the central discourse of climatic stability supplanted concerns with climate change. In his 1924 textbook *Climatic Laws*, climatologist Stephen Visher presumes geographic distribution as the basic boundary around the project to establish "laws of climate." Not unlike the "laws" of economic science and the "laws" of "social physics" for early positivist sociologists, the laws of atmospheric circulation had often eluded scientists, who even in Visher's time looked at astronomers and other physicists with outright jealousy.[5] Yet what about climate was law-like? Except for Visher's first stated law, that "climate changes with the nature and effectiveness of solar radiation," the remaining forty-nine laws dealt with geographic (as opposed to temporal) distributions. For Visher, along with his students and contemporaries, the fruits of Blodget's (1857) declaration of positive climatology had achieved paradigmatic status. What Blodget had to make explicit, Visher could by and large take for granted. The major shift that took place within science and government led to a basic shift in the very meaning of climate and what it meant to govern society in relation to climate.

The coproduction of climatic stability in part explains why it became so difficult for climatologists in the United States, along with those who came to inhabit industrial-capitalist climates, to consider climate change beginning in the mid-twentieth century (Baker 2017; Henderson 2014). Although this issue is pursued in chapter 5, it is already clear how "climate change" could upset scientific practices and government interests insofar as they operated on the assumption of an unchanging climate. The achievement of plans to govern climate with reference to stable times and spaces was a historical development, rather than the outflow of an a priori unchanging climate alone. Thus, to alter the logic of climatic stability—which later climate scientists would try to accomplish—would entail shifting the basic logic of climate knowledge, and with it, the logic of meteorological government. Climatologist Hubert Lamb (1959:299) therefore narrates that a shift occurred over his career during the first half of the twentieth century that came to denigrate climatology as "the dry-as-dust book-keeping branch of meteorology" because climatologists treated climate as "normal" and "static."

Before turning to the rise of climate change science in the mid-twentieth century, however, it is first important to trace a set of historical developments parallel to those addressed in this chapter. The following chapter focuses on weather as a problem of political economy. A focus on weather events and weather prediction reinforces the preceding account of efforts to govern climate under the logic of capitalist society. Rationalizing weather reinforced the basic view among meteorologists of climatic stability insofar as climate signified the unchanging foundation against which they could understand, predict, and evaluate weather fluctuations. By making weather economically legible, meteorologists enabled market society to rationally incorporate weather patterns and extreme events, even as the process relegated climate further into the "normal," unchanging background.

4 Economic Rationalization of Weather

RISK, PREDICTION, AND "NORMAL"
WEATHER, 1870–1930

Chapter 3 demonstrated that climatology, commercial agriculture, and government administration developed together in an expanding, postbellum United States. In the process, the logic of climatology reoriented the meaning of climate within science toward a geographically delineated and temporally stable object. While climate change theory gained new momentum in geologic debates about long-term glaciation cycles, and while concern persisted about human-caused deforestation and aridification of dry and semiarid lands, climatic stability was the functional paradigm in U.S. climate science. However, it would be remiss to approach ideas and practices regarding the stability or instability of climate without attention to the more proximate causes and consequences of how climate knowledge is made. Sociologist of science Paul Edwards (2010) has called contemporary climate change science a "vast machine" held together by a complex material and social infrastructure that connects people, national governments, satellites, computer processing centers, mathematics, and much more. Knowledge, as scholars in the tradition of actor network theory generally hold, is *assembled* though such networks. Take away the network (or, the vast machine), and the ideas and the science fall apart too. In this chapter, I take this basic principle and use it to consider how ideas about the stability of climate were built and maintained as well as fragmented, challenged, and always partial in their reach. What *else* had to be stabilized? As I emphasize, infrastructure had to be maintained, time had to be standardized, meteorological observations and reporting had to be disciplined, and authority to interpret weather patterns had to be professionally monopolized. In short, the cacophony of atmospheric motion and many elements connected to it had to be rationalized.

To trace the rationalization of climate and weather knowledge, this chapter explores how meteorologists worked to assemble a vast array of material elements and human actions, made legible in standard units. Constructing a nationwide *data infrastructure* of weather observation was not simply a logistical issue. It succeeded through meteorologists' capacity to link their professional goals to state and economic interests. In turn, weather prediction served to rationalize weather as a set of economically legible elements, especially relevant to the domains of insurance, finance, trade, and public infrastructural investment.

Moving forward, the first issue addressed in this chapter is the difficulty government-centered meteorologists faced in securing the telegraphic circulation of weather observations, beginning with the origins of the U.S. Signal Service around 1870. Such a challenge involved a meteorological *technopolitics* (cf. Hecht 2011), marked by struggles between those who would stabilize the meteorological data infrastructure and those who, for one reason or another, would subvert it or deny its significance to weather knowledge. The second issue addressed in the chapter is the social organization of "standard" time across geographic space, an invention that permitted meteorologists to develop a standard view of weather and an account of climatic stability. Third, the chapter focuses on the formation of *meteorological consumers* that adopted formal discourse regarding normal versus abnormal weather. Finally, the chapter shows that weather reporting and forecasting served to *economize* weather events, that is, made them fictitiously commensurable with dollars and cents. Weather was experienced in new ways as extreme and damaging, through which weather risk and "disaster" entered public consciousness and economic rationality (Bernstein 1998; Steinberg 1991). In contrast, for meteorologists, government, and the public, climate became the stable statistical-geographic background to a progressive, if tumultuous, socioeconomic drama.

TECHNOPOLITICS OF THE METEOROLOGICAL NETWORK

The 1870 legislation authorizing a national weather service initially mandated the development and distribution of storm warnings. The Signal Service of the War Department held the responsibility for organizing and issuing such warnings. Tracing the work of the Signal Service helps to show that, despite a political context of demilitarization and federal government withdrawal (especially after Reconstruction), climate science, practices of government, and atmospheric processes coproduced one another.

From Military Signals to Meteorological Data Infrastructure?

After the Civil War, "signaling," or the practices and techniques of coding, making, displaying, and interpreting visual and related signals (chiefly for military communication), tied the fate of the Army Signal Service to the possibility of instituting national weather services. At this time, Chief Signal Officer Albert Myer, whose method of signaling had become standard for Union forces, published a revised *Manual of Signals*. Myer was professionally invested in maintaining the organizational existence of the Signal Service, which his superiors threatened to decommission. Given this uncertainty, Myer mobilized his ties to War Department officials and meteorologists to defend the importance of a corps of men trained in standardized signaling techniques: "The actions of the late war, in which victory or defeat has sometimes hung upon the transmission of a signaled message, have rendered it certain that military signals will be used in the future military and naval operations, of our arms. They will be employed in the continued Indian warfares in the Interior" (Myer 1866:v). Myer feared the loss of training facilities, trained signalmen, and the network of signal officers and materials that upheld not only his career but also the very relevance of signaling to government.

The issues that Myer faced in the postbellum military bureaucracy arose in relation to a previously separate problem in the burgeoning shipping industry. With a dramatic increase in rail and Great Lakes shipping tonnage, individuals connected to industrial production and trade experienced weather in new ways—specifically, as a *risk* to the circulation of

WEATHER SIGNALS

ADOPTED FOR GENERAL USE BY THE SIGNAL SERVICE ON AND AFTER MARCH 1, 1887.

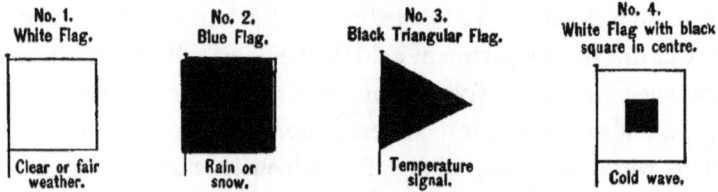

Figure 8. "Weather Signals" resembling Myer's military signals and promoted here for use and display by railroad companies. *Source:* U.S. Signal Service (1887:99).

capital and of agricultural and manufactured goods. Weather became a serious business matter. Astronomer (and later chief meteorologist of the Signal Service) Cleveland Abbe and scientist Increase A. Lapham had been actively seeking funding from corporate entities, including the Chicago Board of Trade and the Bureau of Lake Underwriters, to establish a storm-warning system for the hazardous Great Lakes shipping routes. To this end, Lapham published his treatise "Disaster on the Lakes," calculating that for 1868–1869 alone there had been losses of over $7 million, 530 people, and 231 ships.[1]

Myer was able to link the problem of costly storms to the tools of military signaling (as depicted in figure 8). He also argued that the Signal Service was the government entity that could take up the problem that meteorologists were seeking to coalesce their scientific work around— namely, how weather patterns and storms form and move (Fleming 1990). Economic historian Erik Craft (1999) argues that private corporations, shipping underwriters, and boards of trade were reticent to provide weather services for clients or the public because the costs might outweigh private benefits. Yet producers, distributors, traders, investors, and meteorologists alike increasingly viewed weather as a fundamentally economic problem.

In 1870, Congress authorized the Signal Service to function primarily as a weather service, initially on the Great Lakes and Atlantic Coast shipping routes, before extending such service to other areas and including regional forecasts relevant to specific crops and transport industries across the country. Government investment in formal weather services, Craft (1998, 1999) calculates, resulted in immediate increases in market efficiency, especially for Great Lakes shipping and, by extension, the financial facilitation of agricultural markets.

Bricolage Bureaucracy and Action-at-a-Distance

In the 1870s, meteorologists working for the Signal Service designed methods for collecting meteorological data and disseminating the storm warnings they were now mandated to provide. Such methods called for unprecedented coordination among people, meteorological instruments, and information dispersed across the country. In general terms, a fundamental struggle for those in government involves the mundane capacity to implement orders across space and time. Bruno Latour (1987) calls this capacity "action at a distance." Such actions characterized Signal Service meteorologists' power over storms. To understand why this is so, let us consider the basic structure of weather observation, reporting, and forecasting.

In the early 1870s, the Signal Service began to utilize a diverse network of people, called "observers" who took instrument readings at established times and transmitted them, via telegraph, to the Signal Service Central Office in Washington, D.C. There, signal officers and their staff interpreted the diverse weather reports and created the daily weather "synopsis and probabilities," later called the "forecast." The forecast for the upcoming and following day could then be transmitted back through the telegraph network to diverse organizations, including especially local newspapers; regional boards of trade and exchange; railroads; and scientific, commercial, and farmers' organizations. Time was of the essence. The forecast window was short, given a relatively poor understanding of storm patterns. So if the network was faulty, then storm warnings and forecasts would be late and useless.

The difficulties in securing the data infrastructure that comprised the meteorological network in these terms were technical as well as social. Let's start at the periphery of the network. Keeping in line the geographically dispersed human weather observers called for a kind of soft power.

Weather observation in the Signal Service (and later the Weather Bureau) was a patchwork social process, forming something of a "bricolage bureaucracy." The weather observers' assemblage consisted of a range of informants who often held divergent interests. For example, a standard retrospective review of a month's weather patterns may provide a list of observational sources, like the following published in the Signal Service's *Monthly Weather Review* in April 1875 (U.S. Signal Service 1875a:1):

> The present weather review expresses the main features of the April meteorology, as deduced from the following reports and records:
> Reports from 88 Stations of the United States Signal Service
> Reports from 10 Stations of the Canadian Meteorological Service
> Reports from 267 Volunteer Observers
> Reports from 1 United States Naval Hospital
> Reports from 27 United States Army Surgeons
> Records furnished by Private Observers, Marine Logs and the Press

Notice that most observations were recorded by volunteers. Farmers were the most common "voluntary observers" (later classified "cooperative observers"). However, unlike army surgeons, volunteers did not always perform the systematic observation demanded by Signal Service forecasters in Washington.

To shape observers' behavior and maintain a network of weather observation, Signal Service agents strategically engaged local economic interests and existing organizations. For example, an early Signal Service (1872) manual, *Instructions to Observer Sergeants*, detailed how signal agents needed to first learn to construct effective instrument housing to assure "their protection from local influence." After securing a recording station, each agent was ordered to "as soon as practicable, put himself in communication with the board of trade, chamber of commerce, board of underwriters, and such other bodies as may desire to co-operate with this office in the efforts to make the service useful." Forming "meteorological committees" within existing such organizations was to "be urged as a matter of special importance" that could support long-term, continuous communication with the central Signal Office (U.S. Signal Service 1872:6). Subsequent records indicate vast success in such recruiting efforts, judged by the growing list of participating cooperative organizations listed as hosting weather stations in annual reports of the chief signal officer. Yet the initial work in the field by signal agents remained tedious, especially since until the 1880s the Signal Service lacked extensive educational resources that could standardize communication of meteorological knowledge and techniques.[2]

With crucial but limited bureaucratic control, national weather services were backed by a political ideology of civic democracy. "Cooperative" volunteers at times comprised over 90 percent of weather station personnel (Marvin 1896:556). Observers were typically portrayed as "public-spirited persons [working] for the benefit of their communities" (Calvert 1931:iii). Yet administrators needed to forge a delicate balance between a vision of voluntary civic participation and a bureaucratic structure that could sustain expert rigor and social control.

It is telling that meteorologists who initially sought to establish the official national weather service argued that the Department of War was the logical home for a storm-warning system. As the congressional resolution forming the weather service stated, "Military discipline would probably secure the greatest promptness, regularity, and accuracy in the required observations" (reprinted in Raines 2011:43). This appeal to discipline had long marked a rationale that military organizations could best provide systematic weather observations, both in the United States and other countries (U.S. Signal Service 1872:399; see chapter 2). In 1884, for example, an unnamed Signal Service historian (U.S. Signal Service 1884:3) argued, "An economic feature of the Weather Bureau is that it is a *military service*, and disciplined observations could not have been secured by a civil corps." Later retrospective accounts by meteorologists also attributed the evolution of weather services to "the application of discipline of military exactness" that coordinated observers, offices, and meteorological infrastructure "into an efficiently working machine" (U.S. Weather Bureau 1908:xvii).

For meteorologists to generate storm warnings and forecasts, the "machine" had to run continuously and efficiently. The meteorological order represented or mapped by the Weather Bureau was preceded by social order and secured by power relationships. The Central Office in Washington was especially significant, for it operated as a national amalgamator of information. As one example, the 1883 *Regulations for the Signal*

Office depicted strict attention to bureaucratic organization at the Central Office. Any communication "not necessary to the proper discharge [of] duties" was strictly prohibited. Office rules forbade the use of offices for visitors or "entertainment"; "unnecessary conversations," "private letters," and newspapers were "strictly prohibited" during office hours (U.S. Signal Service 1883). Thus, administrators at the Central Office created a space of formal discipline, which through its minute level of social control powered the larger "machine" of weather observation.

How the Signal Service Central Office spatially organized its national offices demonstrates a similar attention to order as a prerequisite to producing successful storm warnings and forecasts. By 1880, 110 enlisted signal service members staffed twelve units, or "rooms," of the Central Office (Raines 2011:54). These rooms included one for correspondence and records, a telegraph room, a property room, a printing and lithography room, a room devoted to the *International Bulletin*, an instrument room, a map room, an artisan's room, a station room, a fact room, a study room, and a central library (U.S. Signal Service 1878:429). Through the rationalization of space, the Central Office organized the action-at-a-distance that held the data infrastructure together and synchronized the work of the observers and forecast distributors.

Peripheral Visions: Making Official Weather

From the Central Office to local sites of weather observation and reporting, governmental order faced challenges in making representations of weather "official," that is, flowing from an exclusive and impersonal authority. Although instruments could be standardized and telegraphic communication systematized, human observers, like the weather they were observing, did not always behave as expected. One range of problems stemmed from observers' consistent temptation to forecast or predict the weather. Reflecting this concern, the 1872 *Instructions to Observer Sergeants . . . on Duty at Stations* emphasized that *"observers must confine themselves strictly to the instructions issued from [the Central] office, and will not, under any circumstances, publish, or cause to be published, forecasts or predictions of the weather"* (U.S. Signal Service 1872:12). Creating one's own forecasts was not authorized: doing so would amount to a deliberate circumvention of official representations of weather.

Meteorologists had an especially difficult time securing order at the peripheries of the observational network. As one example, in the 1873 *Annual Report of the Chief Signal-Officer*, a field officer first reported possibly faulty results of river-gauge measurements, then raised the issue of a signal station in Cairo, Illinois, where the Ohio and Mississippi Rivers converge:

> The station has not been inspected since the date of last report. Sergt. Thomas L. Watson was relieved February 18, 1873, for neglect of duty, and was succeeded by Sergt. David Harnett, who left the station without authority, and was reduced [in] the ranks while absent. He was subsequently apprehended, tried, and found guilty of desertion, and is now serving out his sentence for that crime. Sergt. E. Garland was ordered to succeed him, and is now at the station. His reports are not promptly rendered, and it is intended to relieve him unless a change for the better is made. (U.S. Signal Service 1873)

Capacity to control even the most "disciplined" state agents also proved difficult. Indeed, physical examination at the Signal Service training program, at Fort Myer, in Washington, D.C., included identifying physical characteristics of staff, evidenced by meteorologist Henry J. Cox's recollection of enrolling in the Signal Service: "I found this precaution was taken so that I could be more readily identified should I desert from the Corps" (Cox 1991). Lack of promotional opportunities within the Signal Corps only made efforts to sustain a stable corps more difficult (Raines 2011).

Unreliability and limitations of human observation and labor also led to their rather steadily being replaced by self-recording instruments that could partially protect the data infrastructure from the limits and whims of human observers (see figure 9). Meteorologists and associated actors faced challenges in designing and manufacturing self-recording instruments, followed by difficulties with inspecting and maintaining them to ensure valid results. Because of these ongoing problems, coupled with the limitations of human observers, the Weather Bureau purchased and implemented automated meteorological instruments almost as soon as they were invented.[3] Problems remained. Bureau Chief Willis Moore (1903:3) implores observers not to transmit recordings beyond "plain facts" nor to falsify instrument recordings, reasoning that "attempt[s] to fill in missing records in a manner to make them appear as automatic and original, is liable to defeat the aim of scientific observation."

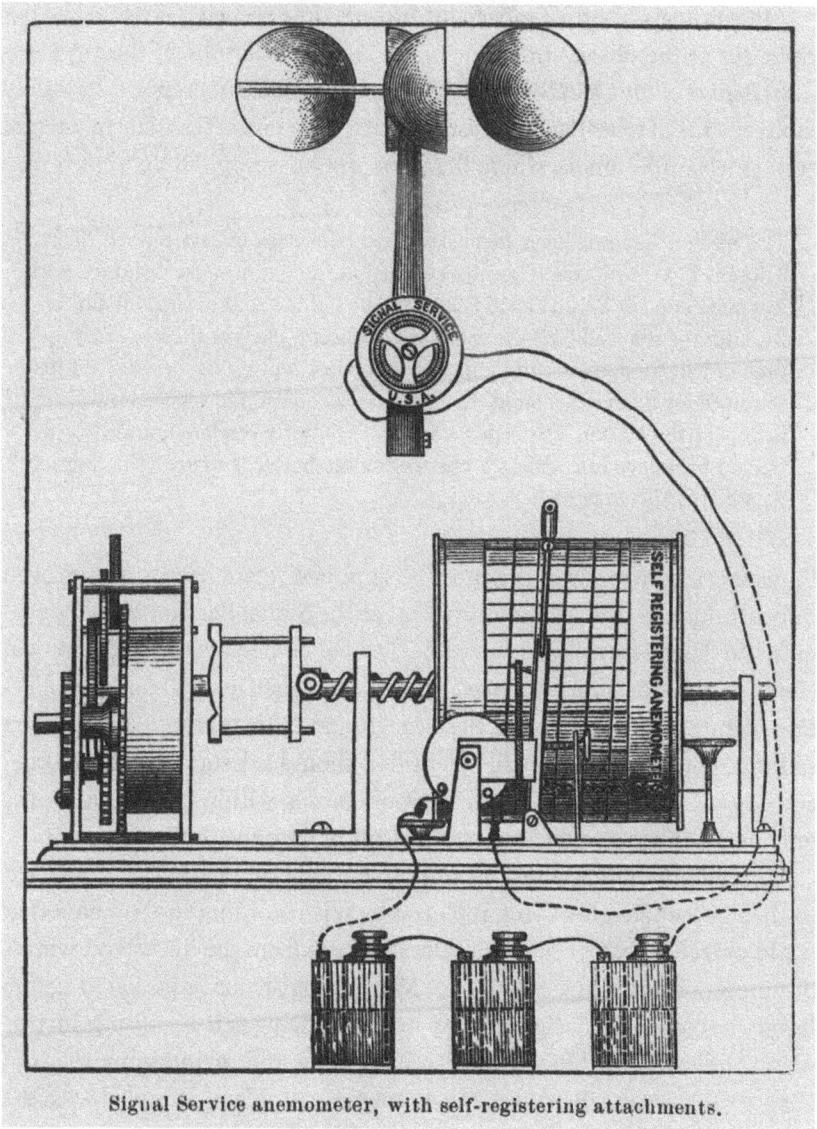

Signal Service anemometer, with self-registering attachments.

Figure 9. Diagram of Signal Service–issued anemometer with self-recording features. *Source:* U.S. Signal Service (1884:16).

If one aspect of technopolitics involved the stable, regular transmission of meteorological data, another took shape at the limits of state authority and meteorological knowledge in the U.S. Far West. Recall that meteorologists in prior generations had connected their concerns regarding climate change and disease to the governmental problem of expanding state territoriality. At the close of the nineteenth century, the sociotechnical network of signal officers and telegraphs similarly linked problems of meteorological observation to the governmental legibility of frontier society. Indeed, their observations were not simply meteorological; those tasked with meteorological observation also provided domestic surveillance communications, linking the meteorological network to a range of threats to socioeconomic order. As James Fleming (2000b) has shown, weather observers were charged with reporting to Washington on the status of railroad strikes and military confrontations with Native tribes.

The telegraphic lines themselves could become a site of political contention and technopolitics, especially for Native Americans seeking to counter U.S. facilitation of white settlement and appropriation of land from native people. As the *Annual Report of the Chief Signal Officer* in 1887 (U.S. Signal Service 1887:143) stated about Indian warfare and protection of settlers in the Southwest: "There can be no constructions more important for holding a frontier or protecting the first steps of advancing civilization than the telegraphic lines." The reporter outlined military tactics to protect war parties, military signal posts, and commercial outposts from Native American attack. Often, as the report stated and as Raines (2011) identifies more broadly, Native Americans strategically cut telegraph lines. For U.S. officials, the problem could be circumvented through circuitous networking of telegraphic lines and mindful positioning of new lines and military divisions, as well as a capacity to deploy manual (nonelectric) signaling in cases of telegraphic network interruption or failure. Therefore, the 1887 reporter (U.S. Signal Service 1887:143–44) expressed the fundamental links between the "connection of military posts," the "incidental protections the stations [provide] frontier villages" to advance commercial interests, and "the meteorological information" also tasked to the service. One Signal Service report declared, "As an engine of civilization, the frontier telegraph rivals the railway, enabling the Government to throw an aegis of protection over the rapidly expanding wave of western emigration," which the reporter argued was "facilitating, no doubt, the sale and settlement, as well as the material development of the public lands." (U.S. Signal Service 1884:24–25). In broad terms, then, meteorologists worked to make "official" weather by calling into being a technopolitics of infrastructure, comprised chiefly of an array of humans and nonhumans variously aligned with the goals of meteorologists and allied state actors.

STABILIZING TIME AND SPACE

The material stability of the infrastructure was but one dimension of the larger process of rationalizing commercial activity through weather and climate knowledge. Meteorologists' efforts to stabilize time and space also encountered challenges but were equally important to a rationalized view of weather.

Until the mid-1880s, the United States was in no sense characterized by "standard time." Time, for the purpose of recordkeeping, scientific activity, or social coordination, had previously been established by local observatories and later developed around the demands of "railroad time," itself initially a company-specific patchwork of "times" (Bartky 2000; Zerubavel 1982). This situation presented a serious problem for meteorologists, whose capacity to organize action-at-a-distance relied on temporally standardized measurement across space and time-sensitive data transmission. Only through reimagining time in relationship to weather could observations make a national view of weather patterns and, as I show, of the temporally "fixed" nature of climate.

The problem of standardizing meteorological observations across the United States was an immense challenge, with various proposals for efficient and uniform systems of time on offer in the U.S. and internationally (Abbe et al. [1879] 1880; Dowd 1884; Bartky 1989). Professor Maury (1880), writing in *Popular Science Monthly*, captured the difficulty involved for meteorologists.[4] He reported that "painstaking and indefatigable observers" were being "systematically vitiated" because of inconsistent observation methods and "the more fatal lack of uniformity in the hours of observation." Maury compared the "synchronous" method, dominant in prior decades, in which each observer reported weather conditions at "his own *local* time," to the "simultaneous" recording practices structured by what Maury (1880:291–92) termed "actual time." Maury personified the problem of "local" versus "actual" time with reference to the maps that meteorologists produced to represent storms: "A weather-map based on such non-simultaneous reports, instead of faithfully mirroring the sky overhanging a continent, necessarily gives it rather a *wry* face" (1880:292). Despite this challenge, Maury argued, synoptic visions of the globe were coming into view in a novel way, bringing forth "clarified conceptions of its massive yet orderly machinery," which Maury discussed metaphorically as a "great steam apparatus" that must be "viewed as a whole" for "its operations be clearly understood" (1880:305, 310). Time, as Maury and colleagues argued, had to be torn from its local and experiential contexts and reconfigured objectively with respect to nature's "machinery." But how could meteorologists accomplish such a feat?

One challenge for achieving standardized time involved shaping public attitudes to better align with the sort of diachronic time that organized meteorologists' worldview. Meteorologists' efforts articulated with various political, economic, and scientific projects to establish time uniformity, which, as Ogle (2013) shows, proved slow, arduous, and hardly universal across scales and locales. The progressive aim, as Abbe (1893:38) put it, was "to effect a complete reform in the popular mind as to the desirability and possibility of greater uniformity and accuracy, and especially to do away with the old conservative ideas as to the need of a local time." To rectify this issue on a formal level, Abbe participated in the General Time Convention of Railroad Officials in 1883, which aimed to employ the mean time of London's Greenwich Observatory to standardize time across railroads in the United States. The effort was successful, yet as Abbe (1893:39) recollected, "It was still necessary to induce towns, cities, states, and people to give up the old local and adopt the new standard times," a task that meteorological and time-standard committees advanced by mobilizing the popular press to "prepare the public for the change that was to come" (1893:39).

The problem of aligning weather observation and related practices through a novel system of time was not a specifically American one. Meteorologists pursued time standardization projects at an international scale on the basis that understanding national weather would require knowledge of the atmosphere outside state-territorial borders. As early as 1872, Albert Meyer had urged Congress to help facilitate "a grand chain of interchanged international reports, destined with a higher civilization to bind together the signal service of the world" (quoted in U.S. Signal Service 1884:19). As he reasoned, "the atmosphere is a unit, and to be understood must be studied as a unit. . . . we must extend our investigation far beyond our territorial limits." That same year, Dutch meteorologist Christophorus Buys-Ballot (1872:17–23), in preparation for what would become the Vienna International Meteorological Congress, highlighted that national and regional differences in instrumentation and time presented serious challenges. Subsequent Congresses ultimately settled on international meteorological standards for simultaneous observations, while also establishing a stable entity, the International Meteorological Committee, to oversee the system.

In 1875, publication of the *Bulletin of International Simultaneous Meteorological Observations* commenced, including in the United States. The U.S. meteorological infrastructure provided substantial contributions, featuring more "primary" (high-quality) stations reporting regular observations than any other country (International Meteorological Committee 1879:44). The breadth of meteorological observation was nonetheless limited, privileging especially the territorial limits of Euro-American states and imperial territories, or those parts of the Northern Hemisphere "so far as they have been placed under meteorological surveillance," to use the words of one meteorologist (U.S. Signal Service 1884:19).

In 1884 Cleveland Abbe represented the U.S. Signal Service at the International Meridian Conference in Washington, arguing that establishing standard time was critical to coordinating international meteorological activities. On the global scale, meteorology depended especially on colonial networks that held not only common interests in standardizing time, but also the infrastructural capacity to coordinate geographically disparate events, instruments, observers, and international standards.

Beyond shaping public opinion and aligning international data networks, a further challenge was how to ensure that technologies and people upheld the standardized time once it was formally established. In 1893 Abbe recollected his often-frustrated previous efforts to standardize time across the Signal Service. He referenced "the great annoyance caused by the uncertainties in the local standard of time used by our voluntary observers" (1893:38). Despite clear directions posted on Signal Service documents, Abbe claimed, observers had not yet internalized a national standard time. The task of making time standard was likewise frustrated by difficulties in maintaining standardized instruments. Officers frequently monitored, tested, and reconfigured instruments and recording practices throughout the entire Signal Service network, "the object being to assure ourselves that all regular stations throughout the country were employing a uniform standard, accurate to the nearest second" (Abbe 1893:38). To aid this effort, a kind of laboratory was set up at the Central Office to calibrate time across the network. The American Watch Company produced a contracted clock, kept "in a hermetically sealed brass case in order to obviate the influence of changing atmospheric pressure and any possible magnetic influence on the steel pendulum." For additional protection from influence of "outside" weather, Abbe recalled, "A pier was specially prepared in the subbasement of the War Department in a room whose temperature variations were slight and were partly annulled by a self-controlling heating arrangement" (1893:39; see also U.S. Signal Service 1884). Only by these inventive means could meteorologists stabilize a "standard" time.

Once it was successful, synchronizing time provided opportunities for climatologists to revise existing accounts of climate and reconstitute diverse geographically local descriptions into a more uniform, national one. Alexander McAdie, for example, reduced and statistically interpolated local meteorological records for the period from 1877 to 1889 (McAdie 1891). He reported "corrections" for "time and locality" in his "Tables of Corrections to Hourly Temperature Readings to Reduce to True Mean Temperature" (McAdie 1891:x). Thus, McAdie effectively had to reconstruct national time in order to hold time constant, resulting in a new national model of climatic space. Such was the kind of work necessary to achieve "actual" (or synchronous) time with respect to weather and climate.

Thus, meteorologists' success in stabilizing the times of weather and the spaces of climate, from local to global scales, depended on the powers that could coordinate the world based on one or another standard "unit." Fabien Locher (2009) outlines a roughly parallel process in the case of French and international European storm warnings, which, as in the

United States, had begun in the early 1870s. It became possible to speak about "the weather of France," Locher shows, when it was represented on storm maps and reproduced though routine storm warnings. The politics of meteorological infrastructure thus gave way to new views of the weather. In particular, "the weather" materialized as an extralocal, national object of knowledge, government concern, and public experience.

MONOPOLIZING METEOROLOGICAL CONSUMERS AND PUBLICS

A prerequisite for widespread application of predictive knowledge is trust and authority in whomever can legitimately speak about the future (Anderson 2005; Pietruska 2017; Andersson 2018). Authority is often won through struggle. Just such a struggle unfolded within the domain of weather prediction. Translating official predictive information into its public use was a long and arduous process, not an automatic one. To monopolize the authority to report and predict weather, government meteorologists had to invest significant resources in shaping how various public users consumed weather information. They also had to contend with alternative forms of predictive authority, namely popular weather forecasters (also called weather prophets) and folk weather knowledge.

Beginning in 1872, the Weather Bureau had provided "synopses and probabilities" regarding anticipated weather. At that time, official statements began by offering a "synopsis" of recent and current weather in various locations, to which meteorologists attached the "probabilities," or forecasted weather, for the upcoming day or two. In official registers, the probabilities were later followed up with "facts" that could compare observed weather to the previously posted "probabilities" and hence verify their accuracy. In the 1870s and 1880s, meteorologists did not regularly call these activities "weather prediction" or "weather forecasting." However, Weather Bureau authority clearly penetrated many facets of social and economic life.

A report to the chief signal officer from Robert B. Fulton, director of the Mississippi Weather Service (U.S. Signal Service 1889:109), provided an initial example of state success in forming a public around climate: "The study of the climatology of the State by this service has been appreciated by the public, and the results worked out have been freely used by public speakers and the press in setting forth the agricultural and other advantages of this State." Fulton argued such services permitted market specialization in Mississippi industry. Cold-wave warnings successfully tailored to the decision-making of farmers helped them to protect early season fruits and vegetables and advance regional capacity for supplying seasonal products to northern markets. Insofar as "the public is being educated to appreciate and understand the [Signal Service] aims and its methods," Fulton (U.S. Signal Service 1889:109) reportedly witnessed growth in routine public understanding of meteorology as a government service.

On the national level, the meteorologist and director of the Weather Bureau's Climatological Division, Frank Bigelow (1900:7), claimed, a "system of mutual support" provided the "greatest value in establishing the weather service firmly among the necessary adjuncts of our modern life." He cited how agricultural societies, boards of trade, and industries successfully performed "missionary efforts in their respective communities" and "instruct[ed] them in the utility of the national service" in order to "interest the people in meteorology" (1900:7). Widening print media publication of daily Weather Bureau probabilities meant that a wider public, not only consumers of specialized warnings and crop-related "precautions," became consumers of official weather information. At the subnational level, state weather services held as their mission "to educate people up to an appreciation of [the Service's] importance," and hence to "lead them on to become their own weather prophets," assisted by official predictions "sent on ahead of the storm" (Glidden [1895] 1897:4). Meteorologists who facilitated public use of official forecasts thus advanced a general understanding of climate and weather that could be internalized in public culture roughly consonant with Weather Bureau meteorology.

Although weather prediction remained partly decentralized through state-level Weather Services on the one hand and international meteorological ambitions on the other, "national weather" at this time emerged as an object of observation, professional science, and public consciousness. By the 1880s the Signal Office had organized formal meteorological education for officers, complete with standard curriculum and certification exams. Textbooks, primers, and popular scientific coverage of meteorology reflected the work of the Weather Bureau. According to the report of Bureau meteorologist Frank Bigelow (1900:83), "meteorology is extending rapidly throughout the common schools of the country as a required branch of instruction for every child." As he explained, "these changes have certainly resulted from the persistent propaganda of publications emanating from the Weather Bureau during the past thirty years." Bigelow (1900:84) argued that the provision of daily forecasts had created "obedience of navigators to the storm warnings" and a "growing dependence of the railroads" on cold-wave warnings that threatened perishable goods. Routine forecasts, he continued, caused "gradual improvement" in the "great agriculturalist's care" to heed the frost warnings and embrace information on safe crop zones delineated by "normal or abnormal temperatures and rainfall."

Having assessed the widening use of meteorological science in public life, the meteorologist and textbook writer Douglas Archibald presented in 1897 an increasingly common narrative. The study of the atmosphere, he argued: "Is even now only just emerging from the stage of myth and speculation into that of fact and certainty. This desirable result has been chiefly attained by the disuse of vague speculation and the application of the known laws of physics" (Archibald 1897:5). Archibald articulated a common view that meteorology was transforming from a domain of myth and speculation into one rooted in the practical application of scientific laws. Notwithstanding such positive assessments and widespread public support for a national weather service, around the turn of the twentieth century, Weather Bureau meteorologists held unstable and contested jurisdiction over predictive weather knowledge.

The prospect of long-range weather forecasting was especially contentious because it required making predictive statements about weather that, to no small degree, could be incorrect. Only in 1908 did the Bureau officially begin issuing "weekly outlooks," at first experimentally and then in 1910 in a standard form of general "forecasts" (still primarily oriented to agricultural interests). As Katharine Anderson (1999) shows for Victorian-era forecasting broadly and Jamie Pietruska (2011) demonstrates for the Weather Bureau specifically, official meteorologists had to contend with competing cultural authorities, most prominently popular forecasters who rejected claims that Bureau-centered knowledge was the only kind of forecasting useful to farmers, merchants, and the wider public.

The historical legacy of weather prophets who had rejected "official" meteorology loomed over the prospects of long-range weather forecasting. Henry Vennor, for example, had become famous in the late 1870s for successful storm predictions using nonstandard methods, and he used this credibility to publish widely read almanacs and newspaper weather predictions (see, e.g., Vennor 1877). Vennor and others like him both exploited and helped to unsettle the epistemic authority of the Weather Bureau, the officials of which remained circumspect about long-range forecasting techniques and outright despised public interest in popular prophets (Somerville 1979; Pietruska 2011). Almanac writers, like Bureau officials, benefited from the capacity to easily circulate print media and to sell and publish monthly and seasonal forecasts (Sagendorph 1970). Like government bureaus, popular writers maintained networks of correspondents and observers who formed the basis of forecasts and could allegedly verify successful seasonal predictions. These writers built expertise by developing proprietary techniques that could be mastered by the individual forecaster or that required insights into a specific geographic region. Almanacs, compared to government services, moreover, readily drew upon folk weather lore that resonated with subscribers. Non-Bureau weather prophets also successfully drew upon controversial (yet hardly rejected) knowledge, including for example the effects of the moon and sunspots on weather (Garriott 1903:30–34). Popular forecasters creatively synthesized science, folklore, and regional knowledge, which often made it difficult for Bureau officials to demarcate their own scientific research from what Moore (1904:3) derides as "pseudoscience."[5] Prediction was still labeled prophecy, and prophetic authority was anathema to government meteorologists invested in rational-bureaucratic approaches to weather. The instability of charismatic prophecy (Weber 1978:1114) could be resolved by forecasters' appeals to science, correspondence networks, and other tools akin to those of formal government bureaus.

As one rejoinder to the problem of popular forecasting, Weather Bureau meteorologist Edward Garriott published a government bulletin titled *Weather Folklore and Weather Signs* in an attempt to inform public audiences of the relative accuracy or danger of lay weather knowledge (Garriott 1903). He compiled hundreds of ancient proverbs and common sayings from around the country and measured their value. Using the language of science, he thus drew boundaries between valid facts and what he deemed useless "folklore." Meteorologists recognized that science could not abolish folk knowledge. Instead, Garriott reasoned that by engaging folk knowledge, meteorologists could strategically overcome the reality that "today, fakirs and charlatans in the various professional and scientific fields, astrologers, fortune tellers, and long-range weather forecasters command, in civilized communities, a lucrative following" (Garriott 1903:29). The Bureau thus publicly put traditional knowledge to the test, hoping that under evaluation, folk beliefs could conform with scientific accounts of weather. This outcome could then bolster the position of the Bureau over its forecasting competitors.

Bureau bulletins and reports sometimes intervened to warn the public against unofficial, especially long-range, forecasts. For example, New Jersey Section director Edward McGann opened the January 1905 report of the Bureau's Crop and Climate Service with an article titled "Unreliable Weather Forecasters." He states: "About this time of the year farmers are wont to receive almanacs from various sources and in them find forecasts the weather of each month of the year. Dire are some of [the] predictions" (McGann 1905:3). To such "fraud" he responds: "It is the opinion of the leading meteorologists of the world that the public interests are injured by the publication of so called long range forecasts. . . . The persistent efforts of certain men to foist their predictions upon the public for personal gain have reached such proportions that it is deemed advisable fairly and temperately to counteract the influence of those who it is believed are preying upon the credulity of the public."[6] Pietruska (2011) shows how the Weather Bureau's reluctance to embrace official long-range forecasting efforts stemmed from officials' desire to strongly separate predictive science from superstitious or self-serving prophesies. The "rambling" of "fake" forecasters and rainmakers, as Weather Bureau officials had labeled them (Garriott 1904; Moore 1905), persisted as a problem in the early twentieth century. Local oracles stood ready to trumpet their own expertise over the sometimes inaccurate official forecasts, while the Weather Bureau remained reticent to generate long-range forecasts that held greater uncertainties.

Within efforts to stabilize official visions of the weather, the epistemological battle around prediction can be compared to how the Signal

Service had treated local weather knowledge just two decades earlier. For example, in 1883 Signal Officer Henry Dunwoody edited a large volume, *Weather Proverbs*. In a tone like his contemporaries' ethnological writings, Dunwoody and contributing authors described qualitative "popular weather prognostics" (Dunwoody 1883:9), which, they held, may have assisted Bureau prediction beyond a one- or two-day timeframe. Dunwoody especially drew upon eclectic methods of assessing popular plant and animal behaviors as signs of upcoming weather. He worked to reinterpret these methods in the language of science. As an examination of Dunwoody's work reveals, the Signal Service's understanding of popular prognostication relied on circulars that officers had distributed, imploring citizens to share local knowledge. Dunwoody then reported their results to enhance Weather Bureau forecasting efforts, concluding: "Many of these sayings express in a crude form the meteorological conditions likely to follow, and have resulted from, the close observation on the part of those whose interests compelled them to be alert in the study of all signs which might enable them to determine approaching weather changes" (Dunwoody 1883:5). Dunwoody's strategy was to deploy the power of the Bureau to integrate local "weather prognostics" that Bureau officials believed to be available throughout society. This strategy compares sharply with meteorologists' actions in the early twentieth century, when the relationship of government to local "fake" knowledge reoriented toward monopolizing official representations of weather.

How did the Weather Bureau succeed in providing credible, long-range predictive knowledge? Only by rationalizing uncertainty, through the introduction of forecasts and delicately qualifying their predictive capacity with appeals to science, did weather forecasters eventually succeed in facilitating weather prediction beyond a one- or two-day period (Pietruska 2017). Under the direction of Bureau Chief Willis L. Moore, the Weather Bureau began issuing official long-term forecasts in 1908. Public understanding of weather forecasting as a legitimate effect of scientific advances helped meteorologists secure their role as spokespersons for future weather. Although the Bureau's authority over prediction remained open to criticism, by the time Willis Milham's 1912 textbook *Meteorology* is published, the age of "speculation," he optimistically argues, is finally ending (Milham [1912] 1918:3). Even though popular forecasters and public users continued to criticize meteorology as a profession, the Weather Bureau had garnered authority to police meteorological knowledge. For example, one federal statute, approved by Congress in 1909, made "counterfeiting weather forecasts" punishable by a $500 fine and/or ninety days' imprisonment. Through legal and other strategies, meteorologists and the Weather Bureau enacted a professional closure on meteorology as a predictive science and a domain of official weather services.

Beyond eschewing the work of alternative weather prediction and securing a monopoly on long-range forecasting, the Weather Bureau intervened directly in public consumption of weather information during the first decade of the twentieth century. For example, in the March 1909 issue of the *Monthly Weather Review*, T. C. Maring discusses a bid for an "aesthetic" but "accurate" model "weather kiosk" (see figure 10; see also Marvin 1909). The kiosk was designed to facilitate public understanding of what Maring (1909:90) calls "street weather" and to make visible the otherwise hidden mechanisms of weather-recording instrumentation. Other demonstrations of Weather Bureau practices, for example at publicly attended events like the 1915 Panama-America Exhibition in San Francisco, complemented routine services and provided meteorologists with an opportunity to display Weather Bureau expertise (see Alter 1915:453).

Although the success of official weather was hardly totalizing in its scope, the subsequent effects of meteorologists' rationalization of weather were far-reaching. As one example, in 1915, T. Morris Longstreth wrote a lengthy weather forecasting guide for a popular audience, called *Reading the Weather*. On the one hand, Longstreth grounds practical predictive techniques in science. He opens with a reflection on meteorologists' success in having "overwhelm[ed] the old, buttressed superstitions," specifically whereupon "at last Science established some sort of Weather Bureau in 1870" (Longstreth 1915:i, iii). On the other hand, the author takes for granted that the daily life of his intended audience is often neither impacted nor guided by official weather forecasts. Thus, Longstreth reasons, popular knowledge is called for and is plausibly endangered by the "quickening hand" of science (1915:i). Science and nonscience, for Longstreth and a larger public, did not need the boundary lines that Weather Bureau officials had constructed. Questioning the sole legitimacy of official weather forecasting, Longstreth compares the "anticyclone" of the

Figure 10. Weather Bureau kiosk in Washington, D.C., in 1923, providing public display of real-time instrument recordings. *Source:* Library of Congress (1923); see also Marvin (1909:90).

"weather man" to common observations, and he notes that long-held beliefs about the behavior of pigs before a storm show that "the barnyard antedates the barometers as forecasters" (Longstreth 1915:20, 25). Although science and lay knowledge, for Longstreth, both testified to "our well-ordered atmosphere," his writings demonstrate that the tools of forecasting were not in the exclusive domain of official forecasters, opening the door to Longstreth's goal of cultivating a popular "weather-wise" disposition: "An unconscious desire, a little conscious knowledge, and a good deal of experimentation with the cycle of the days, and you have a weatherman" (1915:64). Popular culture and official forecasting persisted as alternative, though compatible, bases for predictive weather knowledge. The advent of official forecasting did not absolutely monopolize or rationalize weather knowledge, although popular knowledge had come to reflect the basic principles of a rationalized meteorological order.

NORMALIZING WEATHER

As we have seen, the Weather Bureau network—comprising observers, telegraphy, instruments, and the Central Office in Washington—provided meteorologists with opportunities to gain a novel official power to represent, map, and predict atmospheric patterns. Developing this power entailed integrating a bricolage bureaucracy and data infrastructure into weather reports and forecasts and monopolizing authority to create and interpret weather and climate information.

How does this historical instantiation of meteorological government relate to the dynamics of climate knowledge? Recall that the fundamental stability of climate was, by the 1870s, largely taken as given among scientists, thanks to the "positive climatology" heralded by Lorin Blodget and others in the previous two decades. My argument here is quite simple: weather prediction based on statistics and probabilities increasingly relied on an assumed, stable climate. When climatologist Elias Loomis (1868) wrote in his widely circulated and frequently updated *Treatise on Meteorology* that weather prediction was "impossible," he qualified his judgment: "The climate of a country remains permanently the same from age to age. . . . Assuming, then, the established constancy of climate, we can predict beforehand the probable character of any month of the year," however imperfectly (Loomis 1868:157, 158). If climate was considered to change, then the idea of weather prediction based on synopses of historical climate records would crumble. But knowing future weather was simply too important, and rational anticipation of tomorrow's events was fast becoming the central occupation of the meteorological profession.

Meteorologists' successful *normalization* of weather thus represents the critical link between, on the one hand, the goal to make official weather, and on the other hand, the stabilization of climate. Establishing the binary—normality/abnormality—of climate and weather was a necessary step toward rendering climate a stable background. But actors' focus on weather as normal or abnormal was less embedded in debates about climate reality. Rather, it was concerned with how anticipating future weather could facilitate the rationalization of economic activity. Official weather and climate discourse were thus primarily oriented toward economic consideration of what was "normal" and "abnormal."

Examples, drawn from the primary, widely distributed professional publication of the Weather Bureau, the *Monthly Weather Review* (henceforth *MWR*), from 1870 to 1920, reflects meteorologists' and their consumers' understanding of capricious (abnormal) weather set against (normal) climatic spaces. Typical *MWR* and other Weather Bureau reports tabulated recent weather events on a vertical axis of geographical areas and a horizontal axis of weather parameters represented by "departures from normal." Reports, most typically on monthly timescales, designated such departures numerically in terms of "excess" and "deficiency," quantified as either absolute values or proportions (see figure 11).[7]

Monthly Weather Review reports adhered to differentiating a "normal" value against any "abnormal distribution." For example, the April 1877 issue provided an introduction to the month's weather across the United States, stating: "It is a most remarkable fact, as numerically shown, that, in every district of the United States east of the Rocky Mountains, the April temperature has been extraordinary low. . . . The only exception this abnormal distribution of temperature is on the Pacific coast" (U.S. Signal Service 1875:4). Appeals to abnormal and normal weather permeate interpretations of records, including those with comparatively short time series. In August 1878, for example, one observer reported: "A comparison with the averages for August, during the past seven years, shows that the temperatures have been from one to two degrees above normal throughout the Gulf and Atlantic States . . . but have been about normal in the Ohio, Mississippi, and Missouri valleys. On the Pacific coast, the monthly mean . . . is six degrees below average: at San Francisco it is about normal, and at Portland, two degrees above" (U.S. Signal Service 1878: 580). As another example, a standard Weather Bureau (U.S. Department of Agriculture 1896:555) tabular summary of relative humidity indicated, "Normals are for a period of eight years, except for Los Angeles and Wichita, which are for seven years." Although some reports may have been works in progress (relative humidity was a newly systematized measure), these evaluations indicate the central discourse by which meteorologists generally

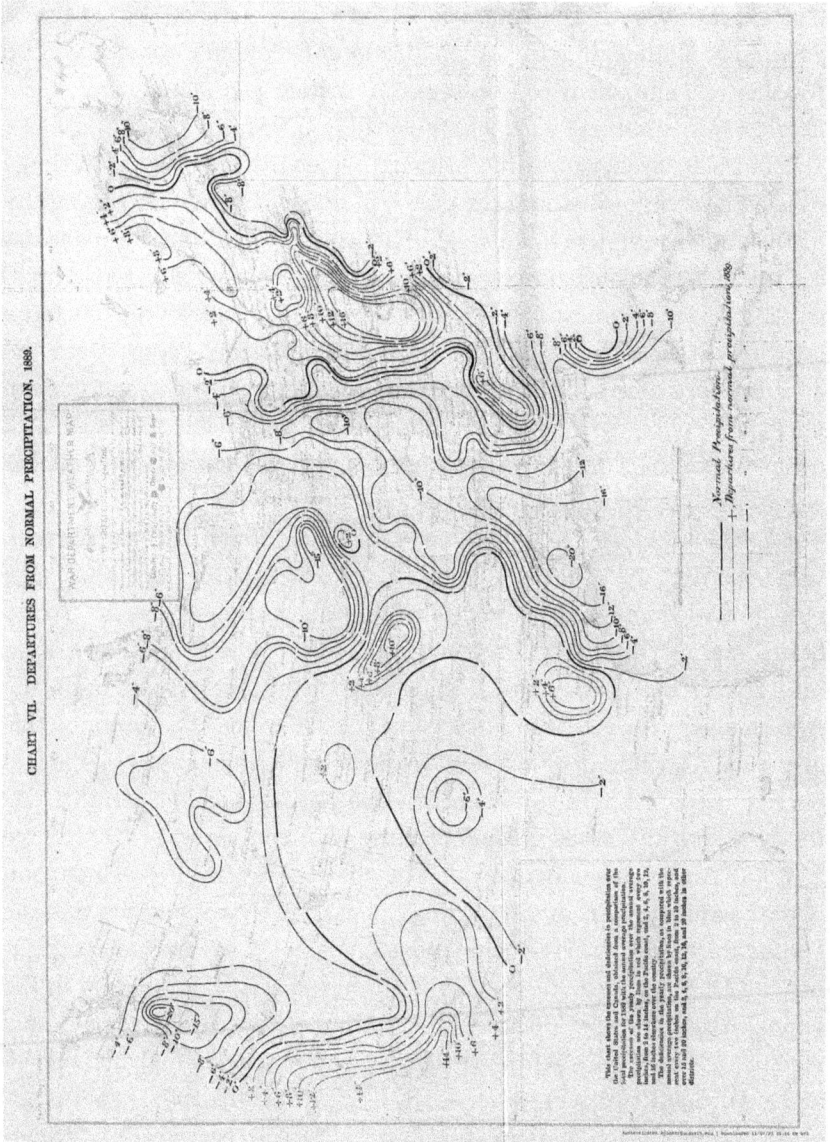

Figure 11. Graphical depiction of U.S. showing "Departures from Normal Precipitation" recorded for 1889. *Source: Monthly Weather Review* (1889: chart 7).

represented weather by reference to "deficiencies" and "excesses" of "normal" values, even in cases that clearly belied such designations.

Another relevant case regards the treatment of Puerto Rico, which did not make systematic Weather Bureau meteorological observations until 1898. Section Director Oliver Fassig reports in a 1911 article, "The Normal Temperature of Porto Rico," that Weather Bureau data confirm a depiction of the island as holding an "equitable," "comfortable and healthful" climate. Fassig defends the evaluation, stating: "Carefully made daily temperature observations extending over a period of five years in the Tropics ... will yield an average annual value which is within a fraction of the true normal value" (Fassig 1911:299). The tendency toward normal representations of weather and climate was strong enough to evaluate a climate with reference to a five-year time series that could approximate the "true normal value."

In later decades, normal/abnormal weather was reified further, especially insofar as "climatological sections" (see figure 7 in chapter 3) were presented as units of analysis. For example, a "condensed climatological summary of temperature and precipitation by sections" would typically represent average values, "departures from the normal," and extreme values, stratified by geographic section.[8] Thus, meteorologists' attribution of normality matched the basic representations of climatic stability.

An important implication of meteorologists' efforts to represent normal weather was the capacity to evaluate specific events with reference to statistical normalizations. In the "Introductory" section of the *MWR* April 1875 issue, as an early example, the author observed: "The most noteworthy peculiarities of the weather are as follows: (1) The extraordinary and almost universal continuation of the cold weather, (2) The frequency, lateness, and destructiveness of the April frosts, (3) The lateness of the rivers and lakes in opening to navigation, [and] (4) The unusually high range of the barometer" (U.S. Signal Service 1875a:1). Meteorologists' use of evaluative language helped them bridge their technical expertise in constructing official weather reports with public experience in various localities. Bridging the official with the local was an important objective, because it could reinforce epistemic authority regarding weather knowledge.[9]

In the *MWR*, meteorologists "normalized" weather though quantitative reductions of observer reports, although the accompanying scientific reportage of events and patterns routinely narrated community experiences of the impacts of weather events. In practice, such reportage involved interpreting meteorological records with reference to an informant's sense of damage or "severity," an evaluation that extended beyond numerical representations of "excessive" and "deficient" values. For example, to address abnormal rainfall in the August 1876 (U.S. Signal Service 1876: 6) issue of *MWR*, in a special section on "Drought," the author cited "deficiency" and "excessiveness" alongside observer reports of "cattle suffering," "streams dry and wells low," and other local experiences of severity:

> *Droughts*—Fla.: Mayport, ground very dry, season unusually sickly.
> Ill.: Riley, rain needed.
> Maine: Standish reports drought very severe, streams dry and wells low; West Waterville, driest August ever recorded.
> Mass: Amherst, vegetation scorched
> Texas: Corsicana, cotton crop greatly injured, and stock suffering from want of water.

Based on statistical and evaluative normalization of weather, meteorologists and those who came to rely on their expertise could incorporate weather events into economic calculations. The stability of a normal climate could then form a foundation for evaluations of risk, damage, and threats to economic productivity.

RISK AND THE ECONOMIC RATIONALIZATION OF WEATHER EVENTS

Meteorologists helped to establish normal and abnormal weather events in their relationship to economic rationality. People had for centuries recognized that their ability to reliably anticipate weather had serious economic consequences. In economic terms, at their inception in the 1870s weather warnings and climatology were profoundly successful in reducing losses and securing profits in agriculture, trade, finance, and shipping industries (Craft 1999). The economic import of the increasing rationalization of weather entailed a growing concern for making individual events legible in economic terms.

Just as stable climate "zones" had become markers of commercial investment strategies and government administration, weather "normals" and extremes became a language for evaluating economic risks and opportunities for profit. Henry Dunwoody (1894), who had organized a decade of state-level Weather Bureau reports to Washington, helped to articulate the central idea that the economic value of weather services, when compared with projected losses in the absence of such services, was so great as to be "fittingly expressed by the word 'incalculable'" (Dunwoody 1894:124).

Claims to incalculability notwithstanding, many interests became pinned to making weather events economically legible, that is, economized and subject to calculations about dollars gained and lost. Those interested in doing so could then collect on insurance and damage claims; financialize shipping, trade, or commodity markets; and defend the importance of weather services to the national political economy. Dunwoody (1894:123) promoted the Weather Bureau's capabilities in these terms: "The Weather Bureau can show in the case of the tropical hurricane of September 24–29, 1894, that 1,089 vessels, valued at $17,100,413, remained in port. In the hurricane of October 8–10, 1894, 1,216 vessels, valued at $19,183,500, heeded the warning. . . . But for the warnings these vessels would probably have gone to sea, and it is but fair to presume would in such event have met with disaster." Weather Bureau officials—ever in need of justifying their congressional appropriations and public benefit—coordinated with boards of trade, emerging weather-related insurance companies, and financiers to make events commensurable with economic costs, losses, and benefits.[10]

By economizing weather events, meteorologists could facilitate economic expansion by rationalizing anticipations of weather that impacted market relations. They could also regulate disputes that involved *unanticipated* weather events. By the turn of the twentieth century, "disaster" carried a clear socioeconomic meaning. Economic calculation of weather events consisted of comparing "abnormal" events to "normal" patterns and thereby shaping evaluations of experienced events. The power to interpret extremes could then shape appeals to the state by private entities or democratic publics to protect against possible future events that were simultaneously anticipated but unknown (Steinberg 1991; Levy 2014). Governing thus entailed new techniques for understanding, valuating, and managing weather as *risk*. In the 1904 *Yearbook of the U.S. Department of*

Agriculture, meteorologist Henry Cox addresses the shifting legal definitions of nature and weather: "The act of Providence, the legal term being
actus Dei, is indeed the favorite argument heard in many trials whether it
be the overflowing of a sewer through extraordinary rainfall or the cessa-
tion of building operations on account of prolonged wet weather or severe
cold. In all such cases the weather man is needed" (Cox 1904:307). The
"weather man," who could take the authority of official weather knowledge
into the legal domain, undercut prior assumptions about weather as natural, divine, or a matter of circumstance.

Forms of "weather insurance" were also proposed in the early twentieth
century (see Reed 1916), expanding upon prior instantiations of seasonal
insurance risk in weather-impacted industries like shipping and government disbursement of crop and related insurance subsidies (Cronon 1991;
Pietruska 2017). As the arbiter of weather events, including their costs,
benefits, predicted probabilities, and records, "the weather man" replaced
providence as the manager of the unanticipated future.

Supplementing God and circumstance, meteorologists helped to organize public experience of normal and extreme weather. Tragic disasters,
which made clear the persistent, even increasing, vulnerability of human
life to catastrophe, most notably brought this situation into view. In such
cases, the alleged stability of climate "normals" confronts unexpected
and unanticipated events. For example, in the 1900 Galveston hurricane
(among the most costly and deadly disasters in U.S. history), the national
public considered the Weather Bureau and especially local meteorologist Isaac Cline partial saviors for issuing a storm warning. The warning helped residents anticipate the storm's significant flooding and severe
weather. Residents of the island settlement of Galveston, along the Gulf
Coast of Texas, were especially vulnerable to hurricane-induced storm
surge. However, the local public and media also blamed Cline and the
Bureau, insofar as Cline had rested confidently on the city's safety from
danger. In the decade prior, Cline had vocally criticized unfulfilled plans to
build coastal fortifications to protect the burgeoning city's wealthy properties from tropical cyclones—an expert position that failed him miserably
in 1900 (Larson 1999; Pielke et al. 2008). More generally, publics came to
view weather and climate as domains that could, by the official discourse,
be rationalized, predicted, and anticipated. Rational treatment of extreme

weather as risk thus deepened the relationship between meteorology and the state, now faced with predicting and managing disasters.

Public engagement with official climate knowledge routinely involved meteorological experts settling matters of fact related to risk, responsibility, and economic losses. As Weather Bureau chief Marvin (1920:567) argues in the case of the Bureau's "Climatological Services" Division: "The value of this work is incalculable. It affects and benefits the entire people. . . . In New York City alone the weather records are brought into court by personal appearance of a Weather Bureau official more than 500 times a year. Several thousand certificates are issued annually over the seal and signature of the Secretary of Agriculture for court use. . . . The economic value of the climatological work of the bureau is enormous." Despite the increasing use of official knowledge, however, weather events continued to frustrate efforts to reliably economize risk. An example can be found in Hoffman's (1901:24) proposal to sell tornado insurance to state and municipal governments. Proposed "wind storm insurance" rates, for example, charged per $100 of city property were set at 20¢ per year in Indiana, Illinois, Michigan, and Wisconsin, and raised to 25¢ per year in Missouri, Kansas, Iowa, and Minnesota (Hoffman 1902:36). The business has yet to effectively transpire in a "scientific and profitable manner," he claims, because probabilistic statistics of tornado events (being geographically local and holding large year-to-year differences in damage costs) do not easily match the financial risk of insurance provision as evaluated by possible underwriters.

As Hoffman's scheme reported, crude attempts at setting values followed political-geographic boundaries as proxies for risk categories. Regardless of the limitations, meteorologists and those who relied on official climate knowledge increasingly incorporated weather events into strategies to measure, arbitrate, and economize weather risk.

CONCLUSION

From the 1870s to the 1920s, weather—represented through a discourse of normal/abnormal phenomena and with reference to a "stable" climatic background—became a basically rationalized element of economic activity, governmental administration, and public consciousness. This development built on the capacity for climatologists and related experts to align their science with social actors interested primarily in commercial development and state administration, an alignment that produced a stabilization of climate and the rationalization of weather observation and prediction. This power over weather provided meteorological science a formal position within the state bureaucracy, without which infrastructure and knowledge about continental weather would have been impossible to establish. Through the institutional arrangements between formal government and the sprawling network of meteorological data infrastructure, meteorologists succeeded in forging weather that was both national and official. Weather and climate could then enter economic government. Meteorologists coproduced evaluations of weather risk by translating atmospheric events into the language of economic costs, benefits, losses, and financial risks. Although meteorologists may not have faced the moral resistance to the economization that Viviana Zelizar (1978) outlines for life insurance in the latter nineteenth-century United States, they contributed to the basic social transformation of the era, namely the rationalization of American capitalist society.

In the context of an industrializing United States, the stabilization of climate and the economization of weather fit hand in glove. Through official discourse about normal weather, climatologists and meteorologists could carve up climates in numerous ways geographically, but temporally, this practice was dependent upon rendering climate as stable—the normal background. Establishing the stability of climate and the normalization of weather was not simply a matter of statistical aggregation of data concerning an objective atmosphere "out there" in nature. Indeed, others advocated alternative methods for understanding, measuring, and forecasting weather and climate. Local weather prophets had little use for the national view of weather. Indeed, it threatened their claims to knowledge. And medical climatologists were less concerned with understanding climates as geographically delineated and temporally stable. Such an understanding would have entailed relegating components of their work to medicine and the human sciences.

To forge a stable climate required the development of complex social relations and material associations. This forging process is visible in the politics surrounding meteorological infrastructure—what I call techno-politics—including diverse efforts to discipline weather observers; compete against alternative weather forecasters; coproduce state territorial governance; and designate official time, administrative space, and reported weather.

In the following chapters, which comprise part III, I turn to the transformations through which climatic stability came unhinged as a core logic of meteorological government. Chapter 5 accounts for the rise of renewed concern, within science and government, about the *instability* of climate. I address how historical developments at the intersection of scientific disciplines, beginning around 1930, took hold of a new vision for governing climate in subsequent decades.

PART III Climate Crisis and the Politics
of Climate Expertise

5 The Climate State and the Origins of a Climate Science Field, 1930–1980s

I can well recall the warning of Dr. John von Neumann
that weather manipulation, not the intercontinental
ballistic missile, would be the ultimate weapon for
the protection of the free world.

—Senator Alan Bible (U.S. Congress, Senate 1965–1966:2)

Global warming was not discovered. Rather, climate change became a central matter for science and government because of how distinct but overlapping social struggles came to integrate atmospheric research into state- and war-making. Resolving these struggles, this chapter shows, provided distinct but compatible rewards to actors in the climate science field and the state, a process that potently brought forward scientific and governmental concern over global climate dynamics (inclusive of, but hardly reducible to, global warming) by the 1960s. In this chapter, I conceptualize the actors and activities involved in the governmental apprehension of global climate as a *climate state*.[1]

To be sure, meteorologists and state actors were not done with the rationalization of climate and weather that an era of "positive climatology" had ushered in nearly a century earlier. Indeed, up to the 1970s, atmospheric scientists routinely participated in weather and climate "modification," a development that demonstrates how broader efforts to understand atmospheric dynamics were frequently shaped by ambitious state and imperial goals to control those dynamics technically (Harper 2017; Schubert 2021). Yes, scientists initially helped to extend the rationalization of

161

climate to the aim of predicting and controlling climate. However, they simultaneously pursued their own agendas of understanding climate dynamics, including anthropogenic climate change. As I show, "climate crisis" as it began to take shape in the 1980s can in part be understood as a crisis within the logic of meteorological government, rather than only as a crisis of the climate system itself.

GLOBAL CLIMATE DYNAMICS AND THE CLIMATE STATE

Historians of science have addressed the relationship between climate research and the political economy of the mid-twentieth century, with a special emphasis on the Cold War in national and transnational contexts (Doel and Harper 2006; Dörries 2011; Hart and Victor 1993; Weart 1997). In the U.S. context, scholars have found that Cold War military priorities durably transformed American science (Leslie 1993; Wang 1999), especially as military patronage redirected geosciences toward military applications (Doel 2003; Hacker 2000). Post–World War II science gained an expanding position in U.S. government. On a different front, meteorologists' ambitions about global measurements were rewarded through international professional efforts to create what Edwards (2006) has called "infrastructural globalism," which supported a more thoroughly *global* climate science that could later also support a globalizing political economy. I have already addressed how the state monopolized weather and climate knowledge over the course of the late nineteenth and early twentieth centuries. Climatology and meteorology, broadly speaking, had become state-centered sciences. As I show, however, by the 1950s and 1960s the logic of climate research became more autonomous from the state compared to the previous decades, even as climate researchers were also institutionally embedded in government. By "autonomy" of a climate science field (following sociologist Pierre Bourdieu), I mean the field formed a more recognizably hierarchical order that produced a clear set of positions and orientations among participants. Bourdieu conceptualizes a field as a patterned space of social relationships and individual positions that recognize shared rules of practice that become embodied in individuals' practical actions and dispositions (what Bourdieu terms *habitus*). Social action within any field involves participants drawing upon their experience to accumulate a field-specific form of "capital," which functions through accumulation and exchange with other forms of capital. Scientific prestige, like economic capital, can serve as a kind of social currency. "Scientific capital," then, can be defined as resources and competences recognized as conferring scientific authority (Bourdieu 2004; Hess 2006, 2011; Hong 2008; Ruget 2003; Panofsky 2014). Distinction among scientists in a field, or between science and other social domains, may inspire "boundary-work" to differentiate legitimate from illegitimate activities (Gieryn 1983). This conceptual approach is useful insofar as it helps to show the social bases upon which legitimate knowledge—say, "good climate science"—gets negotiated and established.[2]

As scholars studying expert fields have shown, how experts and intellectuals are able to intervene and change society is often related to how social fields are structured, with autonomous fields and knowledge brokers often garnering resources to influence typically more powerful political and economic domains (Eyal 2013; Medvetz 2012; Eyal and Buchholz 2010; Sapiro 2003; Mudge and Vauchez 2012). For example, intellectual debates among academic economists, as Yves Dezalay and Bryant Garth (2002) show in the context of twentieth-century Latin American state formation, have significantly shaped national and international politics and finance (see also de Souza Leão and Eyal 2019).

A major problem with treating climate science as a social field is that science inevitably involves technologies and nonhuman objects that do not behave like people (Law 1992). We have already seen how human struggles, such as those to create a standardized network of meteorological observations across North America, tied in with material struggles featuring infrastructure, instruments, and the weather. This basic perspective can be applied to how mid-twentieth-century climate change science related to U.S. state-making through a process of "matched" struggles. Struggles may be considered matched when actors with previously unrelated, undeveloped, or even contradictory concerns construct "hinges" that link actors, objects, technologies, and issues across social fields. In his study of professions, Andrew Abbott (2005:255) frames hinges as issues that secure professional jurisdictions by providing distinct yet compatible rewards. For example, he analyzes medical licensing as a hinge between the medical profession and the state that emerged by providing "payoff not only for either the doctors or their 'irregular' competitors, but also for some political group against *its* political competitors." In other instances, Timothy Mitchell (1998) argues in the case of economics, as does Bruno Latour (1988) for French microbiology, that scientists often gain power by enrolling diversely interested actors and attaching them to scientists' professed ability to solve *those actors'* problems. By tracing matched struggles, it becomes possible to trace and reconcile how the U.S. climate science field emerged as an autonomous source of power by virtue of the embeddedness of this field in broader networks of power, especially within the state.

DISCIPLINING THE ATMOSPHERE?
AMERICAN METEOROLOGY AND STATE-MAKING

"Climatology" and "climate science" today may be publicly considered interchangeable terms (Barry 2015; Nuccitelli 2015), yet through the mid-twentieth century climatology remained rooted in descriptive statistical methods that "climate scientists" would later find inadequate. Climate science relied on different credentials, skills, and orientations to climate, anchored especially in climate modeling. Those who came of age as climatologists in the 1930s and 1940s encountered an academic situation in which the study of climate had been limited to a subspecialty among geographers and a strongly subordinated concern among meteorologists (Bryson 1997; Landsberg 1941; Leighly 1954). In William Koelsch's (1996:529) terms, studies of climate dynamics "were stillborn." Drawing boundaries between "climate science" and "climatology," meteorologist Peter Lamb (2002:4) recalls that the new "climate science," largely advanced not by climatologists but by meteorologists, represents nothing short of a "scientific revolution." By expanding the ways in which researchers constructed climate, the "revolutionaries," Lamb concludes, "rescued the moribund 'climatology' of the first two-thirds of the twentieth century."

In the decades prior to the 1930s, when climate mapping was the central activity of professional climatology and weather forecasting was the primary occupation of meteorologists, questions about climate change (including anthropogenic global warming) had been raised by a dispersed array of researchers primarily outside meteorology and climatology. Only with hindsight have scientists and historians argued that early scientific investigations into climate change initiated a shared trajectory with studies of global warming that would begin in the mid-twentieth century (Bolin 2007; Weart 2008). Claims like Lamb's about the scientific status of climate dynamics and the professional location of its "revolutionary" spokespersons are central to the various moves of professional boundary-making that I now trace, initially for meteorology in the late 1920s.

Overturning Weather Bureau–Centered Meteorology

Atmospheric scientists (Bolin 1959, 1999; Phillips 1998) and historians (Cox 2002:179; Harper 2008) identify the Swedish scientist Carl-Gustav Rossby as a founder of "modern" American meteorology, carrying Scandinavian-based developments in meteorology, often ascribed to Vilhelm Bjerknes, to the United States (Friedman 1989). Exploring Rossby's career trajectory helps to show how a new form of atmospheric research emerged because Rossby and his network were able to successfully match struggles within meteorology to those of state actors in the United States, particularly during World War II.

Rossby completed a Norwegian college education in 1918, then went to Bergen, where he studied under Bjerknes at the Geophysical Institute. There, researchers developed theories of atmospheric patterns, especially "polar fronts," using the novel method of what came to be called air-mass analysis, rooted in Bjerknes's "primitive equations" of atmospheric motion (Friedman 1989). In Norway, meteorology was a prestigious scientific vocation based in state-supported weather services and university-based research programs (Phillips 1998). As American meteorologist Horace Byers (1960) remarks, "New characteristics of the atmosphere were revealed nearly every day."

Across the Atlantic, meteorology in the United States lacked such academic loci. Climate research remained strongly tied to regional agricultural and other commercial interests. Few options for careers, training, or credentialing in atmospheric research existed outside of the Weather Bureau. These were the circumstances when Rossby arrived from Norway in 1926 with a fellowship to work at the Weather Bureau. In 1927 a conflict between Rossby and Bureau personnel prompted him to leave for the Massachusetts Institute of Technology (MIT), where, in 1928, he organized the first U.S. meteorology department at a research university. Rossby had allegedly frustrated Bureau superiors by providing aviator Charles Lindbergh with an unauthorized weather forecast using air-mass analysis. This Bergen method conflicted with Bureau practices, derided by meteorologists a decade later as amounting to a "guessing science" (Koelsch 1996; Rossby 1934). The dispute between Rossby and the Bureau later became the subject of a mythic tale among Rossby's students. One of them, Byers (1960:252), held that "the Bureau was headed by unimaginative administrators who had no interest in Rossby's scientific brilliance but rather found the young Swede, with his schemes for revitalizing meteorology in the United States, a great nuisance." A line of conflict had emerged.

By attempting to affirm the legitimacy of both the imported Bergen meteorologists' methods and his own training, Rossby adopted a "subversion strategy" (Bourdieu 1975:31) to reorient, or subvert, the basic manner in which individuals conducted meteorological science. At the time, other practicing meteorologists were content with "conservation strategies" that valorized knowledge production established through conventionalized means. Their typical accounts of weather patterns consisted of mapping observations and drawing statistical probabilities based on forecasters' interpretations of past and current maps and tables. This method had guided Weather Bureau forecasting over the previous several decades. Those involved in forecasting often lacked college degrees, and they gained meteorological skills primarily through apprenticeship (Harper 2008). U.S. academic scientists at the time, compared with those in Bergen, for example, had less contact with forecasters or climatologists. Meteorologist Jerome Namias, recalling his position as a "salesman for the Bergen School," interprets the situation as one of "scientific backwardness" that could only be overcome by geophysics-trained "upstarts" located in a "hotbed of resistance" (Namias 1983:746, 734, 741).

Under these circumstances, senior Bureau meteorologists rejected alternative forms of professional meteorological practice. At the outbreak of World War II, Bureau meteorologists remained insular in organization and ambiguous in epistemic authority concerning the kinds of atmospheric dynamics involved in various war theaters. Bureau forecasters

were in a relatively weak position compared to the Bergen-school meteorologists, who had already gained access to university departments and
administrators better positioned to train military meteorologists. Given
increased aviation and the waging of World War II, progress in meteorology was an urgent, high-stakes issue. This context proved fertile ground
for remaking meteorology as a science.

Mobilizing Meteorology

Charles Tilly (1975:42; 1985) has famously argued that "war has made the
state, and the state makes war." Meteorologists' involvement in the state
before and during World War II is an important example of this process.
In the 1940s, U.S.-based meteorologists and state officials followed one
another into the unprecedented messiness of war. In the case of military
meteorology, individuals like Rossby proved adept at solving military
problems while simultaneously institutionalizing their visions for meteorology as a geophysical discipline.

Worldwide aviation in World War II required new forms of meteorological knowledge, including high-altitude meteorology and detailed predictions of precipitation, winds, cloud cover, and other poorly understood
weather elements (Bates and Fuller 1986). The geopolitical and military
struggle to undertake air-based missions in new battle scenarios matched
some meteorologists' goals of studying atmospheric circulation dynamics, especially in the upper atmosphere and in areas of the globe that had
long been understudied by the international meteorological community,
especially what Byers (1960:267) labels "the neglected tropics." Furthermore, mobilizing meteorology simply required an enormous increase in
the number of practicing meteorologists.

The issue of how to establish meteorological training became a hinge
connecting war planners to those seeking to make meteorology into an
academic geophysical discipline. During World War II, between seven
and ten thousand Americans were trained as meteorologists and another
twenty thousand as observers and technicians (Koelsch 1996). Kristine
Harper (2008) calculates a 1500 percent growth in the number of meteorologists from 1941 to 1945. What came to be called the "Big Five" institutions for meteorological training included the University of California–Los

Angeles (UCLA), the University of Chicago, Caltech, MIT, and New York University (NYU) (Allen 2001; Byers, Kaplan, and Minser 1946). Scientists from Scandinavia, who taught Bergen School theories and methods, directed all programs other than Caltech's. Rossby, the first established Bergen meteorologist in the United States, often facilitated the recruitment of Norwegian colleagues to these programs.

The meteorologists, methods, and training programs that Rossby assembled helped to rapidly reconfigure scientific and other resources to create a new, "military meteorology." The government project to expand meteorology had tremendous effects on meteorological training generally. Rossby's pathbreaking MIT program was replicated first in 1940 at the University of Chicago, which established a formal department in 1942 under the direction of Rossby's MIT student Byers. Harry Wexler, like others of Rossby's MIT students, worked as a military meteorologist, then taught at the Chicago department; he was among the first at the Weather Bureau to have a PhD in meteorology. Helmut Landsberg also worked with Rossby on military training programs at Chicago while seeking to elaborate a geophysical basis for a "physical" (as opposed to descriptive) climatology (Landsberg 1941).

The emerging network of academic departments permitted both a newly university-based meteorology and state efforts to integrate science into war-making. Rossby and other European meteorologists, many of whom had fled to the United States in the late 1930s, established the University Meteorological Committee (UMC) to organize military training programs beginning in 1940 (Byers 1970). The UMC provided professional training and instruction, ultimately helping direct government contracts and war-time data collection that matched its and department directors' visions for meteorology.

Rossby and his colleagues imported new forecasting and training techniques while actively shaping the interests of state officials. Especially significant was Rossby's relationship to Francis Reichelderfer, who became the Weather Bureau chief in 1938. Previously a U.S. Navy officer, Reichelderfer had participated in organizing the initial military training programs at MIT. Reichelderfer and Rossby had much to gain from one another, based on their trajectories to that point. On the one hand, Reichelderfer had invested his career in reorganizing the military weather services (Namias 1991). Rossby, on the other hand, stood to maintain a dominant position in meteorology if war could provide a means of expanding Bergen-school, academic meteorology. They engaged in matched struggles—Reichelderfer against existing military and Bureau organization and Rossby in support of an academically centered discipline.

Bergen-school meteorology linked science and the state insofar as its tools and organizers helped to make a world that, to be governable, required meteorological capacity to translate otherwise unwieldy atmospheric phenomena into weather forecasts and atmospheric knowledge relevant to war. Recall that the meteorological data infrastructure in the nineteenth century linked the problems of state territoriality to meteorologists' goals to trace storm patterns. Similarly, in the 1940s a novel infrastructure took shape because of the link between meteorologists and the state. First, state territoriality extended vertically into the atmosphere, especially through aerial warfare. Second, Bergen-trained meteorologists' vision during World War II traveled transnationally. For example, Cushman (2005) has shown that "Pan-Americanism" linked Bergen meteorologists' goals with U.S. and Latin American state interests in the region. Third, Rossby and his colleagues' role in mobilizing meteorology succeeded in strategically organizing new meteorological practices within university programs that could then reproduce efforts to study atmospheric dynamics utilizing mathematics, physical principles, and later, digital computing.

POSTWAR METEOROLOGY: STATE-SCIENCE COPRODUCTION THROUGH NUMERICAL WEATHER PREDICTION

As World War II ended, state officials sought to secure national scientific dominance (Steelman 1947). They wanted to institutionalize the research and development efforts that emerged during the war. Franklin D. Roosevelt poignantly captured a risk-based science policy in his letter to J. Robert Oppenheimer regarding the progress of atomic bomb development at Los Alamos National Laboratory: "Whatever the enemy may be planning, American science will be equal to the challenge" (reprinted in U.S. Atomic Energy Commission 1954:98). The experience of World War II left many scientists with newfound resources and prestige directly related to the larger project to secure U.S. hegemony in the international political economy.

Science remained "the endless frontier," as science policy architect Vannevar Bush (1945) declared. Scientists and policymakers succeeded in organizing that frontier in government through a discourse of "basic science," constructed in opposition to both "pure" science allegedly trapped in abstractions and the "applied" science of technical development (Bush 1945; Calvert 2006; Eisenhower 1954a, 1954b). "Basic" science was to be value-free. However, in Bush's account, it furthermore upheld unique American values of freedom and democracy that resulted in innovation and economic prosperity. Basic science invested the future of government directly in the work of scientists.

The rise of basic research, while still embroiled in a vacillating politics of federal science policy, allowed elite meteorologists to explore the "endless frontier" in a way that promised relevance to national security while also settling the boundaries of their discipline (Needell 2000; Sapolsky 1990). Rossby and his colleagues feared that postwar meteorology might lose its status as a science. As president of the American Meteorological Society (AMS) from 1944 to 1945, Rossby undertook reform efforts that reveal his vision to secure the boundaries of the field. Before World War II, the AMS was open to anyone who paid $3.50 annual dues (Harper 2008:84). Rossby called the AMS "the National Geographic Society of meteorology," suggesting it was geared toward amateurs and unfit to serve a more professionalized field (quoted in Harper 2008:85). In 1944 Rossby reorganized AMS publications in order to advance meteorology's reputation as a science, notably by founding the *Journal of Meteorology*.

Boundary-making in meteorology likewise shaped the Weather Bureau. Meteorologists lamented that the Bureau was suffering an acute "personnel problem" because of conflicts between senior officials and the new, academically trained recruits (Advisory Committee on Weather Services 1953). As Byers (1960) recollects, the kind of meteorology that initially made Rossby *persona non grata* at the Weather Bureau" in 1927 had by 1945 become common within many university departments and weather services.

Rossby and colleagues retained a focus on weather forecasting that could make basic science relevant to emerging military and commercial priorities, especially in aviation. In the late 1940s, Rossby and colleagues' research increasingly turned toward numerical weather prediction (NWP), an entirely new kind of weather forecasting that combined atmospheric fluid-dynamics theory with mathematics and digital computing (Harper 2008; Nebeker 1995). By tying progress in forecasting to physical theory as well as mathematics and computing and by centering meteorology in specialized university research programs, NWP assembled critical resources that later would become central to climate modeling research efforts.

In 1946 Rossby and mathematician John von Neumann negotiated a contract, administered by the newly organized Office of Naval Research (ONR), for what became called the Meteorology Project. This project held different meanings and goals for various participants. It succeeded to the degree that the meanings supporting weather prediction linked together the actions of participants. For Rossby, the project provided inroads into what he considered the central problem of meteorology—"general atmospheric circulation"—which entailed representing atmospheric dynamics in a system of mathematical equations (Rossby 1959). For government officials, securing accurate, longer-range forecasts was the major justification. Von Neumann, however, was most interested in developing an electronic computer, because the differential equations that meteorologists were formulating matched his technical goals in computing (Harper 2003). Von Neumann also emphasized "weather modification" as an important aspect of NWP scientists' promises. Overall, by the early 1950s the fact that multiple but compatible meanings of "weather prediction" were in play facilitated the technical apparatus that produced numerical atmospheric models. Basic NWP research thus survived the decade of work required for its methods to become operational in forecasting.

NWP transformed weather forecasting (Harper 2008; Nebeker 1995). Most important for the present analysis is how NWP and the disciplinary transformation of meteorology affected the genesis of a postwar climate science field. Through NWP efforts, the atmosphere became a new kind of object for meteorologists and, by extension, other scientists and their audiences. It became a global geophysical system that could be rendered calculable, in addition to being observed and mapped. Individuals working to develop NWP did not formally construct and run climate-change models at the time. However, meteorologists and those recruited from computing, mathematics, and physics organized an epistemic and institutional basis upon which an integrated complex of competing atmospheric models, climate theories, and professional positions could be built.

Considering several alternative scenarios suggests that some avenues of climate science may have become foreclosed during this period. First, NWP meteorologists faced the challenge of making equations of atmospheric processes compatible with complex, delicate computing technologies (Charney, Fljortoft, and von Neumann 1950). As Haigh, Priestley, and Rope (2016) show, successfully building a computer (von Neumann's ENIAC computer in particular) could not be taken for granted. Succeeding in these material struggles proved decisive for meteorologists' capacity to reliably construct atmospheric models.

Another area of struggle involved meteorologists' ability to defend their work as politically justifiable, especially significant to the ambitious research that characterized atmospheric and weather modeling in its early stages of development. Here, the link between NWP and weather control remained an important gamble for state officials and lawmakers. Faith in control led to budget appropriations and support that fostered basic atmospheric research (like NWP) into the 1960s.[3] Von Neumann's influence critically sustained the discourse linking basic research to weather modification (Fleming 2010; Harper 2017). In the last publication of his life, Rossby (1959:50) favorably reviews von Neumann's vision of weather control as a central component of meteorology. Despite failures in modification programs, the fervor and promise regarding weather control shaped possibilities for atmospheric research. Without connecting political to scientific struggles, numerical modeling may not have succeeded when it did.

As a third alternative, meteorologists could have failed to connect with other physical sciences. Meta-theoretical and methodological innovations are particularly relevant in bridging scientific fields (Fujimura 1992). The development of the discipline of economics is a telling comparison. It is unclear how "the economy" could become an object of science and government if economists had not borrowed the methods of physicists in the 1870s (Mirowski 1989), developed econometric analysis in the 1930s (Mitchell 1998), and integrated mathematics and computing thereafter (Mirowski 2002).[4] Similarly, in meteorology, weather forecasts until the 1930s had little to do with academic institutions and professionalized disciplines. Academic engagement with physical sciences opened the door for those in other scientific fields to recognize meteorologists because they represented atmospheric dynamics in the languages of mathematics and physics. Had physicists been able to continue to write off meteorological practice as unscientific, or had atmospheric analysts not quantified their work in numerical models, it is unclear how a field of interdisciplinary actors in later decades could or would have invested immense financial and professional resources in explaining atmospheric circulation and climate dynamics.

The process by which "the atmosphere" gained a new meaning within meteorology and for the state had unanticipated effects on the trajectory of climate science. Most significant is the rise of climate modeling.

Overall, the struggles connecting meteorologists to state actors around World War II effectively transformed the foundations of atmospheric science. Rossby's hybrid position straddling science and government linked the tools and logics of state agencies with those of academic science. Military meteorological training and NWP connected academic meteorologists to war-planners, generating opportunities for all involved. For military officials, battles could be won. For others, accurate forecasts and weather modification could secure a more governable future. For leading scientists, training and numerical modeling formed a new "climate" for their work.

FROM OCEANOGRAPHIC TO CLIMATE STATE

Climate science is an interdisciplinary field, marked by direct appeals to cross-disciplinary engagement (Schneider 1977; Weart 2013; Coen 2020). Analysis of oceanography over roughly the same period just traced for meteorology shows the broader geoscience-state relations from which climate science would later emerge.

Making Waves: Roger Revelle and the Transformation of Oceanography

Oceanographer Roger Revelle helped pioneer physical oceanography in the United States, integrate oceanic and atmospheric research, and later construct "the greenhouse effect" as a public issue in the 1970s. Similar to how Rossby's career linked science and government, Revelle occupied a distinct position from which to shape academic science, the postwar science policy bureaucracy, and the U.S. Navy.

Revelle is a towering figure in U.S. oceanography and climate science, credited with numerous major discoveries (Malone, Goldberg, and Munk 1998; Morgan and Morgan 1996). From the mid-1930s to 1950, he was variously an overlooked researcher aboard navy ships, a navy commander, and a director of classified weapons research and oceanic expeditions. During World War II, Revelle negotiated contracts for Scripps Institution of Oceanography (which he later directed) through the federal National Defense Research Committee (NDRC) and the navy. In 1942 the Navy Bureau of Ships charged Revelle with "formulating a wide-ranging program in oceanographic research applied to wartime needs and translation of the results into naval terms," chiefly sonar techniques and weaponry (Malone et al. 1998:6). Oceanography thus developed through war-making, reflected in drastic changes at Scripps in the early 1940s, when, as Rainger (2001:342) calculates, half of each incoming class was uniformed cadets.

Revelle's involvement in establishing oceanographic research connected war-making to state bureaucratic concerns over how to integrate academic research in government. Before and during World War II, proposals for a national-level science program remained unsettled. Military contracts allowed Revelle and others like him to benefit from military-academic alliances (Hamblin 2005:xx, 57). Uniquely, he insistently promoted naval research contracts as platforms for scientific progress, even as colleagues shunned what they believed to be anti-scientific military surveillance and control (Mukerji 1989; Wang 1999). After World War II, the navy was sharply divided between those who favored in-house research focused on direct military operations and those favoring university-contracted basic science.

In 1946 the ONR adopted principles of basic research to support geophysical research. Harvey Sapolsky (1990) and John Hamblin (2005) describe ONR's pivot toward basic science as an outcome of navy officials' rivalry and as an explicit strategy of investing in the work and prestige of scientists as a way to compete with the other military branches. Not until the "Sputnik crisis" in 1957 did scientists effectively secure expansion of the National Science Foundation, the eventual locus of much government-funded basic science (England 1982). Although located within the U.S. Navy, ONR presented a break from centrally controlled military R&D (Sapolsky 1990).

Revelle proved adept at drawing together naval science and oceanography. Through naval contracts awarded to Scripps, Revelle garnered military surplus ships, instruments, and access to oceanographic data around the world. Within the field of oceanography, Revelle's work was particularly oriented to building "physical oceanography" as a discipline. Revelle fulfilled his adviser Carl Eckhart's pronouncement in 1950, "At the present time it is one of the responsibilities of the Scripps Institution . . . to define the limits [of] oceanography, and to stimulate the formation of a unified profession" (cited in Rainger 2001:342). Like Rossby's success in elevating the Bergen School over other approaches in meteorology, Revelle's leadership at Scripps privileged physical oceanography to the detriment of marine biology, consistent with a more general pattern in which military-contracted research favored physical environmental sciences over the biological environmental sciences (Hamblin 2005:12, 24; Rainger 2000, 2001; Doel 2003).

Physical oceanographic research allowed state officials to integrate oceanic processes into military operations while grounding national security in a discourse of basic research. By the mid-1950s, such research yielded opportunities for oceanographers to pursue problems of oceanic circulation and climate, which would become central issues in climate change research.

Changing Tides: Investing in Climate Science

Reflecting on the oceanographic side projects that he had conducted during classified sea expeditions in the 1950s (formally devoted to nuclear weapons testing), Revelle states, "We felt we were revolutionizing the world because of what we were finding. . . . It's quite obvious now that 1948 to 1965 [was] one of the great ages of discovery of the earth" (Revelle 1975). Capitalizing on the "revolutionary" potential of existing oceanography-state alliances, in the mid-1950s Revelle and his collaborators began to conduct research on global warming through analysis of atmospheric and oceanic carbon dioxide (Keeling 1960; Revelle and Suess 1957). Their findings became seminal works in the science of global warming. Others in prior decades had linked carbon dioxide to climate change.

However, the structure of bureaucratic and scientific fields, mutually invested in facilitating national scientific dominance, had created a social
world in which immense resources and many careers might be oriented to
scientific explanations of climate change.

With Revelle as an elite spokesperson, in the early 1960s oceanographic
research provided grounds for climate to emerge as a central issue at the
intersection of science and government. Within government, Revelle—
whom Nathaniel Rich (2018) has labeled "the most distinguished of the
priestly caste of government scientists"—was able to insert concerns about
global warming into federal government reports (Revelle et al. 1965).
Revelle and his colleagues' research efforts would eventually intersect
with meteorologists' atmospheric-circulation research to provide the first
global climate models (Manabe and Bryan 1969). Diverse scientists, organizations, and technologies coalesced around the construction of such
models. Accounts of climate change increasingly had to depend on them,
not least because they formally integrated equations of atmospheric and
oceanic circulation. Ultimately such models would attempt to incorporate
the entire physical earth system, aptly called earth system models (Dunne
et al. 2012; see also Coen and Jonsson 2022).

THE CLIMATE SCIENCE FIELD

An autonomous climate science field took shape, beginning in the 1950s,
through the capacity of climate scientists and their allies to shape the national security state. As the Cold War unfolded, the strategic focus on satellite and ballistic weapons technologies, conflicts over airspace sovereignty,
the space race, and weather modification programs all reformed power and
science in relation to the atmosphere. "Earth" and the skies above it were
emerging as novel objects of standardized measurement and abstract analysis. At the same time, they were becoming a new object of especially global
military-geopolitical strategy. Scientists continued to build credibility by
claiming that basic research fortified key pillars of government—economic
development, national security, and international scientific superiority,
each understood in increasingly hegemonic terms. Furthermore, international scientific collaboration connected U.S. foreign policy to the goals

of geophysical researchers and other states. In particular, collaboration between U.S. and USSR climate scientists ultimately shaped durable international partnerships. The US-USSR Agreement on Environmental Protection was formed in 1972, built on transnational research exchange in decades prior (Doose 2021). For example, as Jonathan Oldfield (2016) shows, Soviet climatologist Mikhail Budyko's work on climate change, specifically his 1956 text *Heat Balance and the Earth's Surface*, among other works, was incorporated into debates about global climate internationally, including in the U.S. (see also Budyko 1969; Oldfield 2018).

Cold War climate science was hardly a pure intellectual exercise among an incipient cadre of climate theorists. Atmospheric scientists' participation in U.S. weather modification efforts in the 1950s and 1960s shows how they engaged Cold War politics in service of their positions in the emerging climate science field. Through scientific advising, state officials argued that national defense demanded modification programs. Navy commander Paul Jorgenson testified before the U.S. Senate (U.S. Congress, Senate 1966:33) that "we regard the weather as a weapon," a perspective echoed by politicians such as Senator Alan Bible, who claimed, "I can well recall the warning of Dr. John von Neumann that weather manipulation, not the intercontinental ballistic missile, would be the ultimate weapon for the protection of the free world" (U.S. Congress, Senate 1966:42). In response, National Science Foundation (NSF) director Leland Haworth (1966:91) testified that progressing beyond the "infant art" of previous modification techniques demanded investment in "basic" atmospheric research. Some scientists, still working within the discourses of "modification" and "basic" science, raised issues of anthropogenic climate change, which, in a turn of phrase, they categorized as *"inadvertent* modification" (National Science Foundation 1965). Climate "modification," whether intentional or inadvertent, illustrates that scientists and state actors established a mutual discourse of climate change.

The "First Wave" of Climate Scientists

Many climate researchers in the 1950s and 1960s had professional origins in the academic departments that Rossby, Revelle, and their colleagues had recently developed. Those pioneers were followed by a distinct *first wave* of climate scientists. Some in this wave had been cadets or technicians during World War II, but a few were military officers. Many earned PhDs in the physical sciences in the late 1940s and early 1950s. Notable figures in this first wave included Jule Charney, Joseph Smagorinsky, and Bert Bolin, all of whom continued to organize climate science through the 1960s and beyond.[5] Smagorinsky (1991:31) recalls that Norman Phillips, among other atmospheric modelers, emerged as one of "a new breed of young turks never before connected with the field of climatology." Like Charles Keeling and Hans Suess, who worked with Revelle at Scripps, these first-wave researchers engaged government agencies less directly. Support for their work was more likely to come from NSF grants than from military contracts. As Smagorinsky notes, moreover, they did not primarily hail from professional climatology. As a result, they advanced novel professional orientations.

In one first-wave career trajectory, Jule Charney entered the UCLA military meteorology program in 1941. There, he studied under Jacob Bjerknes and Morris Neiburger, the former recruited by Rossby to UCLA and the latter trained by Rossby at MIT. As a mathematician who encountered Rossby's work, Charney focused less on the war effort than on developing equations of atmospheric circulation (Lindzen, Lorenze, and Platzman 1990; Phillips 1995). In 1946 Rossby introduced Charney to von Neumann, who had just begun planning the NWP Meteorology Project. Such networks facilitated Charney's advances in NWP and, later, in climate change modeling. Most significant was the famous "Charney Report," often considered among the first consensus statements on anthropogenic global warming (Charney et al. 1979). Such career trajectories in climate science became more widely possible beginning in the 1950s. In effect, participants increasingly organized their work around general circulation models (GCMs), that is, models based on an intricate system of standardized observations, mathematical equations of atmospheric dynamics, and digital computation that reconstruct atmospheric patterns and allow scientists to project climate dynamics (Edwards 2000; Phillips 1956).

The International Geophysical Year:
A "Hinge Event" for Climate Science

By the mid-1950s, those working in atmospheric science and government faced a problem, namely, how to organize a standard view of the global atmosphere amid worldwide geopolitical struggle and contention. First-wave climate scientists, for their part, wanted standardized data to be collected over vast, uncharted areas so that GCMs could be developed and tested against the "real" atmosphere (Edwards 2010; Weart 2008). Recall that in in the extension of meteorological observation networks throughout in imperial domains, including to "new" U.S. territories, had succeeded by meteorologists connecting their problems, including controversies about climate change and disease, to the problem of exploring, mapping, and governing a "civilized" territorial order. In the latter part of the twentieth century, a similar problem unfolded between first-wave scientists' ambitions to build a global science while engaging U.S. global geopolitical strategy (Miller 2004).

Within these circumstances, scientists effectively organized the meteorological and oceanographic branches of the International Geophysical Year (IGY) of 1957–1958. Overall, the IGY is estimated to have included sixty thousand scientists participating from sixty-seven countries (Collis and Dodds 2008). Participants exploited new technologies, especially military rockets and satellites, to gather atmospheric data, and they assembled an international observational network and global data centers to organize vast stores of data (Sullivan 1961). The IGY was a key moment in the coproduction of the Cold War state and a U.S. climate science field. It was what the sociologist Andrew Abbott (2005:260) calls a "hinge event," linking individuals in bureaucratic and scientific fields by infusing their efforts with disparate but compatible meanings. For U.S. officials, IGY gave foreign policy an objective, humanist face. As IGY planner Lloyd Berkner (1954:575) states: "Tired of war and dissension, men of all nations have turned to 'Mother Earth' for a common effort on which all find it easy to agree." Yet as Allan Needell (2000) has documented, the Dwight D. Eisenhower administration supported IGY based on a largely secret primary interest: developing military technologies. Various U.S. government agencies proffered different justifications for their involvement, backed by a taken-for-granted discourse that basic science could enhance national security. Administrators who straddled the concerns of geoscientists and government officials, such as NSF director Alan Waterman, proved particularly instrumental in constructing the mutuality of diverse interests served by the IGY.[6] National security and scientific internationalism coalesced around a program of rational mastery. Basic science, international

Figure 12. IGY commemorative stamp issued May 31, 1958.
Designed by Ervine Metzl. *Source:* Geophysical Year (n.d.).

integration, and technological advancement could perhaps raise humanity
to new heights. This discourse is well illustrated in figure 12, an IGY image
that adapted Michaelangelo's famous Renaissance fresca representing the
touching of God and the biblical Adam. Since the eighteenth century, meteorologists had held visions of an unfolding relationship between rationalized environmental knowledge and a rational, civilized social order. In
the 1950s, IGY represented the apex of such a positive vision.

IGY provided an unprecedented opportunity for those invested in
achieving accounts of global atmospheric dynamics. IGY-supported work
in atmospheric modeling and global warming includes Revelle and Suess's
(1957) "Carbon Dioxide Exchange between Atmosphere and Ocean and
the Question of an Increase of Atmospheric CO_2 during the Past Decades"
(see also Revelle 1958). This work demonstrates how researchers utilized IGY data and opportunities to focus on rising carbon dioxide and its
possible effects on global climate. Beyond specific scientific findings, as
Edwards (2006) has shown, the pursuit of global knowledge that under-
girded IGY became an administrative and technological foundation for
global climate modeling, what Edwards calls "infrastructural globalism."
Efforts to secure global data thus amounted to a formative set of coordinated research practices that marked a climate science field oriented to
expansion and to recognition as a basic science.

Making Boundaries, Fielding Climate Science

Around the time of IGY, climate researchers in the United States worked to institutionalize the field, notably through the National Center for Atmospheric Research (NCAR). Founded in 1960, NCAR linked atmospheric researchers to state allies, especially Alan Waterman and National Academy of Sciences (NAS) president Detlov Bronk, for whom directing such research had become a governmental priority.[7] In a high-profile report entitled *Research and Education in Meteorology* (NAS 1958), scientists argue that meteorology retained a frustratingly inferior position within science. In the mid-1950s, 90 percent of meteorologists were government employees, and George Platzman argued that meteorology continued to suffer from what he called the "trade school blues" (quoted in Weart 1997:331; Mazuzan 1988). Concerned scientists and policymakers organized the NAS Committee on Meteorology, directed by Rossby and Lloyd Berkner, and including Rossby's former colleagues Horace Byers, John von Neumann, and Jule Charney. Just as previous generations of meteorologists had launched campaigns to link their professional goals with public concerns regarding weather, the committee held several conferences in 1956 to garner widespread elite and public support for atmospheric research. Such research, they reported, was "on the threshold of a truly exciting and productive era—one in which man's understanding of his environment is about to increase at a rapidly accelerating rate" (NAS 1958:6). The committee's major report pronounced the "inescapable conclusion" that "our nation's best economic interests require a more active attention to the meteorological problem" (1958:4).

Discourse about "the meteorological problem" held that national economic growth and security demanded an elite center devoted to atmospheric science. With NSF funding, NCAR was established in 1960 in Boulder, Colorado. It is one of the few centers that continues to generate global climate models.

Other centers that stabilized U.S. climate science had similar origins. The reputation of NWP had yielded opportunities for participating researchers, opening possibilities of expanding atmospheric research in the mid-1950s, including in the Weather Bureau. In one key development, Weather Bureau meteorologists, including Francis Reichelderfer and Harry Wexler, worked with the Air Weather Service and Naval Aerological

Service to establish the Joint Numerical Weather Prediction Unit (Fawcett, Hubert, and Stickles 1956). During the 1950s, Harry Wexler invested Bureau resources into researching climate control (Fleming 2011; U.S. Advisory Committee on Weather Control 1957), and in 1955 Joseph Smagorinsky exploited new opportunities within the Weather Bureau to open its General Circulation Research Section. Smagorinsky (1983) recollects: "Reichelderfer undertook to fund the whole [Research Section] from Weather Bureau resources ... even though it was still considered somewhat improper for the Bureau to involve itself in research." In 1963 the Bureau's Section reformed as the Geophysical Fluid Dynamics Laboratory (GFDL), which, like NCAR, remains a critical site that makes "global climate" legible and computationally possible. Under Smagorinsky's directorship in the 1960s, GFDL-affiliated scientists effectively bridged oceanographic and atmospheric sciences and began to develop some of the first coupled oceanic-atmospheric climate models—a breakthrough in climate science (Lewis 2008).[8]

By around 1960, as a result of the efforts of first-wave climate scientists, those who secured global data and institutionalized climate modeling were reshaping the possible orientations, audiences, and networks of future climate scientists. The climate science field became hierarchically organized around climate modeling practices. Attracting the engagement of physicists, chemists, mathematicians, and computer engineers, climate science integrated an increasing number of physical and chemical parameters and those who could speak for them scientifically.

Missing the Wave?

first-wave climate scientists reorganized climate science, others missed the "wave." They continued to participate in the field, although primarily as outsiders and, later, as critics. These individuals included nonacademic meteorologists and climatologists who lacked relevant credentials, the resources of the field's pioneers, and field-specific career investments in explaining climate dynamics. Many of these individuals were not interested in climate change. Although climate modeling generated field-specific scientific capital, which was the currency of prestige, peer recognition, and scientific discovery, for some researchers the general product—global circulation models—did not (yet) generate the scientific capital and career opportunities of weather prediction, modification, and descriptive or agricultural climatology.

The rise of global warming as a public issue began to change the terms by which climatologists related to and evaluated their first-wave contemporaries. By the 1970s, some prominent climatologists, including those with training in geophysics, were voicing concern that climate science—especially research into anthropogenic global warming—might no longer be acceptable as climatology. For example, with the rise of global warming as a hallmark scientific issue, Helmut Landsberg (1972, excerpted in Henderson 2014:73) comments: "Alarming tales have been spread, many of them by persons whose standing as climatologists may well be questioned. And just as the competence of a cardiologist in neurosurgery may be doubted so may the judgment of atmospheric physicists or dynamicists in climatology." Landsberg, like geographers who had treated climate as a statistically normalized average of recorded weather, was clearly experiencing the boundaries of climate knowledge shifting beneath his feet. Climatologists of the mid-twentieth century likely did not envision that physicists or mathematicians would come to be the most prominent spokespersons for climate. The case of James Hansen is telling. Trained as a physicist, Hansen worked in physical climate modeling beginning in the 1970s. His work provided a platform for consistent interventions in politics beginning in the 1980s, leading to many hailing him as a "climate hero" and "the leader of climate science in a warming world" (Block 2008). The meaning of climate science in these designations would have been foreign to climatologists working before the 1970s. The "first wave" was reverberating through science. It follows that the new figure of the climate change scientist would upset the established settlement regarding what it meant to implicate science in efforts to *govern* climate.

CLIMATE SCIENCE OFF ITS HINGES?
THE STATE-SCIENCE DILEMMA

From the 1960s to the 1980s, professional investment in explaining climate dynamics came to maturity. Emerging concern with global warming notwithstanding, state actors shored up investment in climate research by strongly orienting toward the rational mastery of climate. Many believed that problems with unanticipated weather and climate changes could be "fixed" through technical intervention in the atmosphere, as Fleming (2010) shows in the case of climate control. At the same time, scientists who had left behind the World War II era only to gain a degree of autonomy in research institutions were beginning to put forth alternative narratives about the basic *instability* of climate. They came to formally represent the incapacity of science to solve (mathematically, much less technically) what Lorenz (1963) and others found to be the complex, chaotic nature of atmospheric dynamics. Climate scientists who began to consider the possible dangers of anthropogenic global warming produced a *state-science dilemma* beginning in the 1970s: the U.S. Cold War legibility project that would have rationalized climate and led to atmospheric mastery risked failure. In Allan Schnaiberg's (1980) terms, climate science transformed from "production science" (which harmonized with the postwar political and economic order) into a more reflexive "impact science" that made legible the discord, uncertainties, and contradictions of industrial capitalism.

From the 1940s through the 1960s, many state actors had stood to gain from alliances with scientists, precisely because scientists' resources and prestige fitted their own struggles, whether geopolitical, ideological, or more narrowly bureaucratic in nature. In this light, when climate scientists intervened in formal politics, beginning in the 1970s, to raise concern about global warming, their actions represented a historical inversion of prior science–state relations. Seeing like a *climate* state, to extend James Scott's (1998) terms, had initially entailed making atmospheres legible, even predictable and controllable. Yet by the 1970s those actors whose very expertise coproduced a climate state came to envision risks associated with possibly catastrophic climate futures.

Global Warming Consensus and the Rise of Climate Skepticism

Not long after the institutionalization of climate research, some scientists engaged in larger efforts to remove climate dynamics and climate science from governmental concerns. In other words, they participated in actions that would *disintegrate* efforts to govern climate. Particularly significant is the rise of the climate change countermovement, which articulated a broadly neoliberal position regarding both the state and climate change. The origin and structure of the climate science field meant that some scientists peripheral to the field could fuel the rejection of the scientific consensus on global warming in the 1970s and 1980s.

Many scholars have traced the politics of global warming from around 1980 to the present, finding that public "skepticism" toward climate science has been manufactured and led to partisan polarization in the United States regarding climate science and policy (McCright and Dunlap 2000; Oreskes and Conway 2010; Dunlap and McCright 2015; Carmichael, Brulle, and Huxter 2017; Antonio and Brulle 2011). The countermovement against climate science and environmental movements, these studies show, strategically blocked climate policy from the 1970s onward. Did the structure of the climate-science field play a role in this countermovement?

An answer to this question can begin by outlining the fragile consensus achieved in the late 1970s among climate scientists, who established that anthropogenic global warming was a matter of scientific fact and a public issue of grave concern. The structure of political and climate science fields shaped who could be legitimately involved in settling the scientific controversy, who could continue to contest its settlement, and how the contestation played out before a growing public audience.

The relative autonomy of the climate science field allowed participating scientists to recognize the coherence of climate change research, which in turn enabled scientists to collectively pronounce stronger claims to scientific consensus about global warming and its impacts on climate and society. In 1979 the NAS published *Carbon Dioxide and Climate: A Scientific Assessment* (Charney et al. 1979), presented to the NRC's Climate Research Board and known as the "Charney Report" (after Chairman Jule Charney, discussed earlier). The study's authors were all men broadly representative of the first wave of climate scientists—either trained or employed in one of the major academic departments for climate science (including MIT, UCLA, and Stockholm University), affiliated with the NCAR, and/ or mathematicians or physicists. Together, these authors put forth "incontrovertible evidence" of fossil-fuel-based warming (1979:xii). They frame their consensus by stating that their conclusions "may be comforting to

scientists but disturbing to policymakers" (1979:xiii). The report finds
that present warming would have significant global social impacts in the
twenty-first century, such that a " wait-and-see policy may mean waiting
until it is too late" (1979:xiii). In their view, the controversy is settled.

It is remarkable that climate scientists in the 1970s already had secured
scientific consensus around "natural" versus anthropogenic explanations
for global warming. Even so, the structure of climate science—like politics, always a site of differentiation—cannot be reduced to consensus formation. Some championed climate researchers' status of having achieved
consensus and becoming "armed with predictions," as Senator Timothy
Wirth framed hearings before the U.S. Senate's Energy and Natural Resources Committee in 1987 (U.S. Congress, Senate 1987–1988:5). Yet, the
first wave generated new ripple effects once politicians, energy companies,
and some scientists carried scientific controversies about global warming
into the political domain.

When climate scientists initially made policy recommendations in the
1970s, they did so at a time when environmentalism was politically powerful and strongly oriented to national regulatory strategies of governance
(Brulle 2000). Already by 1979, the oil giant Exxon had devoted over half a
million dollars to studying carbon dioxide impacts on climate, having recognized "a good probability that legislation affecting our business will be
passed" (cited in Rich 2018). Although environmental protection was not
initially a strongly partisan issue (recall that Richard Nixon, a Republican, established the Environmental Protection Agency [EPA] and signed
the Clean Air Act), the larger politicization of environmental policy soon
made the "greenhouse effect" an issue owned principally by Democrats.
By the time Ronald Reagan was elected president, policy consideration
of global warming was still in its infancy. Al Gore, who began bringing
legislative attention to global warming in the late 1970s, did so as a fresh-
man congressman, with little recognition in the Democratic Party. With
the help of policymakers, climate scientists continued to advance a strong
case for energy planning and climate policy to halt global warming. However, the political landscape changed. As Naomi Oreskes, Erik Conway,
and Matthew Shindell (2008) explain, by the early years of the Reagan
administration, appeals to scientific uncertainty, reticence, and doubt had
begun to replace earlier legislative consideration of urgent regulation.

But why would scientists—the ones who could speak authoritatively about global warming—join industrial and partisan interests to participate in *deconstructing* consensus regarding the seriousness of global warming? As we have seen, some climatologists, including Helmut Landsberg, were critical of climate scientists' claims about global warming on the basis that "climate science" undermined previously conventional constructions of climate (Henderson 2014). Myanna Lahsen (2013) finds that the most powerful critics of climate science hailed from fields whose participants were not centered on GCMs or other basic tools that had organized the careers of first-wave climate scientists.[9] By virtue of experiences accumulated from the 1940s to the 1950s, many older generation scientists (especially physicists) placed extraordinary faith in technoscientific progress (Kelves 1995). The rise of the environmental movement and its successes during the Nixon administration had challenged aspects of that faith.[10] And although climate science included many trained physicists, climate scientists had come to align more visibly with "environmental" causes that contrasted sharply with the defense-centered legacy of the physical sciences (Moore 2007).

When scientists intervene in politics, contention among them over what constitutes "good," credible science is often accentuated (Gieryn 1999). Such was the case when climate scientists began to seriously engage policymakers throughout the 1980s. Charney and colleagues, who had developed their consensus report on global warming, could not police the boundaries of "climate science" once actors peripheral to the field began to find new audiences in the conservative movement, including media outlets, think tanks, and the government agencies transformed under the Reagan administration. Oreskes et al. (2008) show that, in the early 1980s, some scientists—especially the Scripps director and former Manhattan Project physicist William Nierenberg—leveraged the role of science advising to challenge the scientific consensus that had already been established about global warming. Nierenberg chaired a major report of the National Research Council regarding carbon dioxide and global warming (National Research Council 1983). The report, which was buttressed by an optimistic worldview that society could adapt to global warming, claimed that debate and caution were required to understand climate change. Undercutting the seriousness of the risk that Charney and other climate scientists had established years earlier, Nierenberg and the report received significant media and political attention thereafter. He and fellow physicist colleagues went on to establish the George C. Marshall Institute, a conservative think tank that strongly supported anti-environmentalist views, including climate skepticism. Nierenberg directly criticized research using GCMs as "politically motivated" and too uncertain to have any bearing on national policy, a position that found a ready audience in the wider "Reagan Revolution" (Lahsen 2008:213–14).

In the political landscape of the 1980s, then, peripheral actors in the climate science field were able to help organize a cautious, if not skeptical, approach to global warming science and policy. Industrial elites and conservative politicians could thus reasonably exploit interpretive spaces in which climate change skepticism remained plausible in allegedly scientific terms (Oreskes and Conway 2010). Robert Antonio and Robert Brulle (2011) show that many formal organizations comprising the climate change countermovement (CCCM) formed by the end of the 1980s. Organizations in the countermovement came to obfuscate climate science consensus in part by drawing rhetorical boundaries around "true" science that excluded climate science as uncertain, "post-normal," or politically motivated.[11] Rhetorical boundaries around truth and certainty beleaguered the communication of climate science claims that relied on climate modeling and simulations, in particular, because these techniques involve interpretive flexibility that may be attacked as not representing the "real" climate (Lahsen 2005; Edwards 2001, 2010). As a result, the vociferous ideology of climate skepticism had the enduring effect of polarizing public opinion about both scientific credibility and climate change.

The conservative disintegration of meteorological government offered one path out of the state-science dilemma. The enduring power of climate denial to empower fossil fuel companies, undercut public understandings of science, and block climate policy presents an important trajectory of the climate state under neoliberalism. Neoliberalism has broadly paralleled the fragmentation of the environmental movement and involved environmental policymakers shifting toward market mechanisms and the basic tenet that access to natural resources must be minimally regulated (Brulle 2000; Parr 2013). Like economic planning and regulation, the specter of federal climate policy infringed upon the growing ideology among economic and political elites of the 1980s and 1990s that regulation could not solve social and environmental problems.

Conservative and neoliberal challenges to the climate state, which not inconsequently involved scientists who were peripheral to the climate science field, is clearly not the whole story.[12] Climate scientists made significant strides in organizing government, including through the U.S. Global Change Research Program (USGCRP). Beginning in 1989, the USGCRP has organized expertise within thirteen government agencies to regularly produce the National Climate Assessment, modeled roughly after the assessment reports of the IPCC. Climate expertise continues to proliferate and is well organized within a multitude of governance structures tasked with understanding, mitigating, and adapting to the impacts of climate change. Efforts to strategically anticipate climate impacts have therefore allowed relations between climate experts and state actors to partly obscure the dilemma. Climate policies and climate science serve to bring the global issue of climate change down to the level of bureaucratic organizations. Of course, both the capacity and legitimacy of climate policy remain open to political contestation, presenting fertile ground for new efforts to govern climate change.

CONCLUSION

The "destabilization" of climate that became consolidated by the 1970s unfolded through a transformation in meteorological government. There was a basic shift in how scientists and state actors came to treat climate as a category. Beginning in the late 1920s, some meteorologists worked to upend weather forecasting and effectively shift the institutional locus of atmospheric sciences to universities. A similar disciplinary restructuring unfolded within oceanography. The reformulation of these disciplines helped to transform the basic institutional and epistemological foundations of climate research. Most significant, scientists started to consider "climate" as a global geophysical system that could be known in its dynamics and changes. It was no longer the "stable," time-invariant category that climatologists knew through statistical tables and maps.

The political economy that marked the 1940s and 1950s had a sweeping impact on the geophysical sciences altogether, but two phenomena are especially significant for the rise of climate science: a new position of climate research in the state and the autonomization of climate science.

Regarding the first issue, meteorologists and oceanographers built jurisdiction in the state, providing opportunities for a select interdisciplinary elite to recenter atmospheric research around numerical modeling, physics, and computing. Examining the state-science relationships that characterized the mid-twentieth century period demonstrates that climate science emerged not despite—but rather because of—the embeddedness of science in the state. Scientists succeeded by linking their efforts with struggles among actors in the bureaucratic field, including over how to organize war-making and how to secure hegemony via scientific dominance. Whether ultimately successful (as was the case with NWP) or unsuccessful (e.g., large-scale climate modification) in their efforts, scientists and state actors coproduced a "climate state" that invested significant resources in understanding the dynamics of global climate. Broadly speaking, climate science succeeded through the work of those who could assemble geophysical phenomena and speak for how those phenomena should shape government. This capacity was especially rewarded in the geopolitical circumstances of the Cold War.[13]

Regarding the second issue, autonomization, by the 1960s climate scientists had achieved a degree of independence, thanks to state-supported research centers, a new science policy bureaucracy, and academically centered training programs. The field centered on actors possessing scientific capital, reflected especially in achievements in constructing GCMs. In this context, climate researchers meaningfully investigated anthropogenic climate change. Furthermore, they had already secured a basis for expert interventions into politics. The "autonomy" of climate science was a historically and materially contingent phenomenon. It did not rest on the validity of scientific discoveries or findings alone, but rather on how scientists could enroll allies, ranging from computing technologies to state actors. To invoke Naomi Oreskes's adage again: facts do not speak for themselves; people need to speak for facts. The facts of climate change were first developed among a first wave of climate scientists and assembled through a "vast machine" (Edwards 2010) of centralized modeling practices and a global array of instruments, monitoring technologies, and computing centers.

A "dilemma" ensued, insofar as the discourse of mastering climate dynamics faded in the face of concern for global environmental crisis. This challenge was visible as early as the 1980s, when climate scientists intervened in politics to unequivocally state that global warming should spur government consideration of regulating the burgeoning fossil-fuel-based economy.

No such radical reconfiguration of science, society, and climate change materialized in the 1980s. Nor has one transpired since. In November 2017, more than fifteen thousand scientists signed a "Warning to Humanity" that broadly outlined the "collision course" of the climate-society relationship that, barring radical change, will produce "vast human misery" (Ripple et al. 2017:1026). Radical social movements have organized on the principle that climate crisis can only be averted through revolutionary change. In the face of climate denial on the one hand and scientific claims that policy efforts may amount to "too little, too late," what new possible configurations of meteorological government may arise? "Climate security"—the strategic apprehension of climate change impacts through practices of national security—is a central contemporary configuration, the examination of which lays bare how efforts to govern climate futures have taken shape in recent years.

6 Governing Climate Futures

ENVIRONMENTAL SECURITY
AND SECURITY TECHNOLOGIES

Scientists, environmentalists, and climate policy advocates in the United States applauded the USGCRP when it released its first national assessment in 2000, entitled *Climate Change Impacts on the United States: The Potential Consequences of Climate Variability and Change*. A decade after USGCRP began, and despite environmental setbacks during the George W. Bush administration, the sweeping national assessment represented an important shift in efforts to govern climate change: it focused attention squarely on climate change *effects* and not its causes. To *adapt* to climate change and respond to its short- and long-term impacts emerged as a mainstream policy goal. (The word "adapt" appears 227 times in the report.) In the years that followed, organizations, national governments, experts, activists, and communities continued to converge around a governmental logic centering on strategic anticipation of and preparation for impending climate risk, perhaps even disorder and catastrophe (Felli 2015).

In the early twenty-first century, many calling for adaptive action amid climate crisis intended to support social transformations that would halt global warming while equitably preparing for local impacts. Both the immediacy and inequality of those impacts gained traction in new scientific studies, a focus on climate change adaptation in climate policymaking, and social movement activism regarding climate justice and equity (Roberts and Park 2006). In 2002, environmental groups worked transnationally under the auspices of the International Climate Justice Network. Anchored in Global South rights and livelihood struggles, the network's organizations drafted the Bali Principles of Climate Justice and pressed for their incorporation into climate policy, particularly via the *United Nations Framework Convention on Climate Change* (UN 1992). Although adaptation and equity principles, including the "common but differentiated" responsibility for global warming between rich and poor nations, had been present in the first UNFCCC (UN 1992:5), activists only achieved enduring political visibility for their concerns through large-scale activism surrounding the 2009 Conference of the Parties to the UNFCCC (COP 15) (Chatterton, Featherstone, and Routledge 2013).

But the growing discourse of climate crisis took shape under the weight of other social forces. One such force is the U.S. national security state. How did the logic of *national* "climate security" form, given that environmentalists and scientists had long emphasized climate change as a quintessentially *global* environmental problem? The following two chapters take up two interlocking pieces of this question, both involving "climate security" as a body of knowledge and domain of expert practices. The first piece is the development of security technologies that freshly enabled strategic governance of uncertain climate change futures. I define "security technologies" as the array of scientific products; surveillance, data-gathering, and monitoring activities; and decision-making techniques that rationalize future uncertainty and facilitate strategic action based on anticipated future security environments. The second piece, taken up in chapter 7, is the social organization of think tanks and the novel capacity among participating actors to translate, and thereby shape, climate science and security policy.

By examining these two phenomena, the chapters reconstruct "climate security expertise." This expertise is centrally defined by efforts to render possibly catastrophic climate futures governable in the present through practices and technologies of national security. Other scholars have analyzed the securitization of climate change (Thomas 2015; Trombetta 2014; Oels 2005, 2012; Bettini 2014; MacDonald 2013, 2021). Ole Wæver (1995) had introduced securitization as a general concept to identify policies that, in the name of security, preempt democratic deliberation or restrict rights by enacting emergency measures (see also Buzan, Wæver, and de Wilde 1998; McSweeney 1996). Whether and how climate change has been securitized is debatable (see Oels 2012; Trombetta 2014). What remains to be understood are the sociohistorical origins of actors involved in advancing or constraining climate security discourses, what their stakes are in relevant policy debates and issue frames, and what kinds of technical and scientific resources they draw upon to advance one or another vision of making climate change governable through security.[1]

To address these questions, it is useful to consider climate security expertise as a social field of practice. This calls for exploring how a range of actors, organized groups, technologies, and logics emerged in ways that made climate change legible as a security issue, thereby populating a new climate security field. Climate security expertise can furthermore be treated with reference to how material technologies help to crystalize social relations and form a unique dimension of symbolic power. Loïc Wacquant and Aksu Akçaoğlu (2016:57), following Pierre Bourdieu (1993), discuss symbolic power as "the capacity for consequential categorization, the ability to make the world, to preserve or change it, by fashioning and diffusing symbolic frames [and] collective instruments of cognitive construction of reality." Efforts to establish visions of how uncertain, long-term future environments might be governable reflect symbolic power at work.

Achieving a "consequential categorization" of the world that enables future risks to be actual targets of government action involves simultaneously social and technical capacity. Analysis of infrastructure and scientific tools in prior chapters has already demonstrated their role in power relationships. Technologies, whether telegraph lines or financial instruments, are material "crystallizations of socially organized action" (Sterne 2003:367). They in turn hold a unique capacity to shape practices by virtue of their material constitution. As Andrew Pickering (1993) argues, human action is always changed, or "mangled," by nonhuman forces. Early global climate models, computing technologies, and data infrastructure during the Cold War helped comprise and institutionalize climate research as a scientific field, but these technologies also provided tools for "seeing" the world anew in ways that became politically consequential. Might climate security expertise in recent decades likewise provide ways of "seeing" the climate future? If so, how does the practice of security designate *what* is to be secured, by whom, and through what mechanisms?

Tracing the sociotechnical world of climate security expertise helps to identify a troubling contradiction regarding claims to govern the climate future via national security. Climate security experts have come to express a shared vision: They seek to *depoliticize* climate change, to remove it from public contestation and the play of partisan forces. Yet this vision has entailed a very narrow and fundamentally political view of the future, namely one that anticipates how to protect the nation-state, in its present form, even amid projected global climate crisis and social disorder.[2]

To empirically trace the processes by which climate security in these terms has come about, the chapter proceeds as follows. First, I outline the history of formal governmental institutions of national security in the United States, with an emphasis on the post–World War II national security state and post-9/11 reorganization of security strategies and institutions. Hampered by the successes of the climate change countermovement and its power in government, those who expressed growing concern over environmental risk embraced these strategies and institutions through the issue of "environmental security." Second, I discuss how climate experts came to focus on two substantive issues: climate-induced migration and climate-induced conflict. I then trace the development of individual security technologies that enable the legibility of these risks as real, governable threats.

ENVIRONMENTS FOR NATIONAL SECURITY

Although the first decade of the twenty-first century involved a significantly new direction for climate security with regard to global warming, efforts among experts and government officials to apprehend the security implications of global climate change predate that period. Earlier events set in motion the institutions and proximate factors through which "environmental" security came into being.

Around the end of World War II, the United States was the richest nation-state on the planet, having advanced economic growth through wartime industrial production while avoiding the more catastrophic damage done to European, Soviet, Japanese, Chinese, and other populations and the reduced economic productive capacity caused by military and social upheaval. After the war, U.S. officials advanced the basic governmental strategy by which superior American economic and military power would both protect American national security and transform the world through a U.S.-centered international capitalist political economy (Leffler 1992). The central institutions of the national security state, beginning with the National Security Council, the Central Intelligence Agency (CIA), and the National Military Establishment (founded by the 1947 National Security Act), organized security policy as an anti-communist global geopolitical containment strategy (Leffler 1992; Stuart 2008).

Geopolitical hegemony under the Truman Doctrine involved assessment of the resource capabilities and productive forces of Soviet and Third World states. In this political-economic context, Cold War military priorities both advanced earth/atmospheric sciences and drew upon scientific credibility to establish the governability of an American-led international order (see chapter 5). In the course of institutional rationalization of national security, U.S. officials considered how environmental change would impinge on geopolitical struggle. By the early 1970s, some national security experts began to evaluate future climate risk. The CIA first addressed global climate change as early as 1974. The agency commissioned studies to project the influence of global climate fluctuation on food production, which risked disrupting future U.S. national security interests domestically and around the world, for example, by causing "mass migrations" across borders (CIA 1974:6, 8) and altering the economic assets of the USSR (CIA 1976). Although not at that time attributing climate change to anthropogenic warming—but rather to natural fluctuations or even accelerated global "cooling"—officials who heeded the projections of climate scientists were clear: "The implied political and economic intelligence questions resulting from climatic change range far beyond the traditional concept of intelligence" (CIA 1974:1).

By the 1980s, and especially with the conclusion of the Cold War, "redefining security" became both a strategic challenge and an institutional dilemma for the national security state (Ullman 1983; Matthews 1989). Part of redefining security in the wake of global Cold War included governmental recognition of the environmental toxicity of war; the catastrophic risks attending military-industrial buildup; and the international dimension of Third World famines, conflicts, and other disasters of the 1970s and 1980s. By the late 1980s, anthropogenic global warming captured public and governmental attention as a threat to world order, creating what sociologist Sheldon Ungar (1992) labeled a "social scare." The spread of this "scare" depended on climate scientists' capacity to make interventions in politics by linking contemporary events (in the case of 1988, severe U.S. drought) to the "greenhouse effect," therefore providing a narrative for unfolding events that began to capture media and political attention (e.g., U.S. Congress, Senate, 1987–1988; Shabecoff 1988). In the wake of this scare and despite the rise of climate skepticism, some national security officials trying to redefine post–Cold War threats to hegemony viewed what they began to call "environmental security" as an increasingly important issue (U.S. White House 1991, 1993; Butts 1993; Levy 1995).

At the same time, multilateral trade and governance bodies advancing globalization, including for example the UN Environmental Program, took up "environmental security" under a banner of sustainable development that could allegedly solve environmental crises through economic growth (UN Brundtland Commission 1987; UN 1993). Scholars likewise approached environmental security within a broader goal to reorient the "referent objects" of security analysis and policy consideration away from the state (Rothchild 1995; Dalby 2002; Buzan, Waever, and de Wilde 1998; Barnett 2001). Intellectual and policy activities emphasized "human security" as a paradigmatic departure from Cold-War era consideration of national state security (e.g. UNDP 1994; O'Neill, MacKellar, and Lutz 2001; Webersik 2010; Matthew et al. 2009; White 2011).

Richard Matthew (2002:110) has argued that "environmental security" discourse had "boosted the political capital of certain sectors of the environmental movement" over the course of the 1990s. For example, in 1993 the Clinton-Gore administration sought to link environmental policy objectives to national defense strategy by establishing the position of Deputy Under-Secretary of Defense for Environmental Security. Deputy Under-Secretary Sherri W. Goodman told Congress upon her appointment that in the post–Cold War world, the Department of Defense (DOD) had a mission that included "ensuring responsible environmental performance in defense operations and assisting to deter or mitigate impacts of adverse environmental actions leading to international instability" (quoted in Butts 1993:vi). Based on coordination during the early 1990s between security strategists' objectives, the EPA, and other agencies tasked with environmental regulation, Sherri Goodman (1996:104) later held that the nation had entered "an exciting time for DODs environmental professionals." She affirmed that these "professionals" had overcome the stigma of toxic militarism and won approval from mainstream environmental groups. She cited, for example, the Audubon Society's and the Sierra Club's approvals of the DOD's "environmental performance" (1996:104).

Others in security domains developed concerns about environmental security in relation to the uncertain and contentious future missions of the armed forces in a post–Cold War world. For example, in a classified 1990 study, U.S. Navy officials recognized that addressing global climate change "will be both politically uncomfortable and terribly expensive . . . [and] will necessarily drain off other social programs," a trade-off these officials reasoned to be inevitable (Kelley 1990:9). In a context of demilitarization pressures, the report evaluated, "the Defense Department's budget remains a tempting target" for public scrutiny, meaning that "the Navy must ensure that the policy formulation, planning, and analysis processes adequately address the impact of global climate change before other sectors take precedence in resource allocation matters and public support" (Kelley 1990:2). Still others insisted that environmental change was simply not a military issue. For example, in an article published at the Army War College, "Environmental Security: The Role of the DOD," Kent Butts (1993:1) compared the environment to the "war on drugs," arguing that "while the diversion of U.S. forces to the war on drugs is at least tangentially related to the military mission, an environmental role for DOD is less obvious and has often met with resistance from senior defense community personnel." Tensions in environmental security resulted in haphazard attention to environmental issues in the National Security Strategy during this time (U.S. White House 1991, 1993).

How was such "resistance" to environmental security overcome within government? Denise Garcia (2010:271) has argued that over the course of the first decade of the 2000s an international norm had begun to consolidate around climate change, including formerly "recalcitrant states" (such as the United States), "because of the security threats posed by climate change." Maria Trombetta (2011:142) likewise has found security considerations a major basis for U.S. involvement in international climate agreements, meaning that political gains regarding environment policy have "in many cases been achieved *through* its securitization." Yet the processes through which climate change could be legible to security practices still need to be identified and explained. As I argue, experts came to link environmental disorganization around the globe to other emergent threats to national security.

THE FUTURE IS HERE, CA. 2007

Around 2007 a convergence ensued among experts dealing, in one way or another, with the plausible, radically destabilizing impacts of global warming. Several initially disparate efforts intersected to address climate change as an urgent problem that society must not only work to mitigate, but manage through adaptation, resiliency, and security. Many understood climate change to be decidedly real, possibly irreversible, and holding catastrophic implications for the world as we know it. The sense of urgency was acute, as if time itself had changed.

A few initial points of fact show what comprised the shift out of which climate security expertise emerged. The IPCC (2007a) issued its Fourth Assessment, presenting stronger claims than previous reports to certainty that anthropogenic forcing explained long-term and recent global warming and was already causing dislocating social and economic impacts around the world. Officials and planners across various sectors and levels of government began to embrace "climate change adaptation" (CCA) as a basic policy goal (see, e.g., WUCA 2009). "Climate services" and consultant organizations emerged to address that goal (see Lourenço et al. 2016). Among think tanks entering the fray, the Center for Naval Analysis's (CNA) Military Advisory Board famously calls climate change a "threat multiplier" in its landmark April 2007 report, *National Security and the Threat of Climate Change.*

In short order, "climate security" was beginning to make sense to those concerned with either climate change or national security. The same month that the CNA report was released, the National Intelligence Committee "elevated its efforts and initiated a classified National Intelligence Assessment of climate change," holding that "the time was right to develop a [intelligence] community level product on the national security significance of future climate change" (U.S. Congress, House of Representatives 2008:2; Whitelaw 2008). Private foundations soon advanced partnerships between emerging think-tank organizations, the DOD, and other entities involved in climate security to "bring outside voices to the climate debate" that participants believed had either become deadlocked or was unlikely to protect society against a dangerous future (Sherry Consulting 2009:10). In 2007 and 2008, a legislative bill for a Climate Security Act made its way to the Senate floor, proposing to institute a cap-and-trade program—argued for, in part, through a national security logic in which "global climate change represents a potentially significant threat multiplier for instability around the world." The Senate did not pass the bill.[3]

In the wake of policy failures on the one hand and charges of IPCC conservativism about "abrupt" climate change on the other, responses among scientists, policy advocates, and writers included renewed emphasis on the imminence of catastrophe (Spratt and Sutton 2008; Dyer 2008). Indeed, large groups of scientists soon argued that human society had transgressed several "planetary boundaries," climate among them, and was already eclipsing the "safe operating space for humanity" (Rockstrom et al. 2009). Social movement activists disenchanted with international policymaking processes began to radicalize. Reverberating among radical groups, the choice in the face of a "global emergency" became revolutionary social change or extinction (Extinction Rebellion 2018).

Not surprisingly, then, in 2007, climate science fiction ("cli-fi") solidified as a generally dystopian literary genre (Ullrich 2015). The essentially postapocalyptic understanding of social order—distilled in cli-fi narratives but harmonizing down the corridors of science, religion, social movements, and government—had captured the zeitgeist of the broader fear that society had crossed the point of no return and must prepare for the worst.

DESTABILIZING CLIMATE: MIGRATION, TERROR, CONFLICT, AND DISORDER

Two issues proved particularly salient as climate security garnered a relatively stable field of experts oriented to climate crisis: the climate-migration threat and the climate-conflict threat. Although the issues of conflict and migration overlap considerably, they deserve separate consideration. A survey of these issues demonstrates that they developed through increasingly coordinated social activity, especially dependent upon institutional changes in the national security state after 9/11 and growing concern for imminent climate crisis by around 2007. Once established, these two issues could become objects of novel security technologies. Actors could treat environmental security not only as an impetus to climate mitigation policy, but also as a means for making legible plausible future threats of global disorder. Climate security experts could then emerge as the prophets of a governable order amid impending crises.

"150 Million 'Climate Refugees' by 2050": Disorderly Mobilities and National Security

In the late 1980s and early 1990s, scholars, nongovernmental organizations (NGOs), the United Nations, and national governments raised security issues focused on the intersection of climate change, food crises, and population growth.[4] Particularly significant were "environmental" (and later "climate") "refugees." Terry Kelley (1990:6) argues that sea level rise will have significant impacts on naval missions, warning that "large segments of populations may become environmental refugees, stressing neighboring nations' resource and good will by their forced migration to safer ground." In the context of such concerns, in 1995 Norman Myers, a migration scholar and former British ambassador to the UN, collaborated with the Washington, D.C.-based Climate Institute to publish a report, *Environmental Exodus*. Outlining the dire position of migrants and making predictions through 2025, the authors draw the following conclusion: "For all countries, whether developing or developed, the over-riding objective must be to reduce the motivation for environmentally destitute people to migrate by supplying them with acceptable lifestyles. For developed countries in particular, the prospect will increasingly become a case of 'export the wherewithal for sustainable development for communities at risk—or import growing numbers of environmental refugees'" (Myers and Kent 1995:13). Although widely criticized by other academic migration scholars, Myers continued to advance his career through analyses of "environmental refugees" (see Myers 2005). One migration scholar and critic, Stephen Castles, argues that Myers's and others' "objective in putting forward these dramatic projections [of environmental refugees] was to really scare public opinion and politicians into taking action on climate change" (quoted in Barnes 2013). Myers, as a widely popular expert, provided academic credibility to the work of the Climate Institute, which continued to assert publicly that climate refugees must be a central focus of climate policy action.

Within academic circles, Myers's work provoked studies that aimed to further specify predictions of climate-induced migration. Even so, in the first decade of the twenty-first century, few scholars formed career paths or organizations devoted to climate migration and security. Few demographers and migration scholars were trained to integrate climate data into demographic projections. More explicit climate-migration scholarship therefore developed with a strong orientation to migration policy and climate policy, both increasingly politically contentious issues (Raleigh, Jordan, and Salehyan 2008; McGranahan, Balk and Anderson 2007; Perch-Nielsen, Bättig, and Imboden 2008). These scholars positioned findings on migration and climate change with reference to policy imperatives to address root causes of migration and vulnerability to climate change as a basic state strategy that could mitigate otherwise unmanageable crises. For instance, Rafael Reuveny and Will Moore (2009:476) conclude their study of migration in Southeast Asia: "We think it is imperative that developed countries invest in climate change mitigation and social-economic development in less developed countries if they want to successfully regulate their own borders." In what sociopolitical circumstances, regarding security and migration, did conclusions like these make sense?

Salient views among government officials and experts that climate-related mobility presented a serious security risk is inextricably tied to the larger securitization of migration. Castles and Miller (2009:11) argue that migration was not central to U.S. security strategy through the Cold War, and only around the turn of the century did the "migration security nexus" emerge (Miller 2000; Andreas and Snyder 2000).[5] In short, practices and institutions of national security changed dramatically during the first decade of the twenty-first century and with enduring consequences for how climate change might constitute a security threat. Governments had restricted rights of refugees and immigrants (Buff 2008). The UN High Commissioner on Refugees, humanitarian agencies, and human rights groups have shown that securitization of asylum-seeking led on the one hand to restrictive policies, and on the other hand to more punitive practices of detention and expulsion. Militarized fortifications and surveillance had led to, as Stephen Graham has put it, "ubiquitous borders" policing "dichotomies of safe and risky places" (2010:89, 91). Among the enduring ramifications of the U.S.-led war on terror is the criminalization of migration (Hallsworth and Lea 2011; Donnelly 2013). New governing institutions (especially the Department of Homeland Security) have rationalized preemptive security measures that aim to protect a "homeland" against any number of risks arising from marginal populations. Within the logic of governing "risky" migration, as Gregory White (2011:90) shows, government officials began raising the alarm that mobility resulting from environmental change represented a clear security threat.

A militarized migration system coincided with a deeply politicized climate science and climate policy failure. In 2001, President George W. Bush renounced any intention to abide by the Kyoto Protocol, an international treaty that would have required the United States to regulate and reduce carbon emissions. As Aaron McCright and Riley Dunlap (2003) have shown, the Bush administration's stances on climate change rejected both scientific consensus and national popular opinion. Policy inaction and increasingly partisan polarization among the U.S. population was largely based on the successful conservative countermovement comprised of think tanks that bridged the fossil fuel industry and the Republican Party (see chapter 5; see also McCright and Dunlap 2011). Policymakers and a cadre of experts drew upon the security implications of climate change as a frame that was "central to penetrating positions and persuading desired political change" (Garcia 2010:276). Given the politics of migration and terrorism at the time, climate-migration expertise gained special relevance to those invested in governing climate change as a real-world threat.[6]

Within government, it is no surprise that climate migration gained prominence as a national security threat over the first decade of the twenty-first century. The Department of Homeland Security Climate Change Adaptation Task Force, for example, analyzed environmental change and dangerous human mobility as primary emerging threats, recognizing that the "concept of homeland security has broadened" beyond preventing terrorist attacks "to include the principles of effectively managing multiple risks to the Nation's security" (DHS 2010). The DOD also reflected a new approach to strategically "place emphasis on the ability to surge quickly to trouble spots across the globe" and to move "from static defense . . . to mobile, expeditionary operations" and "from Department of Defense solutions to interagency approaches" for fighting the "Global War on Terror" (DOD 2006:v, vi, vii). The DOD's 2006 Quadrennial Defense Review (QDR) (DOD 2006) discusses neither climate change *nor* immigration, but by 2010 the QDR is clear that "the effects of climate change" and associated trends, including resource scarcity, urbanization, "demographic tensions," and new diseases together "add complexity to the security environment" because their "complex interplay may spark or exacerbate future conflicts" (DOD 2010:iv). The 2010 QDR therefore addresses climate change as an "accelerant of instability" that "may spur or exacerbate mass migration" (DOD 2010:85), which military forces must prepare to manage.

Parallel to DOD officials, the National Intelligence Council (NIC) has acknowledged that security forces in immigrant-receiving countries were exhibiting "increasing concern about migrants who may be exposed to or are carrying infectious diseases that may put host nation populations at higher risk" (NIC2008:14, 16; see also Podesta and Ogden 2007:123; Alperen 2017:64). DHS has projected in a report that climate-induced mobility will lead to the "transmission of infectious diseases" and heightened risk of pandemic (DHS 2010:2). In efforts "to ensure diseases do not enter the U.S.," DHS officials (2010:3–6) report that the homeland is "particularly exposed" to threats from its nearest neighbors—threats including state instability, extremism, terrorism, transnational criminal networks, expanding "ungoverned spaces," and "disorderly population displacements." Employing epidemiological language of exposure for a variety of bordered threats, DHS reiterates a vision of the future that demands an increasingly prepared, militarized border to secure U.S. territory and citizens from the outside world.

Drawing upon the extant literature on climate-migration, the 2008 "National Intelligence Assessment on the National Security Implications of Global Climate Change to 2030" reports: "The greatest concern will be movement of asylum seekers and refugees who due to ecological devastation become settlers" (NIC 2008:16). The NIC (2008:28, 29) projects, for example, that Europe will be "threatened by climate problems from other parts of the world," and the most significant impact "is likely to be migrations" from Africa and the Middle East, to which European militaries will "have to increase their activities in securing their borders and in intercepting migrants." The NIC thus echoes the warning developed by Myers and others of climate-induced migration as a threat external to northern states but one that will impose upon them.

The 2010 DHS *Climate Change Adaptation Report* focuses primarily on threats to the U.S. Southwest and Southeast regions because, the authors argue, these areas are particularly susceptible to "potential cross-border impacts" originating in the "near perimeter": Mexico, Central America, and the Caribbean (DHS 2010:ii). DHS holds that emergent patterns of climate-induced migration to the United States, including "mass migration" and "new normal migration," "may challenge enforcement, processing, and response capacities across government" (2010:2, 5). DHS officials have also expressed concern for the effects of transnationalism, through which "deepening social connections" between immigrants and their communities of origin may "incentivize more cross-border migration" (2010:6). Dangerous public responses among those "with connections to these vulnerable communities," the report reasons, may force government "concessions" and divert resources "away from other priorities" (2010:6). The process could then only deteriorate in an ungovernable fashion: initial displacements will "create 'staging grounds' for immigration to the U.S." in areas that may "then become future 'flashpoints' of conflict" (2010:6). The logic through which social order can be governed thus positions climate change as a primary cause of mobility, which by its nature is dangerous, intolerable, and must be anticipated lest the homeland be left exposed.

"Chronic Security Anarchy":
Climate-Conflict and National Security

Issues surrounding global warming and violent conflict paralleled the development of the climate-migration issue as both an emerging body of expertise and a target of government.[7] As an initial example, economist Ola Olsson frames her analysis of climate change and civil conflict in Darfur in the mid-2000s, relying especially on a meta-analysis of climate-conflict studies that "demonstrate that a one standard deviation in temperature towards warmer temperatures or more extreme rainfall is associated with an increase in inter-group conflict by 14 percent" (Ollson 2016:1).[8] Academic studies and dozens of media reports similarly have drawn upon the meta-study, published in *Science* (Hsiang, Burke, and Miguel 2013), especially by referencing the causal link between warming and conflict. As one of many possible examples, Olga Khazan, in the article "Hotter Weather Actually Wants to Make Us Kill Each Other" in the *Atlantic*, reviews Hsiang et al.'s work as "one of the few studies that points to climate change's likely impact on human society, as opposed to things like melting ice caps" (Khazan 2013). The media network Climate Central summarizes the findings of a different high-profile report on climate change and the Syrian conflict (Kelley et al. 2015), published in the *Proceedings of the National Academy of Sciences*, by stating, "Climate change doubled or even tripled the likelihood of the drought that became part of a cascade of events that have killed and displaced millions, gave rise to Islamic State and left a country in ruins" (Kahn 2015). The Arab Spring, beginning in 2011, more broadly has been associated with climate change impacts on regional conflict (Werrell and Femia 2013). The concluding goal of such studies and their popularizers is clear: to "change the calculus" of the costs of dealing with climate change by demonstrating the disorder unfolding before us (Khazan 2013).

Blaming global warming as a cause of specific conflicts remains controversial among scholars, despite appeals to consensus (Solow 2013; Homer-Dixon 1991, 1994; Burke et al. 2009; with criticisms by Buhaug et al. 2014; Dalby 2013; Livingstone 2015). Academic disagreement notwithstanding, climate-conflict researchers have presented a seemingly tangible body of evidence and expertise that could, experts have hoped, spur preemptive action toward adaptation to climate change and thereby preserve international social order in a context marked by fledgling climate policy action.

Central to the logic that climate change–induced violent conflict is a U.S. security threat is the discourse of global warming as a *threat multiplier*. DHS argues that "the most significant homeland security impact will be *indirect*, resulting from climate-driven effects on many countries and their potential to seriously affect U.S. national security interests" as a "threat multiplier" (2010:i). Scheffran (2008:20) provides a similar narrative regarding the "proliferation of risk," specifically resource competition and forced migration, which can "bring neighboring states into crisis" (2008:22). These "spillover effects," Scheffran (2008:20, 23) argues, can "destabilize regions," increase "migration-induced conflict," and "overstretch global and regional response capabilities" (see also Pumphrey 2008). In their scenario-based study of African and South Asian climate security impacts, Podesta and Ogden (2007:116) present a similar causal metaphor for the climate-conflict relationship through which a "chain reaction" of "dominos" (among them migrant radicalization) would produce "cascading geopolitical implications," meaning crises will be "all the more dangerous because they are interwoven and self-perpetuating."

Once the war on terror began, the relationships between climate-induced migration, violent conflict, and terrorism provided a way to demonstrate how a "threat multiplier" effect worked on the ground. Security experts Paul Smith (2007) and Campbell et al. (2007) argue that climate change will undermine national security objectives in weak states, in which terrorism was projected to thrive. McKeown (in Pumphrey 2008:109) argues that young men will be the most mobile in climate-induced migration flows, producing a specifically compounded threat, given that young male migrants form "the primary recruitment pool for militaries and guerrilla armies the world over." McKeown (Pumphrey 2008:136) draws upon the National Security Council's *National Strategy for Combating Terrorism* (U.S. National Security Council 2006) to argue that climate adaptation policy forms a component of the larger anti-terrorist strategy.

Those concerned with defeating terrorism and blocking migration of terrorists projected that climate change would overwhelm the capacity of "weak" or "failed" states. Failed states, one report concludes, will produce "conditions for internal conflicts, extremism, and movement toward increased authoritarianism and radical ideologies" (CNA 2007:6), a process that will extend civil conflict to uncontained "transnational security threats" (Jasparro and Taylor 2008; DOD 2010; U.S. White House

2010). Despite critics, security experts and officials came to agree on the robust connections between climate change and "failed states," "mass migrations," and "heightened conditions for terrorism" (Garcia 2010:285). In order to secure social order, this discourse suggests, future climate risk must be addressed directly while also being governed via attention to its function as a "multiplier" to manifold existing threats.

SECURITY TECHNOLOGIES: MODELS, SCENARIOS, AND HOT-SPOTS

Experts successfully made climate security a governmental category by bringing the central issues of migration and conflict to bear on existing government agency functions, priorities, and political valences. Yet it would be simplistic to argue that increased scientific concerns during the early years of the twenty-first century, simultaneously blocked by partisan rejection and enabled by military interests, somehow produced "climate security" expertise. In order to understand how experts could integrate uncertain, possibly catastrophic futures into government, basic technical and scientific developments are critical to explore. What kinds of skills, bodies of knowledge, and products could climate security expertise produce to make future risks legible? In short, climate security experts came to skillfully operate between the worlds of climate science and security policy. Climate security experts were clear that climate scientists themselves are typically unable to provide products relevant to security policy. Climate security experts took it upon themselves to advance security technologies that could link together climate science and government. By becoming invested in security technologies, security expertise could come to be organized around technical, allegedly apolitical, tools through which efforts to govern climate must pass.

Environmental security expertise had long relied on technological innovation. Formal organization of environmental security within the Clinton administration included officials intervening in and facilitating military data and research infrastructure, most of which was classified and housed within the DOD. Butts (1993), responding to the formation of the DOD office devoted to environmental security, argues that global warming science could benefit from military geophysical research and global data. The Strategic Environmental Research and Development Program (Butts 1993:7–9) therefore worked to coordinate environmental R&D across the security establishment. Related efforts included the CIA's MEDEA program, established in 1993 as a way to facilitate "bridge building" between the intelligence community's and civilian environmental scientists' data, capabilities, and expertise (MEDEA 1995:49). The MEDEA program included an Integrated Database Management Program that facilitated global integration of "spatial and temporal [oceanographic] modeling, forecast and performance modeling, and decision support modeling," the latter specifically tailored to naval operations (MEDEA 1995:46). In 1996, CIA official John Deutch discussed the success of the MEDEA program, stating, "The Intelligence Community has unique assets, including satellites, sensors, and remote sensing expertise that can contribute a wealth of information on the environment to the scientific community." The effort, he concludes, "will add significantly to our nation's capability to anticipate environmental crises" (Deutch [1996] 2007). Data sharing and related practices in the 1990s set the stage for more advanced, concerted efforts in the twenty-first century to strategically anticipate environmental and sociopolitical events caused by global warming.

From Global Climate Dynamics to Regional Climate "Events"

Refined climate models at finer-scale geographic resolution provide an important way for scientists to project how climate change will impact specific areas. Scientists can then use regional modeling to attribute geographically delimited "climate events" to global warming. Regional climate modeling techniques, when utilized by climate security experts, thus can become a powerful technology through which to govern future security threats. In the IPCC's (1992) First Assessment, scientists provide rough regional climate projections, but by the Third Assessment (IPCC 2007a), scientists have begun to develop impact assessments that integrate GCMs with other data to take into account anthropogenic climate effects at increasingly finer resolutions and more defined regional scales (Hulme et al. 2001; Benestad 2005).[9] "Single-event attribution" studies thenceforth allowed scientists to delimit climate events with reference to global warming (Diffenbaugh et al. 2017; Lahsen and Ribot 2022).[10]

The increasingly mainstream capability to explain specific events with reference to global climate has been crucial for making climate security threats legible. Experts have long recognized that security threats entail high-risk, low-probability events, but only since the mid-2010s have they adopted the use of event attribution analyses. One example is Kelley et al.'s (2015) study, which evaluates the positive impact of anthropogenic global warming on conflict and migration in Syria. Experts and journalists have interpreted the study as a case of a threat-multiplying climate event. What made the study possible and then resonate in this way? Kelley et al.'s study relies on meteorological records as well as model-based interpolation of historical regional climate dynamics. When combined with simulations of future climate scenarios, such techniques have allowed climate scientists to isolate and "attribute" the effect of anthropogenic radiative forcing on events and plausible future trends. In the case of the war in Syria or the Arab Spring, studies link drought, urbanization, and armed conflict as patterned series of events that are likely to increase in the future because of global warming. This kind of evaluation is new for climate science. Significant in regional climate studies is the relatively novel scientific capacity to isolate the impact of anthropogenic climate change on regional-scaled patterns and events—in this case, regarding Mediterranean drought (Hoerling et al. 2012; Seager et al. 2014; Kitoh, Yatagai, and Alpert 2008; Wu et al. 2011; Lionello and Giorgi 2007).

Senior intelligence officers only several years earlier, in 2008, perceived that regional modeling was at that time inadequate but "required" in order to embrace what Fingar (U.S. Congress, House of Representatives 2008:18) calls "strategic climate change." Such modeling inadequacies compounded a problem: security analysis could not integrate "social, economic . . . military, and political models" and was therefore limited to "a scenario-driven exercise and an imprecise science" (2008:19). Finger concludes that "outside expertise" would assist the intelligence community in these efforts (2008:19). Sure enough, over the following several years, climate simulations that featured regional modeling and event attribution augmented the more abstract global climate modeling of the 1990s (e.g., the IPCC First Assessment). Climate-conflict expertise could then emerge on a scientific footing.

Earth System Monitoring and Global Surveillance

Parallel to advances in regional climate modeling, climate state-science partnerships have enhanced synoptic monitoring and surveillance to provide what Josef Aschbacher and Maria Milagro-Perez (2012:4) label "a complete and ongoing picture of the Earth." For example, the European Commission and the European Space Agency have operationalized a Global Monitoring for Environment and Security (GMES) program, which aims to provide "the largest environmental and security monitoring system on the planet" (2012:4). The system has integrated satellite remote-sensing, climate, and other on-the-ground data for use in environmental and "homeland security" decision-making, including monitoring climate change impacts and assisting enforcement efforts enabled by the European External Border Surveillance System (Aschbacher and Milagro-Perez 2012:4; Donlon et al. 2012).

Government agencies justify developments in global environmental monitoring, like the GMES, in part because they provide data inputs to refine earth system dynamic models. Such models are critical to measuring global warming and other large-scale atmospheric patterns. But some scientists also recognize the opportunities that their expertise, in satellite imaging for example, may provide to defense-centered interests given "the nation's commitments to operational early warning, surveillance, and reconnaissance systems" (Vetta 2011).[11] Inevitably, as one space scientist reasoned using the case of Arctic sea ice loss, "demand will increase for security patrols—and for the space systems to support them." Global warming is projected to increase economic and military operations in an ice-free Arctic, potentially upsetting the current balance of power in the region. States with territory or other Arctic geopolitical interests rely on both monitoring and scenario-based views of the future as guides to operations and policy (Strawa et al. 2020; Labe, Magnusdottir, and Stern 2018). As Vetta argues, security strategy requires new kinds of environmental monitoring expertise. Vetta (2011) emphasizes the overlap of experts' skills with national security interests: "Like the national security community, the climate research and environmental monitoring communities need to achieve what their defense counterparts have termed 'persistent surveillance.'" The use of monitoring capabilities therefore projects an ongoing synergy between environmental science and security policy (Brutschin and Schubert 2016; Chen et al. 2022).

Scenario-Planning: From Heuristics to Strategic Forecasting

A related set of security technologies includes methodological innovations to identify "thresholds," "tipping points," and risk categories in highly uncertain and nonlinear climatic and social systems. Scenario-planning, simulation, and dynamic forecasting techniques together function as what government agencies and experts have come to call decision-support tools that make possible strategic anticipations of a range of abstracted "futures" (Moss et al. 2014). Such tools often must be tailored to specific users by providing decision support services (DSS). Climate security experts thereby construct scenarios of climate-society interaction that, like regional climate modeling and environmental-security monitoring, connect climate science to what many experts label the pragmatic needs of security decision-makers.

Early systems theorist Herman Kahn defined scenarios as "hypothetical sequences of events constructed for the purpose of focusing attention on causal processes and decision points" (quoted in Pickett 1992:8). Scenario analysis, and related forms of game- and systems-analysis, emerged within the planning exercises, war games, and futurism especially developed by Project RAND and other institutes during the early years of the Cold War (Kahn and Wiener 1967; Kahn 1976). In the early 1970s, futurists associated with the Club of Rome, including Meadows et al. (1972) and Mihajlo Mesarovic and Eduard Pestel (1974), developed scenario techniques into mathematical "World" simulations of global environmental scenarios. Futurism, at the intersection of science and anticipatory government, had emerged in full force (Andersson 2018).

Since its beginnings in the late 1980s and culminating in the first Climate Change Assessment (IPCC 1992), scientists working internationally through the IPCC have developed and refined projections of earth system trajectories that incorporate and narrate the probable climate effects of policy scenarios. The simple construction of four emission-policy and radiative-forcing scenarios has served as an anchor for climate science. Scenarios provided comparable bases for estimation of phenomena

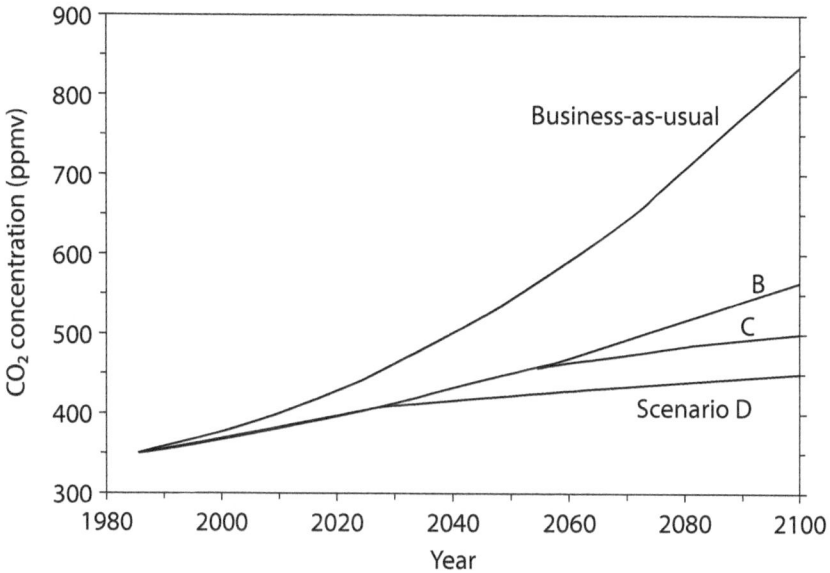

Figure 13. Graph projecting CO_2 concentration over time by comparing emissions reduction policy scenarios. (Original caption: "Atmospheric concentrations of carbon dioxide resulting from the four IPCC emissions scenarios.") *Source:* IPCC (1992:70).

as diverse as sea level rise, aerosol concentration levels, and the costs of mitigation policy over a century-long timeframe. The "business as usual" (BaU) scenario, for example, assumed no change in emissions policy, a doubling of atmospheric carbon dioxide compared to preindustrial levels between 2025 and 2050, and a resulting increase in mean global temperature from 1.5°C to 4.5°C (IPCC 1990:WGII). Policy actions could then be proposed, assessed, criticized, or fought for based on comparison between the BaU and alternative scenarios.

Beginning in the 1990s, scenario-based planning became popular in various environmental management and administrative contexts (Amer, Daim, and Jetter 2013; Raskin 2005). As Simone Pulver and Stacy Van-Deveer (2009) show in a meta-analysis of environmental scenario studies, use of the technique increased dramatically after the 1970s and especially during the 2000s. Among climate scientists, research shifted from using scenarios that had only examined global greenhouse gas (GHG) emissions scenarios, like that pictured in figure 13, to include climate-change impacts to social systems (Mitchell et al. 1999).

Those who sought to develop security technologies regarding global warming readily adopted scenario analysis and scenario planning as tools through which to link projections of climate change to those of sociopolitical dynamics. One highly influential and generally apocalyptic scenario analysis, developed by Peter Schwartz and Doug Randall (2003) and issued by the Pentagon, loosely compiled scientific findings to project how security officials might need to respond to an "abrupt climate change scenario," that is, one after critical thresholds in the climate system had been passed. The scenario projects disrupted food supply and globally unequal social conditions, which will lead to regional conflict, border militarization, and international war. Thus, the authors project, the United States "will find itself in a world where Europe will be struggling internally, large numbers of refugees washing up on its shores, and Asia in serious crisis over food and water." They conclude: "Disruption and conflict will be endemic features of life" (2003:22). Schwartz and Randall (2003:2, 21) are clear on the implication of their scenario exercise: the DOD needs to improve "predictive climate models" and "vulnerability metrics" that can help national security officials "anticipate which countries are most vulnerable to climate change and therefore, could contribute materially to an increasingly disorderly and potentially violent world." Security experts' more sophisticated use of scenarios followed up on this call.

By 2007 more precise, differentiated regional scenarios began to emerge in government and think-tank scenario planning, depending in part on improved regional-scale climate modeling. Campbell et al. (2007:7) stratify alternative warming scenarios by world geographic region. Under what they label "the *catastrophic* scenario," defined by a 5.6°C global temperature increase by 2100, the authors project in some areas "strong and surprising intersections between the two great security threats of the day—global climate change and international terrorism waged by Islamist extremists." Regional scenario construction as a basis for securing "strategic futures" with reference to climate change emerged in various national contexts around this time, for example within the UK Ministry of Defense. As one report states, disorder resulting from climate change will have to be secured not by control, but through rehearsal: "The benefit of strategic futures work is not that it predicts the future, which is unpredictable, or enables organizations to control it. It is about rehearsing possibilities, so one is better able to respond if they happen" (UK Ministry of Defense 2007:1).

The U.S. National Intelligence Committee (NIC) carries forth the same anticipatory logic in its 2012 report, *Global Trends 2030: Alternative Worlds*. In it, the NIC (2012) presents scenarios of regional climate change that have the potential to produce, among possible alternatives, "tipping points" in global social order (2012:i, 62). Subsequent DOD operational training included scenario planning. In a 2015 report to Congress on climate and national security, officials discuss how the U.S. Northern Command, tasked with defending U.S. territories into the Arctic region, had developed "extreme weather-driven scenarios" for use in operational exercises that rehearse responses to climate-induced "catastrophic events" (DOD 2015:11).[12]

Around the same time, political scientists and intelligence officials sought to bring simulation-based conflict forecasting to bear on anticipated sites of climate-induced conflict (see Brandt, Freeman, and Schrodt 2014:945). The methodologies employed in conflict forecasting include statistical and simulation techniques used by atmospheric scientists, although they had yet to orient toward environmental conflict forecasting per se. Thus, one expert review (ASP 2012:12) draws attention to advances in scenario-based crop yield forecasts: "But defense planners *also* need computer simulations that forecast how future warming might influence social and political stability." Security experts are clearly attracted to the possibility of strategically forecasting climate-conflict scenarios and events.

CLIMATE CHANGE "HOT SPOTS"

Alongside region-, sector-, or impact-specific scenario planning, some analysts have bridged security and environmental forecasting techniques by performing "hot-spot" risk analysis. In 2006 the prominent climate modeler Filippo Giorgi developed the Regional Climate Change Index (RCCI), among the first to depict "hot spots" based primarily on data-driven climate models (Giorgi 2006). Identifying hot spots in subsequent studies linked dynamic geophysical processes (e.g., hot spots of sea level rise) with socio-political events (e.g., hot spots of political instability) (see Holland 2013). In its Third Assessment, the IPCC (2007b) likewise begins to label river mega-deltas hot spots of vulnerability and displacement. In one quantitative hot-spot analysis, Roger Torres et al. (2012) constructed an RCCI for Brazil and integrated global warming hot-spot projections into what the authors call a Socio-Climatic Vulnerability Index. Based on a matrix of these two indices, the authors spatially identified "socio-climatic hotspots."

Although often incorporating alternative scenarios over time, "hot-spot mapping" is primarily a technology of spatial representation (de Sherbinin 2013). As a security technology, hot-spot mapping aims to focus attention on specific high-impact zones and emphasize spatial variation in climate-change ecological and social disruptions. The resulting representations are generally intended to help identify security priorities, for example, to prepare for political instability in the case of projected water shortages in Central Asia (Scheffran and Battaglini 2011; NIC 2008).

In effect, scenario planning and hot-spot mapping shape climate security practices by virtue of their power to represent complex, interdependent climate impacts across spatial and temporal scales. More broadly, the historical development of climate security expertise rests on the capacity to link together, on the one hand, scientific techniques for projecting spatially delineated dynamics and events, and on the other hand, governmental priorities to make legible a governable social order in the context of climate crisis. Of course, what that social order might look like would necessarily be a political question, given that climate-induced migration and climate-conflict analyses are structured by historically specific geopolitical interests, not to mention political decisions regarding global warming. Even so, security technologies make futures in part matters of expert legibility.

CONCLUSION

Analysis of security technologies demonstrates how the historical situation for those working in science and government on national security and environmental issues came together in the space of "environmental security." The space was undoubtedly productive, not only by raising concern about global warming in public and government forums, but in generating a new way of seeing future threats tailored to state priorities. Security technologies allow actors to strategically anticipate climate risk around the world and to thereby govern climate change in its *effects*. Security technologies rationalize future-oriented action. The legibility of risky but governable futures has helped actors to augment climate policy aimed at addressing its causes while circumventing the decades-old political problem of the climate change countermovement that aims to deny the "legibility" of climate change causes and consequences altogether. Yet as the following chapter explores, climate security expertise is hardly apolitical. Those who came to occupy the space of climate security expertise may have intentionally (and at times, quite successfully) skirted around the ideological and partisan battlegrounds featuring climate denialists and environmentalists. Yet the resulting visions of the future to be governed are perhaps some of the most potently political visions of all.

7 Future Struggles

CLIMATE SECURITY EXPERTS
AND THE DEPOLITICIZATION
OF THE CLIMATE FUTURE

Fear of, and preparation for, the future rarely occurs in a political vacuum. When it comes to efforts to prepare for climate crisis, the first decade of the twenty-first century was decisive. As outlined in chapter 6, innovations in climate science, public concern regarding climate emergencies and national security threats, and governmental discord in the face of George W. Bush–era climate politics converged. These convergences helped to make "climate security" a space for taking an (un)governable climate-changed future seriously: treat climate change futures as real events and visible scenarios that could be placed on a map and put on a timeline.

Security technologies proved critical to making such futures legible, but at the heart of any form of expertise are human social actors. Like people in many circumstances, experts act in patterned ways and for reasons that are conditioned by their wider historical and social situation. In sociological terms, climate security expertise exhibits a social structure. Within this structure, which I sketch in this chapter as a climate security field, people give future scenarios meaning, contest those meanings, and try to shape others' actions with regard to them. As such, the struggle over the future is an indelibly social process involving both cooperation and competition.

Here I focus on the formation of an organizational ecology primarily populated by think thanks. The term "ecology" signifies that competition and interdependent relationships among organizations shape social action.[1] My approach to understanding think tanks is informed by Thomas Medvetz (2012), who theorizes that think tanks in the United States have come to form an *interstitial field*, structured as a site of exchange between political, bureaucratic, economic, media, and other relatively coherent social domains (see also Stampnitzky 2013; Eyal 2013; Panofsky 2014). In what follows I demonstrate how a range of actors emerged and coalesced from around 2005 to 2010, becoming organized through think tanks nested at the intersection of scientific, political, media, and bureaucratic fields. Climate security experts differentiated their practices and dispositions, even as they found common cause in making climate change an issue of national security. I characterize this process with reference to two axes of differentiation among participants: (1) the *cultural capital* that participants came to draw from scientific, political, media, and bureaucratic fields, and (2) the *translation work* participants learned to perform between these fields, with security technologies frequently serving as important points of contact between them.

The upshot of understanding hierarchy and differentiation among climate security experts' fields is that it can demonstrate distinct but complementary interventions into politics. So the chapter concludes by showing how climate security, on the one hand, advances a mode of governing climate that confronts the inherited legacy of the science-state dilemma and reactionary ideology, especially climate denial (see chapter 5). On the other hand, climate security presents a deeply problematic vector for collective action regarding climate change by imagining "climate crisis" as a situation wherein the nation-state secures some populations while those "outside" the imagined community face the full force of climate disruption.

THE CLIMATE SECURITY FIELD

During the final years of the George W. Bush administration and prior to the election of Barack Obama, many Democrats were reaching for justifications for climate policy. Conservative political forces had helped to increasingly polarize public opinion about global warming (Oreskes and Conway 2010; Dunlap and McCright 2015).[2] Climate scientists had advanced their capacity to project climate-change impacts, even as they faced political pressure and censorship within a Republican-controlled government (Zabarenko 2007). In state agencies, security and defense officials by and large were beginning to consider climate change as a real threat. As the Bush administration proceeded, political pressure among conservatives to deregulate environmental protection coincided with the tremendous post-9/11 expansion of the national security state (Priest and Arkin 2011; Donnelly 2013; Hallsworth and Lea 2001). Many individuals working in government agencies, media outlets, think tanks, and climate science, as addressed in chapter 6, therefore strategically expressed a vision of climate change as a major security threat. Given the political landscape, think tanks provided a context in which such a vision could take root.

By around 2005 to 2010, an array of individuals working primarily in think tanks began to "own" climate security expertise as a set of professional skills, techniques, and identities. Before that time, one would be hard pressed to find a self-defined expert on climate change and national security among either academic scientists, political activists, government officials, or popular writers. Few had yet successfully built a professional career around such issues. This situation changed, especially as individuals drew upon their involvement in new national security institutions, especially regarding environmental security in the 1990s and the post-9/11 security field and the facilitation of security technologies.

The American Security Project and the Origins of the Climate Security Expert

One entry point into the organizations that comprised the field is the 2007 founding of the American Security Project (ASP), a Washington, D.C.-based think tank that its founders described as a "nonpartisan organization created to educate the American public and the world about the changing nature of national security in the 21st Century." ASP was founded by Heather Higginbottom, who had studied public policy at George Washington University and over her career served in various positions within

Democratic administrations, including as Deputy Secretary of State for Management and Resources in the Obama administration. The founding CEO, retired brigadier-general Stephen Cheney, had received military training at the National War College, after which he served in the Marine Corps and became a fellow of the Council on Foreign Relations, a member of the Secretary of State's International Security Advisory Board, and a member of the State Department's Foreign Affairs Policy Board.

In major works, including a 2012 *Climate Security Report* and, in 2014, *The Global Security Defense Index on Climate Change*, ASP outlined the nature of the threat of climate change by drawing upon the credentials and know-how of military staff members: "As a national security institution, ASP knows that there is no such thing as 'certainty' on the battlefield—there is only uncertainty. The same goes for preparing for climate change; the United States must be ready for contingencies" (ASP 2012). As a primarily public-facing organization that has aimed to gather a broad constituency around security issues, ASP has relied less on scientific expertise than on the extensive experiences offered by retired military officials, whom it mobilized for nationwide public speaking events, conferences, and media campaigns. Thus, climate and environmental scientists have not been on staff with ASP, nor have they been a primary feature of the events that ASP has organized. Climate security expertise at ASP, in short, has not relied on scientific capital but rather on the capacity to translate complex future possibilities into public support for climate security policies and strategies.

For ASP, action on climate security has largely been addressed as a public relations challenge. Thus, Cheney has appeared on many major news and television programs—including CNN, Fox News, and the Weather Channel—to represent ASP and speak credibly about the threat climate change poses to national security. In seeking to take nonpartisan stances, ASP spokespersons have generally avoided commenting on any given party platform or record. The problem, in their worldview, is traditional modes of thinking: "The same-old solutions and partisan bickering won't do" (ASP 2012:63). ASP similarly framed one report, quoting board member Senator Gary Hart: "Conventional thinking will increase our vulnerability. Anticipation, imagination, flexibility and experimentation are required to make us secure in an age of profound revolutionary change" (ASP 2012).

In the case of the social organization of ASP, climate security arose as a point of contact between older, retired military officials and politicians who could, with the help of a younger workforce motivated by climate governance (itself a recent and growing specialization among university programs), invest their specific experiences and skills in climate security expertise. As in other think tanks, the rank and file of the organization included interns and adjunct fellows, for example master's-level students or recent graduates from universities in the Washington, D.C., region specializing in international relations, public policy, and security. ASP's program staff has comprised people with similar educational backgrounds. For example, Esther (Babson) Sperling, the former program manager of climate security, received an MA in global security studies from Johns Hopkins University, after which she continued to work on climate security issues, primarily in South Asia. Andrew Holland, director of studies and senior fellow for energy and climate, represents a common alternative path for younger professionals. He entered ASP via politics as a legislative assistant for then senator from Nebraska Chuck Hagel (who also sits on ASP's board of directors). A fellow and member of other think tanks, including Securing America's Energy Future (SAFE) and the International Institute for Strategic Studies, Holland has authored many of the ASP climate security reports, drawing heavily upon the testimony of an established network of military, think-tank, and academic sources.

ASP is internally heterogeneous, but as an initial case it suggests the formation of a novel kind of social actor that exhibits a particular climate security "habitus," or set of practical habits and embodied cultural dispositions. Climate security experts successfully perform a "get it done," pragmatic orientation to climate change, while holding in tension boundaries between science, politics, public opinion, and state institutions. General Charles H. Jocaby, who retired in 2014 as commander of U.S. Navy Northern Command, demonstrates a climate security habitus at work. In a 2015 video interview, Jacoby stood in uniform and discussed how climate change *"can* be considered a politicized issue," but, as he stated with a serious countenance and demeanor: "I'm a soldier. I'm a *requirements* guy. I'm a mission accomplishment guy. And so, for me, it's, 'be in favor of what's happening.' And so, I deal with the facts. Whatever the cause is less relevant to me than the *effect."* Climate change presented, for Jacoby, a "legitimate mission that we readily embrace."[3] Those in the climate se-curity field whose expertise rests on making future climates distinctly legible through projections may be uncomfortable with Jacoby's stated ambivalence toward the causes of climate change. Yet distancing projec-tive science from present reality provides a way for Jacoby to overcome the politics of science by replacing it with "requirements."

Retired military officials, who have occupied high-profile positions in most climate security think tanks, routinely emphasize the nuts-and-bolts pragmatism of military preparedness. Retired Marine Corps lieutenant-general John Castellaw and Army brigadier-general John Adams, during a 2014 event tour with ASP, put it this way: "Even as our comrades on active duty in the U.S. military forces plan for the impact of the rise in sea levels in places like Bangladesh . . . , members of Congress and others continue to deny the obvious. The truth is that climate change is real and poses sig-nificant challenges for our nation's security" (Dobkin 2014). By framing climate change as a matter of military strategy, climate security experts can distance themselves from what they at times see as the incapacity of legislative institutions like the U.S. Congress.

Staking Claims: The Case of the CNA Military Advisory Board

The Center for Naval Analysis (CNA) is widely recognized as being among the most significant think tanks from which climate security expertise emerged. Compared to ASP, CNA is more oriented to government de-fense policy and holds closer working relationships with climate scientists and active security officials. Most significant, the CNA Military Advisory Board (MAB) provided a kind of programmatic manifesto for climate se-curity among defense strategists—its April 2007 report, *National Security and the Threat of Climate Change.* Authored by eleven high-rank retired military officials, the report marked an authoritative and unprecedented consensus among a well-organized group of participants in the security elite. In the report, CNA effectively bridged scientific understandings of the severity of climate risk to the decades of military experience of MAB members. The report addressed climate change as a "threat multiplier" that must be addressed within all sectors of the security and defense com-munities and across all levels of government.

Like ASP staff, MAB members argued that preparedness and bipartisan action must proceed despite uncertainty, important to the political context of the second term of the George W. Bush administration when it released its original report. In one of many testimonials (called "Voices of Experience") that accompany the findings from the MAB's security analyses, retired general and former Army chief of staff Gordon Sullivan concludes, drawing from experiences dating from the Cold War: "We never have 100 percent certainty. We never have it. If you wait until you have 100 percent certainty, something bad is going to happen on the battlefield" (CNA 2007:9). By comparing the risks inherent to battle with the need to act boldly on climate change, military-centered climate security experts have projected a military habitus as relevant to confronting climate security risk.

Peer organizations welcomed the boldness of CNA's claims. Consider the Strategic Studies Institute, a think tank housed at the U.S. Army War College and a center of national strategic research with an aim to "influence policy debate and bridge the gap between Military and Academia" (strategicstudiesinstitute.army.mil). Its academically centered climate security experts and military officials positioned the 2007 CNA report as a foundational document: "They are moving beyond the causal debate and instead saying, 'We represent the security community and we are convinced that we are facing an imminent threat, and we need to be prepared to deal with this threat'" (Butts 2008:135). In a situation in which many politicians and conservative interest groups viewed climate scientists as politically tainted, the personal testimony by military officials represented an alternative point of entry for government actors to address the implications of climate change.

MAB affiliates more generally have drawn upon a capacity to distance military officials from politics and therefore enact a uniquely real-world orientation to climate science. For example, retired U.S. Navy vice admiral and former NASA administrator Richard Truly recounts his coming to terms with the climate threat, emphasizing: "I wasn't convinced by a person or any interest group—it was the data that got me" (CNA 2007:14). Similarly, MAB advisory board member and navy meteorologist James Titley has stated in a 2011 TedxPentagon event, "So, it turns out, conservatives, who may not always listen to scientists. . . . Sometimes a retired senior naval office can get an audience with those people" (Titley 2011).

Titley, wearing his navy uniform and refraining from "starting with a bunch of statistics," recounted his "journey from being a hardcore skeptic" to becoming a climate security expert (Titley 2011). Titley received his PhD in meteorology from the Naval Postgraduate School and worked in various military positions (including as a navy rear admiral). As of 2024, he held a position as a professor at Penn State's Department of Meteorology and Atmospheric Sciences, where he founded (in 2023) the Center for Solutions to Weather and Climate Risk, and he has continued to sit on the board of climate security organizations, including CNA and the Center for Climate and Security. Through these positions, Titley has drawn upon both his long-standing military tenure and his scientific credentials (however peripheral to climate change research per se) to participate in the climate security field.

By channeling the experience and credentials of its affiliated members, CNA has demonstrated the possibility of creating effective working relationships between climate scientists and military strategists—relationships that military-based climate science can then build upon. The testimony of the MAB member Vice Admiral Paul Gaffney II provides one illustration. As former president of the National Defense University and retired former chief of naval research and commander of the Navy Meteorology and Oceanography Command, Gaffney reflects on the history of military research to conclude: "If climate change is, in fact, a critical issue for security, then the military and intelligence communities should be specifically tasked to aggressively find ways to make their data, talent, and systems capabilities available to American efforts in understanding climate change signals" (CNA 2007:23). CNA has since worked to build ties with scientific experts with whom, over the last decade, they have advanced such capabilities.

Those with military credentials, centralized chiefly in the MAB, do not only provide technical experience. They also provide a new language with which to frame climate change. Take for example the concept of climate change as a "threat multiplier" (coined by CNA 2007:6, 44). CNA MAB founder and executive director Sherri Goodman has used the term repeatedly in CNA reports, conference presentations, and media articles and appearances (see Mathiesen 2015). The term goes uncited in a 2009 UN report (UN General Assembly 2009:6) and in academic literature

(MacDonald 2013:44). Defense Secretary Chuck Hagel uses the term to outline the DOD Climate Change Adaptation Roadmap (DOD 2014a:i). The DOD's Quadrennial Defense Review (DOD 2014b:8) emphasizes that climate impacts "are threat multipliers that will aggravate stressors," leading to "conditions that can enable terrorist activity and other forms of violence" (see also DOD 2015). Think-tank reports and subsequent media attention have incorporated a threat-multiplier narrative to interpret political conflicts viewed as portending future global disorder that must be contained (Werrell and Femia 2013; G7 2015:16–17; Browne and Shank 2015; Romm 2015; CNA 2017).

CNA has provided a long-standing base for climate expertise that can facilitate relationships between science and security strategy. Updated in 2014, the MAB report *National Security and the Accelerating Risks of Climate Change* reaffirms for the authors the urgency they stated in 2007, evaluates the progress made by climate security experts, and points toward the need for a more concerted national security effort. The now widespread discourse of climate change as a "threat multiplier" demonstrates how well the organization has generated effects of climate security expertise in the bureaucratic field. In return, government credibility accumulated via consultation and recognition by government agencies has allowed CNA's experts to reinvest in the organization's stock of climate security expertise (see CNA 2017). Thus, for example, former Homeland Security secretary Michael Chertoff and former CIA director and defense secretary Leon Panetta penned and signed the foreword to the MAB's 2014 report, testifying to the "decades of experience" of MAB members and to "CNA's established analytical prowess," which they attest deserves "strong attention from not only the security community, but also from the entire government" (CNA MAB 2014). In effect, climate security expertise provides, through the work of CNA and others, a collective and authoritative voice for climate security within government.

Institutionalizing Translation: Center for a New American Century, Center for Strategic Studies, and Center for American Progress

Some think-tank organizations and experts have lacked elite military credibility, and they instead have focused on how to shape climate science to better align with the security establishment. Several organizations, including the Center for a New American Century (CNAS), the Center for Strategic Studies, and the Center for American Progress (CAP), have institutionalized practices of translation work that could durably link science and climate security.

Despite developments in climate science that by around 2007 could have plausibly informed climate security expertise, initial reports of that time show how tenuous the relationship between science and think-tank capacity was at its inception. In November 2007, CNAS collaborated with the Center for Strategic and International Studies (CSIS) to publish a report like the CNA MAB's, featuring retired military officials and titled *The Age of Consequences: The Foreign Policy and National Security Implications of Global Climate Change* (Campbell et al. 2007). At CAP, John Podesta and Peter Ogden collaborated on climate security studies, some of which is reproduced in the *Age of Consequences* volume.[4] Their work presented numerous case studies of conflict, disease outbreaks, migration, and related national security threats. The authors rooted their analyses in "climate-change scenarios that had been developed by Pew Center Senior Climate Scientist Dr. Jay Gulledge in consultation with other leading experts in the field" (Podesta and Ogden 2007:1). Based on their interpretation of climate change scenarios, they argue that their findings are "not alarmist" regarding security threats, insisting instead that "this scenario may be the best we can hope for." But the individual who was the conduit to climate science, Jay Gulledge, apparently never held a position of "Senior Climate Scientist" (as the report mistakenly claims). An ecologist by training, Gulledge left his academic post in 2005 to pursue climate security expertise, first with Pew Charitable Trusts' climate security program, and then with other organizations. Lack of in-house scientific expertise opened up space for critiques of "alarmism" from right-wing think tanks, for example the libertarian Cato Institute, which turned to criticizing mainstream climate security expertise as a backdoor vector of bureaucratic expansion (Stewart 2020).[5]

It is less important whether Jay Gulledge was objectively a "climate scientist," a designation that contains a fair amount of interpretive flexibility.[6] What is significant is that careers like Gulledge's proceeded by formally and explicitly aiming to link climate modeling and scenario development to issues that mattered to CNAS, CAP, and other organizations

(Mabey et al. 2011). Linking climate science to security analyses and policies in this way marked a significant innovation at a time when modeling and scenario analysis was primarily limited to work conducted for formal assessments, including those of the IPCC. Thus, even those who were peripheral to the climate science field could benefit from taking on the task of translating science and think-tank work regarding security policy. A comparison is telling. Filippo Giorgi, among the most prominent regional climate modelers in the first decade of the twenty-first century, was not directly interested in the security studies that drew heavily upon his work. Because Giorgi had succeeded in becoming an established climate modeler, his career was not invested in the translation work that others performed. On the other hand, Gulledge, in a telling example of investing scientific capital in the field of climate security expertise, collaborated with CNAS researcher Will Rogers to publish one of the organization's major reports, *Lost in Translation: Closing the Gap Between Climate Science and National Security Policy* (Rogers and Gulledge 2010). Constructing a narrative of a cultural schism between scientific production and consumers' "weak demand signals," which separate the scientific community from national security professionals, the authors identify "a clear shortage of 'translators'" (2010:30, 8). Evidently the CNAS and Gulledge's career, among others, had come to invest in social positions that could problematize this supply-demand mismatch: "Today there are too few individuals dedicated to translating information between the climate science and national security communities. Indeed, it is clear that the rapidly rising demand for technical information to support climate-related decisions may be difficult to satisfy in the near future due to a paucity of skilled mediators" (2010:34). Gulledge and colleagues both constructed and benefited from their claims that "skilled mediators" could close gaps between the logic and practice of climate science and national security.

According to documents from CSIS and CNAS, in 2006 the two organizations began assembling an "eclectic" group of experts that "met regularly to start a new conversation to consider the potential future foreign policy and national security implications of climate change" (Campbell et al. 2007:5). As authors Kurt Campbell and Christine Parthemore (2007:13) frame the resulting report: "Our collaboration engaged, for the first time, climate scientists and national security specialists in a lengthy dialogue on the security implications of future climate change. Our eclectic group occasionally struggled to 'speak the same language,' but a shared sense of purpose helped us develop a common vocabulary and mutual respect." The group included a high-profile list of diverse participants, including economist and security strategist Thomas Schelling, Michael MacCracken (at the time chief scientist and board member of the Climate Institute), Ralph Cicerone (an atmospheric scientist and National Academy of Sciences president), and former CIA director James Woolsey. CNAS experts thus took it as their role to organize science-security interactions and capitalize upon the process by becoming among the first climate security organizations.

Institutionalizing translation work paid off for individual experts. For some, it entailed gaining credibility as scientific experts who, although perhaps peripheral in academia, could participate as "translators." As another example, author and CNAS cofounder Kurt Campbell reproduced his status as a longtime national security strategist. In 2009 he drew upon work among climate security experts to publish *Climate Cataclysm* (Campbell 2009) with the Brookings Institution. By 2009 climate expertise was having widespread effects throughout the world of think tanks. Major organizations like the Brookings Institution began to approach climate change through the discourse pioneered by more specialized think tanks. By and large, these think tanks had come together with a common goal of taking climate change seriously and presenting a nonpartisan basis for "translating" climate science into security policy.

THE STRUCTURE OF THE CLIMATE SECURITY FIELD

Having outlined several important climate security organizations and individual experts, I can now more generally characterize the basic components of a climate security field (represented in figure 14). Although to be a climate security expert may universally require pragmatic and nonpartisan performances, experts have different skills, capacities, and orientations and can be distinguished conceptually based on those differences. One axis of differentiation, following Pierre Bourdieu's (1986) conceptualization, represents relative degrees of "cultural capital." In the present

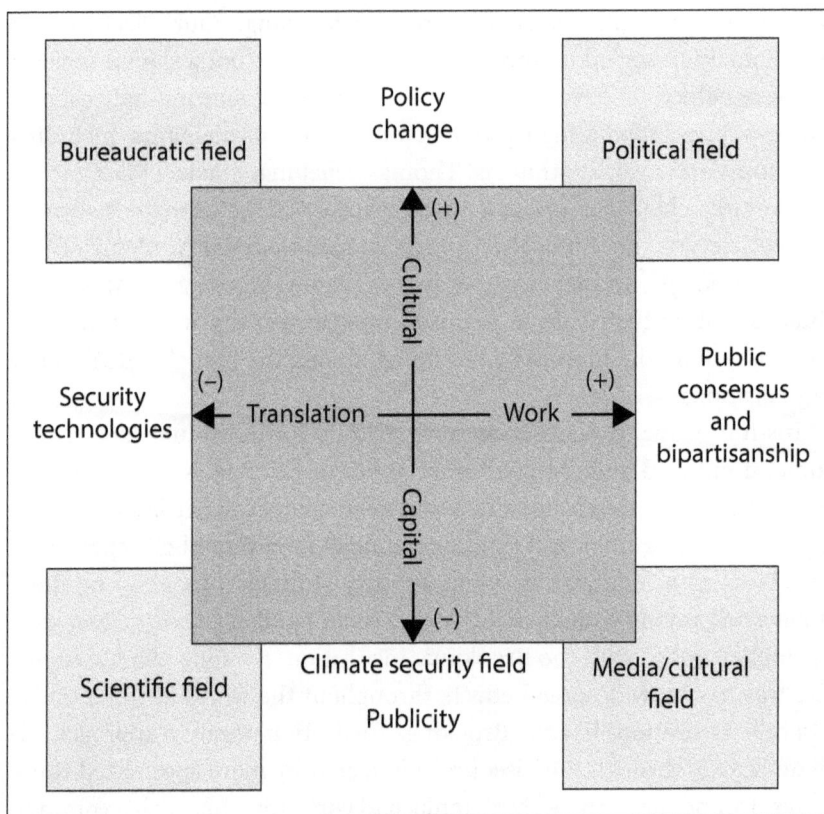

Figure 14. Conceptual field diagram of climate security expertise. *Source:* Created by author.

case, cultural capital is indicated by formal credentials and embodied ways of acting and thinking that explicitly draw from one or another social field and are recognizable to others as significant to climate security. For example, climate science may have its own mechanisms of evaluation and recognition, but for climate scientists to participate in the climate security field, they must perform competences recognizably relevant to climate security. To gain cultural capital, scientists invest in "actionable" science. For academic researchers, these investments may carry professional risk if they detract from established norms around research productivity, but they may also offer opportunities to circumvent the competitive hierarchy of scientific careers. A climate security expert drawing upon recognition from others as being "a scientist" may not possess high standing among other scientists.

Cultural capital is generally field specific. In other words, to be a dominant politician is to master a set of skills and habits routinized for power-seeking purposes, which generally differ from the skills and habits required by a government official seeking to garner recognition as a qualified administrator (Weber [1919] 1946). Cultural capital is also typically institutionalized differently within one field versus another: a military rank holds an ambiguous relation to academic credentials, and one's record of party leadership may or may not be recognized in the same manner as media fame. The climate security expert is a central player in the field to the degree that they can successfully claim possession of a relevant, institutionalized form of cultural capital.

If cultural capital derived from existing fields forms one axis of differentiation within the climate security field, *translation work* forms the second (see figure 14). The climate security expert must skillfully "translate" knowledge and competencies gained elsewhere and invest it in climate security. Experts perform translation work within organizations by producing reports, expert claims, policies, and specific security technologies. These products link together science and government. But translation work must also be recognized, more informally and among participants, as an individual's ability to symbolically transform one thing into another: a political claim into a bipartisan consensus statement, a scientific finding into a security application for a target government entity, a think-tank report into a consumable media platform, and so on. This alchemistic work is difficult, because actors must traverse different fields, for instance science and politics, without making their translation work obsolete. Actors' degree of cultural capital, along with the kind of translation work they perform, relates to the structure of individual think-tank organizations. Organizations overlap differently with scientific, political, bureaucratic, and cultural/media fields.

How actors orient toward the translation work they perform is likewise related to the kind of cultural capital they have invested in the field. Such investment is necessarily tied to their biographical trajectory and the broader goals and scope of each organization. So, although some individuals have come to climate security as retired military officials, others have emerged from existing public policy think tanks. Joshua Busby, for example, authored a Council on Foreign Relations special council report, *Climate Change and National Security: An Agenda for Action*, in November 2007, for which he primarily drew upon interviews with individuals affiliated with CNAS's climate program. Being younger and having completed a PhD in political science at Georgetown in 2004, Busby invested his climate security expertise in academia by contributing to the emerging literature in political science on climate security (Busby 2008). More significant, Busby became involved in projects funded through DOD's Minerva program (designed to enroll academic social scientists in security strategy), including the Strauss Center for Climate Change and African Political Instability at the University of Texas-Austin. As of 2023, Busby remained an active member of the Council on Foreign Relations and a senior research fellow at the Center for Climate and Security (CCS). Working with CCS, he participated in linking think-tank work on security policy to scientific developments in order to explore, for example, tailored climate-political instability modeling.

As another example, the staff, advisers, and products of ASP reflect a moderate degree of cultural capital, drawn primarily from participants' training and credibility in political and government agency affairs. Such cultural capital is successfully translated into media presentations. In this context, claims to climate security expertise are only roughly rooted in specific predictive knowledge, yet they are strongly connected to the public legitimacy of military experience in confronting an uncertain security environment. The organization's ties to media networks and its public efforts to form citizens into a constituency around climate security policy demonstrate that the organization's translation work is less directly tied to developments in climate science and more strongly organized around public pressure for policy action within the security community. For example, the organization's *Global Security Defense Index on Climate Change* (see ASP 2014) provides indices, scores, and maps that compare the degree to which national governments consider climate change to be a threat to their national security. The product does not objectively define climate risk with reference to scientific projections. Rather, it aims to chart isomorphic patterns across countries and thus spur governments (specifically the United States and its allies) to prepare for national security threats more proactively.

The formation of each major think tank that has occupied the climate security field in the last decade generally fits within the conceptual representation sketched in figure 14. It is noteworthy, however, that some individuals participate in the climate security field for ulterior purposes, even if others treat them as legitimate participants. For example, retired admiral Frank "Skip" Bowman was a founding member of the CNA MAB and was subsequently elected as an advisory board member of CCS. Think-tank reports frequently cite his statements regarding climate change. After receiving an engineering degree from MIT in 1977, Bowman worked in the navy, achieving the rank of admiral. By 1996 he was director of the Department of Energy's Naval Nuclear Propulsion Program, and he continued to advance fossil and nuclear energy, serving executive roles in strategic investments (notably, serving on British Petroleum's Geopolitical Committee and as a director with Morgan Stanley Investments). British Petroleum documents cite his "experience of the U.S. and global political and regulatory systems" as important to the Geopolitical Committee's goal to identify and manage "major and correlated geopolitical risk."[7] After retiring from the navy in 2004, Bowman served as president and CEO of the Nuclear Energy Institute, a DC-based nuclear industry trade association. Bowman has more recently advocated ambitious nuclear energy programs through controversial Middle East "nuclear security" projects with a contractor, X-Co Dynamics Inc./Iron Bridge Group (Stein 2017). This career trajectory shows that Bowman's orientation to climate security is institutionally connected to chiefly economic considerations in the energy industry. Yet because long-standing high-rank military experience holds a premium in the field, other people nevertheless uphold him as a climate security expert, which has given him license to provide expert commentary on many scientific and security topics.

LINES OF CONTENTION AND EXPERT INTERVENTIONS

Climate security experts share a common goal to understand and treat climate change as a matter of threats to national security. However, climate security experts also participate in a hierarchy. Despite frequent performances of consensus among climate security experts, actors can contest the structure of the field. Scientific credibility, national security priorities, political stances, and media representations may all come under scrutiny among participants, and they are all subject to change over time.

The Wilson Center's Environmental Change and Security Program presents an important example regarding how such contestation surfaced in the 2010s. The organization has not centrally drawn upon military expertise as a form of cultural capital, and its work, unlike that of numerous other organizations, neither features nor directly acknowledges military officials. The organization has actively rejected discourse of "climate wars" and "climate refugees" (Wilson Center 2016). These substantive positions on climate security reflect the Wilson Center's peripheral position in the climate security field and its alternative orientation to media and state-bureaucratic fields. The organization's leaders openly question media representations of climate security, most notably in their report launch event titled "Beyond the Headlines: Climate, Migration, and Conflict" (Null and Risi 2016).

In relation to federal state agencies, the Wilson Center has primarily partnered with the U.S. Agency for International Development (USAID). Compared to DOD-centered climate expertise, USAID has emphasized "resilience" and "capacity-building," rather than strategic defense secured by military means. Those advancing climate security expertise through the Wilson Center program partially decenter "national security" as the object of what USAID and other officials recognize as a fundamentally "human security" problem (USAID 2012, 2023).[8] Compared to many climate security think tanks, the Wilson Center has also involved international development organizations in its approach to climate security, including the report, *A New Climate for Peace: Taking Action on Climate and Fragility Risks* (G7 2015). The report was commissioned by the G7 and published in collaboration with high-profile, European-based climate security organizations: Adelphi, International Alert, and the European Union Institute for Security Studies. The example of the Wilson Center exposes lines of contention within the climate security field and it demonstrates novel ways for actors to link a contested position within the climate security field with homologous positions within the state (namely, USAID as the development arm of the state).

To summarize thus far, the organizational ecology populated by climate security experts beginning around 2007 has included actors from diverse backgrounds in science, government, politics, and long-standing policy think tanks. Climate security experts organized activity across hitherto uncoordinated social domains. Thus, the formation of a climate security field was structured by interdependence among participants' available resources and sources of cultural capital. However, axes of stratification and hierarchy also emerged in relation to legitimate possession of cultural capital and the "correct" performance of translation work. Even so, the field is relatively settled, having a basis in a set of organizations and recognizable skills, techniques, and competencies. Collectively, the "game" of the field involves making uncertain climate futures legible to security strategists while depoliticizing both climate science and national security practices. The field also has generated effects in each of the primary neighboring social fields in which climate security experts participate: the bureaucratic field, the scientific field, the political field, and the media/cultural field. I address these briefly.

Climate security experts, after accumulating a "vocabulary" (as the CNAS report writers call it) and populating a set of organizations, intervened in the broader fields from which experts themselves had emerged. Such interventions have been especially significant to defense policy and institutions of the national security state. Initial efforts to integrate climate security expertise within the bureaucratic field were strongly rewarded during the Obama administration. For example, experts provided testimony that supported successful amendment of the National Defense Authorization Act that, beginning in fiscal year 2008, required consideration of the effects of climate change on defense facilities, capabilities, and missions. Retired officials involved in climate security think tanks continued to provide expert bases for such consideration, as evidenced by the 2018 National Defense Authorization Act and associated reports (U.S. Congress, House of Representatives 2017:165–70). The NIC, under Director Thomas Fingar, drew upon available experts and think-tank reports in its 2008 report (partly declassified in U.S. Congress, House of Representatives 2008; see also NIC 2012). The CIA's Center on Climate Change and National Security was established in 2009 and publicized an intention to "be aggressive in outreach to academics and think tanks working on the issue" (CIA 2009). Subsequent central intelligence and defense reports, including the Quadrennial Defense Review (DOD 2010, 2014b), drew upon think-tank work to characterize climate risk. Regarding President Obama's comprehensive Memorandum on Climate Change and National Security (issued in September 2016), MAB director Sherri Goodman states: "It's not a new direction, but it is reinforcing and formalizing a direction in which the U.S. government was already headed. . . . That's how you turn concepts into action in the government" (Kahn 2016; see U.S. White House 2016). Executive orders and Defense Department directives (e.g., DOD 2016) firmly integrate climate security expertise within government. To govern climate crisis was to render legible future instability in the climate-society relationship, and hence to anticipate how to preserve social order in its wake.

Climate security experts have also intervened in the scientific field. Many of those participating in the climate security field have continued to refine security technologies as a basis for strategically governing climate change. Experts have likewise sought to change the character of scientific information and its circulation through government. For climate security experts, the required translation work between security capabilities and climate science has paved a two-way street. In one direction, the CNA MAB (CNA 2007:39) has argued that "the military and intelligence communities" should actively share their data and surveillance infrastructure to aid "American efforts in understanding climate change signals," a process it has since helped to orchestrate. In the other direction, climate security experts Marcus King and Sherri Goodman (2011:808–9) have called for "actionable" climate information. To that end, they propose "that national security be woven into future impact assessments conducted by the IPCC," an effort that has partially succeeded (2011:808–9).[9] "Actionable science" similarly frames the science policy directives from the Obama administration regarding climate security (U.S. White House 2016:4, 10). The Biden administration's development of a Defense Department "climate.mil" portal took a similar approach, intended to serve as a "one-stop focal point for scientifically credible, neutral, authoritative, and actionable climate change information" that treats climate change as a wide-reaching "threat multiplier" (DOD 2023). Security-oriented scientific products form one of many "climate services" that tailor knowledge production to specific user demands or formats. Notably, investment in "actionable" science may proceed alongside the other, more famous, layer of political contestation over scientific authority—namely, that which pits a view of scientific truth against its ideological rejection.

Regarding the political field, as early as Al Gore's efforts to advance national-level and multilateral climate mitigation policies, security has been an important vector for Democratic (and at times, Republican) appeals to bipartisan support of climate policy. The future of such efforts remains uncertain, insofar as the moderating interventions of climate security experts have hardly done away with polarization. Examples are manifold. When experts with the CCS were called on to evaluate 2016 presidential candidate Martin O'Malley's claim that climate change helped to cause the rise of ISIS, they emphasized that they "can't speak for O'Malley," insofar as "this is not something that is partisan," but rather relied on other climate experts to argue that climate will contribute to political instability (MSNBC 2015). Climate security experts, however, failed at moderating Republicans' and conservative media's rejection of O'Malley's claims (e.g., Fox News 2015). The conservative countermovement has been generally far more successful at mobilizing political resources to block climate action, including within the security and military establishment and especially with the support of the Trump administration.

Notwithstanding partisan polarization regarding climate change and U.S. national security, climate security expertise has retained its commitment to security expert consensus as a possible strategy for garnering public support. A major political intervention has taken the form of the Climate and Security Consensus Project (CCS 2016) and related statements issued by long lists of high-profile signatories. In 2016, the Climate and Security Advisory Group (CSAG), chaired by the CCS, provided the comprehensive *Briefing Book for a New Administration,* in which they argued that climate change should be a central concern across national security strategies, programs, and agencies. They projected their expertise, identifying as a "non-partisan group of 43 U.S.-based military, national security, homeland security, intelligence and foreign policy experts" (CSAG 2016:3).

Under conditions of extreme polarization, efforts to depoliticize climate security within government have presented experts and their allies in government with persistent, if not intractable, challenges. Climate policy remained split on a partisan basis, despite partial victories of climate security experts. Successful interventions include the Langevin Amendment to the 2017 House Defense Authorization Bill, which required consideration of climate change in defense operations. Yet climate security expertise rests on shaky political grounds. Climate security expertise was positively politicized when, on October 13, 2015, during a televised debate between candidates for the Democratic nomination for the U.S. presidential election, Senator Bernie Sanders answered the moderator's question, "What is the greatest threat to our national security?," stating firmly and plainly: "*climate change.*" Climate change, he argued during his campaign, "is every kind of issue all at once. Its role in exacerbating global conflict and terrorism makes it a national security challenge." He concluded, "Climate change is not just an 'environmental issue,'" pointing out that "[t]he CIA and the Department of Defense both say that climate change is one of the great security issues facing this planet."[10]

To the chagrin of climate security experts, the indirect effect of Sanders's and others' claims was a polarizing one. On December 18, 2017, the Trump administration released its National Security Strategy (U.S. White House 2017). Unlike prior NSS reports, including those issued by the George W. Bush and Obama administrations, Trump's did not mention climate change, much less heed the formal briefs provided to him by climate security experts. What the document did note was that climate *policies* are "dangerous" to the stated goal of "ensuring energy dominance," meaning that "U.S. leadership is indispensable to countering an anti-growth energy agenda," which Trump outlined to include "onerous regulation," implicitly implicating those regarding fossil fuels (2017:22). Trump, who strategically removed climate science and climate policy from government, likewise pulled climate change from his administration's vision of security (Voosen 2018).[11]

Within hours of the NSS release, climate security experts took to media and other channels to make statements on the dangerous implications of Trump's view of national security vis-à-vis climate change. Francesco Femia and Caitlin Werell, the copresidents of the CCS, compared the Trump administration's strategy to prior agency and expert reports, concluding: "To national security and defense leaders, there's absolutely nothing political about climate change. It's a security risk, it makes other security risks worse, and we need to do something big about it" (Werrell and Femia 2017). Although Trump's decision not to address climate change in his administration's national security strategies is hardly surprising, the organized challenge to Trump's NSS by a self-described "climate security" community demonstrates that the field's participants continued to intervene in anticipation of a political administration that could valorize their expertise in government.

Experts at times became the center of struggle between state agencies, the Trump administration, and the U.S. Congress. As one example, Dr. Rod Schoonover, a senior analyst of the Bureau of Intelligence and Research (within the U.S. State Department) and former director of environment and natural resources at the NIC, left his post in 2019 after facing White House and congressional censorship of his report and testimony to the House Intelligence Committee's hearing on the national security implications of climate change. His work, although consistent with mainstream climate security expertise, was ultimately rejected for inclusion in the *Congressional Record* (Friedman 2019; Schoonover 2019). The commentary recorded on the hearing documents is unabashedly skeptical about both scientific consensus and climate security claims. The commentators write, for example, that IPCC "computer models . . . don't work" and are thus useless, and that emerging science on climate "tipping points" represents "propaganda" since the time of Al Gore.[12]

Although nearly all executive-level climate change policy efforts were curtailed by the Trump administration, climate security experts at the time continued to invest cultural capital in translation work, alternatively oriented toward building a broadly cultural vision of climate change as a matter of national security. Experts who maintain ties to media outlets have been especially significant for generating a public constituency for climate security. For example, CCS, established in 2011, has since around 2015 become a central climate security organization largely by consolidating programs of other organizations. Its directors, themselves largely having become experts through the press rather than scientific or policy credentials, aggressively seek to promote media attention to any report, study, or statement regarding climate change and national security. CCS thus maintains a real-time social media, press, and blog presence that did not feature among the activities of think tanks established earlier. More than previously established climate security organizations, CCS experts portray themselves not as an organization that has specific expertise, but rather as a "hub" of climate security experts, understood to be the legitimate spokespersons who can effectively shape representations, especially via press releases and media coverage, about the threat of climate change.

Film production illustrates another avenue for experts to use media to promote a nonpartisan approach to climate security. Consider the documentary film *The Age of Consequences* (2016). Borrowing the title of the 2007 CNAS report of the same name, the film was directed by Jared P. Scott and funded by the Goldman Sonnenfeldt Foundation (a part of Michael Sennenfeldt's larger portfolio of nonprofit and business ventures in energy security). The film drew upon a mix of military sources and climate security experts, especially those involved with CCS. The film's experts portray climate change as an "accelerant of instability" and a primary national security threat. The film represents a strategic public intervention that sharply differs from contemporary major films focusing on climate change. Another film is *This Changes Everything* (2015), based on the book by Naomi Klein with the telling subtitle "Capitalism versus the Climate." The film presents a political-economic narrative that rejects climate security as a relevant frame and centers around social movements that confront the capitalist system and status quo government. On the other side is *Climate Hustle* (2016), narrated by prominent climate-change denialist Marc Morano. The film supports conservative skepticism regarding climate science and policy. *The Age of Consequences* clearly seeks to raise social alarms and resist climate skepticism, yet without the critique of the basic structure of society central to *This Changes Everything* and left environmental groups. Rather, *The Age of Consequences* elaborates climate security discourse into a familiar genre—that of the thrilling action film. In an accessible, public-facing manner, experts link scientific authority to national-military legitimacy and the get-it-done orientation specific to climate security expertise. The production of the film, and the narrative that underlies this cultural intervention, is thus an effect of the orientations and agendas that originated in the climate security field.

As a narrative trope, constructing ours as an "age of consequences" frames the present historic moment in relation to a broad governmental vision, one in which climate security strategies can (and must) protect the country from impending disorder around the globe. From around 2005 onward, climate security experts established their role as spokespersons for this disorder. By occupying an organizational ecology of think tanks, climate security experts have succeeded in translating scientific, political, bureaucratic, and cultural production. Such translations have provided a substantial basis for approaching climate change as a tangible security crisis and an apolitical set of national threats.

Internationally, climate security expertise has steadily gained ground. When North Atlantic Treaty Organization (NATO) secretary general Jens Stoltenberg participated in the COP23 climate summit in Dubai in 2023, he emphasized that NATO was working to shape international norms upholding "green defense" and "environmental security" as pillars of future military strategies. NATO's Centre of Excellence for Climate Change and Security was established in 2021 for just this purpose. The discourse of climate security, along with the international and think-tank organizations advancing it, meant that the Biden administration could readily incorporate it into security and defense strategy. The integration of climate science into security and defense strategies was mandated by President Biden's Executive Order 14008, signed in January 2021, which intended to "[put] the climate crisis at the center of United States foreign policy and national security." Defense Secretary Lloyd J. Austin III praised the order and oversaw the mandated development of the DOD's Climate Risk Analysis report (DOD 2021). DOD's report was published shortly after the NIC released its comprehensive *Global Trends 2040* assessment, which integrates climate change into scenario planning with an emphasis on climate-induced conflict and resultant global insecurity risk (NIC 2021). Political polarization notwithstanding, clearly climate security expertise is finding its place among the tendrils of the national security state.

DEPOLITICIZING CLIMATE CHANGE THROUGH SECURITY?

Those who have participated in building security technologies and making up a new kind of social actor—the climate security expert—have all claimed, in one way or another, to be addressing the problem of climate change: global disorder is coming, and drastic action is necessary to ensure that civilization as we know it may be preserved. The history of climate security is strongly marked by strategic efforts to depoliticize science and security. This effort yielded successes during the Obama administration, was forced onto the plane of political conflict during the Trump administration, and has recovered substantially in the U.S. in recent years given the interstitial structure of climate security expertise.

Climate security experts may ultimately have their day, especially given that scientific projections and international climate policy failure seem to inspire ever more catastrophic visions of the future. As collapse narratives are legitimated in science and culture (Brozović 2023), apocalyptic governance seems all the more rational. Those who claim authority over the question, "What is to be done when the time is too late?," may succeed in future political environments. Such a question may mobilize new political associations that amplify climate security expertise. If so, climate security experts would likely become more central to governmental practices regarding climate change. They would do so based on cultural capital and translation work they have already established.

The transformation in meteorological government toward governing the social impacts of climatic instability is not yet complete, insofar as discourse surrounding climate futures depend on the play of forces within science, politics, and government and the response of those forces to ongoing environmental change. Nevertheless, the (a)political posture of climate security discourse was critical to the establishment of a climate security field. Therein lies the fundamentally contradictory character of climate security expertise: as actors work to depoliticize climate change, they simultaneously advance a deeply political vision of how the future must be governed. In Pierre Bourdieu's (2015) terms, by a kind of "magic," the symbolic power of climate expertise is to develop a consequential categorization, a *vision* and *division* of the social world, that must be anticipated, protected, and brought into being. Based on this implicit symbolic power, the climate security field has provided an effective basis for confronting the partisan grounds of the countermovement characterizing "climate denial," even though the Trump administration proved remarkably successful in suffocating state action on climate change. Some sociologists indeed lauded security and defense experts as "unexpected allies" to those involved in political battle against climate skepticism (Szasz 2016).

Climate security experts may block political movements to disassemble climate policies, which I have characterized as comprising a climate state. Climate security may also offer a successful bridge between climate science and government through the coproduction of security technologies. As with mandating emission targets, inventing geoengineering technologies, or creating a market for carbon credits and offsets, "climate security" appears to make climate change a problem that can be solved in tangible terms by apolitical and technical means. In short, security technologies and related social practices make the climate future, however uncertain or catastrophic, in principle *governable* by virtue of their asserted capacity to represent climate, space, time, and social dynamics in newly legible ways.

Nevertheless, experts who draw upon security technologies, for example dual-use military surveillance and environmental monitoring systems or strategic scenario planning, inform deeply normative visions of future global developments. These visions are inherently political, often involving the projection of an unequal division of power and control over resources that calls for the maintenance of, as examples, U.S. world energy dominance; border security that can manage and exclude migrants (White 2011); defense spending; and the vast network of U.S. military bases, many of which are threatened by climate change (Johnson 2004; DOD 2021). Climate security discourse in this respect reflects on a global scale the quests to govern climate during the eighteenth and early nineteenth centuries, centered as they were around social and geographic delineations that, once understood, could inform practices aimed at maintaining a civilized order. By making global risk and disorder legible through security technologies, climate security projects a political order, in which "outside" risks must be anticipated and contained and those "inside" the territorialized body politic must be protected.

While focusing on projecting a world of disorder, however, climate security discourse may foreclose opportunities to confront the basic social structures that are producing the increasingly unsustainable and unequal world that climate security experts want to avoid. The problem does not appear to be, as the literature on securitization warns, that climate security may enact a "state of emergency," through which climate change is taken out of democratic consideration.[13] Indeed, climate security experts in the United States have been more involved in building engaged constituencies in civil society than in subverting democratic institutions. Nevertheless, the (bio)politics of climate security fundamentally involves an effort to channel knowledge, representations, and anxieties about climate change into a hardening of state security as the natural "adaptive" response to climate change. "Adaptation," as Michael Watts (2015) traces more generally as a form of explanation for action, can obscure the power relationships that are often invested in representing a societal trajectory in quasi-natural terms. The discourse and practices of climate security, however rooted in effective translation between science and policy, remain both the products and the exercise of symbolic power.

The jagged historical contingencies that gave rise to climate security expertise, along with the ongoing contention within the climate security field regarding what counts as "security," are hardly adaptations to a climate "out there" in nature. To believe as much is to misrecognize the basic governmental content of the very category of climate, substituting the mechanics of nature for what are in fact the mechanics of power. Just as in other periods and contexts, however, other ways of knowing, seeing, and governing climate change remain possible. Climate crisis remains open to various ways of treating the future that may embrace or contend with the present configurations of climate, science, and government. The way in which future struggles have taken shape within the field of climate security demonstrates that historically contingent social action can produce new ways of envisioning, anticipating, and acting upon the future.

A genealogy of climate security raises the issue of what could be different. The conclusion takes this up. But it doesn't do so by tinkering with the existing terms of climate science and policy. Rather, it brings in the long view of meteorological government and considers what resources, at once intellectual and political, may be available for reconfiguring climate, science, and government today. The chapter thus pulls together existing, and possibly new, avenues for governing climate beyond the limitations of climate security and outside appeals to technocratic answers for stabilizing climate amid crisis.

Conclusion

LEGIBLE ALTERNATIVES? REMAKING CLIMATE,
RETHINKING CLIMATIC STABILITY

A genealogy of efforts to govern climate has brought into focus three roughly consecutive configurations of climate, science, and government. The first is a matter of *climate change and social order*, the second of *climatic stability* and *economized weather*, and the third an elaboration of *climate crisis* and the specter of governing potentially catastrophic futures. Let us briefly take stock of where we have come before turning to the implications for addressing the pressing cotemporary problems at the intersection of climate science and government.

Part I began by tracing the formation of what those concerned at the time called "rational accounts" of climate and weather, with a focus on North American colonies and then the early United States in comparative context from the 1770s to the early nineteenth century. Two major meteorological controversies shaped the relationship of climate knowledge to government at the time; the first was over the climatic causes of health and disease, and the second was over how American climates and bodily "constitutions" might have been changing. Both controversies had serious implications for how actors evaluated the progress of civilization. Expert positions in these controversies emerged primarily in medicine, political economy, and natural philosophy, domains in which people sought to develop prescriptive tools for making a productive, healthy, and disciplined population stratified by social categories. What emerged within these controversial issues was a diffuse logic of governing climate, which later intersected with U.S. state formation over the first half of the nineteenth century, just as similar controversies were coproduced in other imperial and state formations. Climate knowledge up to the mid-nineteenth century engaged three core components of U.S. state formation: territoriality, bureaucratization, and ideologies of racial stratification. Through involvement in these processes, meteorologists succeeded in producing a meteorological observation network, scientific evaluations of climates' healthfulness and productivity, and a racial climatology that connected empirical climatology with racial thought at the time.

As traced in part II, a different logic of meteorological government emerged vis-à-vis industrial-capitalist climates. In this logic, actors considered climates primarily as geographic-statistical zones, marked not by change but rather by climatic stability. The new professional project of "climatology" from the 1860s to the 1910s put the stability of spatially delineated climates on a scientific basis, primarily through the simple use of exhaustive observation and statistics. During that period climatologists' scientific activities became closely linked to state action, especially government facilitation of commercial agriculture and trade. In turn, scientists and their allies successfully institutionalized climatology within the U.S. Weather Bureau and associated programs within the U.S. Department of Agriculture. "Stabilizing" climate unfolded as a scientific and governmental program keyed not only to the reality of observed regional climates but also to the goals and interests regarding agrarian-commercial and state expansionism. Indeed, Lehmann (2022) and Locher (2009) find in the case of French and German colonization projects that anxieties about climate change arose during this same period, likely because of the politically grandiose but socioenvironmentally fragile and novel nature of such projects.

The stabilization of climate articulated with the formation of modern weather prediction. Weather prediction practices relied on a stable climate as a reference point for normal weather, against which abnormal events could be measured. Making capricious weather events (along with geographically demarcated climates) legible to administrative and economic interests supported the rationalization of American capitalism. Yet the scientific and governmental practices that would rationalize weather from roughly the 1870s to 1920 faced numerous challenges. Meteorologists and state actors struggled first to materially establish the meteorological data infrastructure, and then to secure official meteorological knowledge against alternative bases of knowledge. Official knowledge, taking the form of Weather Bureau storm warnings, reports, and (by the early twentieth century) long-range forecasts, served to rationalize how the public anticipated weather. In effect, discourse about weather in both science and public culture became structured around "normal" versus "abnormal" weather events, against which "climate" stood in as a stable background. The development of "official" and "normal" weather, predictive forecasting, and a climatology centered on "stability" made it possible for social actors to successfully incorporate atmospheric phenomena into rational economic calculations, especially central to commercial agriculture, finance, insurance, and trade.

Part III showed that initially, over the period from around 1930 to the 1980s, the logic of governing climatic stability intensified as atmospheric scientists and state actors envisioned a world characterized by a more predictable, even controllable, climate. Yet in the decades following World War II, the configuration of basic science within government provided a unique opportunity for researchers in meteorology, oceanography, and related disciplines to establish a climate science field. Unlike the climatology and meteorology of prior decades, "climate science" centered on explaining global climate dynamics and, by the 1970s, understanding anthropogenic climate change. Through the 1960s, appeals to climate control continued to feature in broader Cold War visions of control, yet such an approach could not contain the emerging structure of climate science. Nor could it be mobilized to address the increasing legibility of global warming. Control faded. Emphasis on deliberate, technical climate modification became concern for *inadvertent* and possibly uncontrollable human-caused modifications (Schubert 2021).

The result, as of the 1970s, was a state-science dilemma. This dilemma stemmed from atmospheric sciences becoming prestigious in the scientific field but ambiguous as engines of economic production and social progress. In a time of heightened (and not terribly partisan) environmental awareness, climate scientists drew upon their expertise to raise public concern for anthropogenic climate change. Meteorological government in these terms directly undercut any logic that presumed climate to be stable, much less controllable. Legislative and regulatory battles ensued regarding whether and how to govern global climate change, a polarizing process that resulted, on the one hand, in climate-oriented environmental policy, and on the other hand, in the climate change countermovement.

Climate crisis emerged not only as a result of global warming but also through a crisis of governance. Climate crisis is not simply a fact of climate history. It also represents the accumulated history and institution of meteorological government at a given historical moment. Under the circumstances of an unstable climate, the crisis became how to govern a disorderly climate system and preserve social order. Sociologically tracing the implications of this question required following trends in the relationship between innovations in climate science/technology, the social structure of expertise, and the historically specific configuration of the state. As examined in Chapters 6 and 7, an emergent pathway for meteorological government out of the dilemma involved projects regarding climate security, which had roots dating back to the Cold War period but flowered into maturity in the first decade of the twenty-first century. Given political polarization, discourses and technologies of climate security may be resolving some political and scientific issues that have beset attempts in national and international forums to confront climate crisis. Especially through the coproduction of security technologies, climate security emerged as a field of expertise oriented toward depoliticizing climate change and making legible a governable future. In the face of uncertain risks and possibly catastrophic disorders around the world, climate security experts, especially in the United States, are drawing together science and government and providing a vision for securing the power relationships that characterize global society and existing geopolitical arrangements.

The foregoing genealogy holds serious implications for two issues I believe are at the heart of climate science and climate governance today. The first concerns recent efforts at "climate stabilization," which can be approached with reference to the historical juxtaposition of science, climatic (in)stability, and capitalism over the past two centuries. By way of considering this issue, I argue that science regarding climate stabilization should more deliberately incorporate social dynamics—especially social inequality. I call this plausible configuration of knowledge "critical climatology." The second issue is the prospect of confronting the political bases of climate security, a project that does not yet exist but could be developed by drawing lessons from the long history of climate knowledge as a site of power struggles, thereby developing possibilities for alternative configurations of science in relation to politics. Addressing this issue, I focus especially on possible alternative directions for coproducing science and social order as they might be expressed within the discourse of climate justice.

REVISITING CLIMATE STABILIZATION

Those who deny climate change as well as those seeking to optimize global temperature by geoengineering both hold to discourses of climate stabilization. Climate-change skeptics believe that climate is stable, or at the least, hardly influenced enough by human activity to warrant curbing fossil fuel use. Those seeking to deliberately moderate global temperature believe human action can—and must—rebalance, or stabilize, our presently warming plant. The basic problem of climatic stability should be read in relation to the rise of industrial capitalism. Let us pull out the historicized terms of climatic stabilization in light of recent developments.

Over the course of the development of modern capitalism, weather has represented a blessing and a curse. If weather presents a source of place-based wealth, it also represents a potential threat to any project that treats nature as a regular and stable source of material resources. Atmospheric phenomena, to borrow James Dunbar's (1781) eighteenth-century terms again, frequently resist being "recovered from Chaos" by science and formed into a new domain of humanity's "patrimony." Yet for a time, weather and climate were in fact apprehended by rational principles and harnessed to structure the formation of rational industrial-capitalist society. Today, however, global warming is disruptive to both the stability that climate presumably had and the climate knowledge that helped to make atmospheric phenomena legible within modern capitalist society. The modern view of nature as a well-ordered machine, historicized by ecologist Carolyn Merchant (1983) as having wed the scientific revolution to capitalistic innovations, may have partially incorporated "climate" as an element of Dunbar's proposed "patrimony," but only at apparently what has become great peril to a rationally ordered society that might reasonably project itself into a stable future. As the landmark Stern Report (2007) made clear and recent IPCC reports continue to affirm, the economic costs of global warming are projected to be enormous and will likely involve economic recessions and extreme market volatility—developments consistent with James O'Connor's (1998) conceptualization of environmental crisis as a fundamental driver of capitalism's contradictions.

Nevertheless, anti-environmentalists and those aligned with fossil-fuel industries protect the claim that climate has its own relatively stable equilibrium, meaning industrial society can and will flourish (Epstein 2014; Lomborg 2001; Simon 1996). "Nature, Not Human Activity, Rules the Climate," was the title that announced the Heartland Institute–funded Nongovernmental International Panel on Climate Change 2008 report. Skeptics like these argue not only that humans lack the geophysical agency to create the climate changes that scientific research overwhelmingly identifies as occurring, but also that climate is fundamentally stable and naturally suited to capitalist growth. Innovation, perhaps through "natural capitalism," can handle any remaining negative market externalities (Hawken, Lovins, and Lovins 1999).

Such a climate ideology continues to resonate with certain findings from American public opinion surveys. For instance, Leiserowitz, Smith, and Marlo (2010) find that a large majority of those who do not think global warming is happening hold a mental model in which "Earth's climate system is very stable," as opposed to "fragile" or even "gradual to change." This is a striking juxtaposition: in 1857, still in the throes of a meteorology and a public fundamentally concerned with *changing* climates, climatologist Lorin Blodget (1857:481) wrote of climate: "Attached ideas of change to the whole subject is difficult to remove." If Blodget's climatology helped pave the way for an approach to climate suited to capitalistic growth and bureaucratic administration, climate "skeptics" today seek to uphold a similar, if farcical, position.

On the one hand, skeptics and many of their unwitting followers adhere to a myth of a stable *natural* order (that aligns with capitalism in particular). On the other hand, climate-change policy holds to a discourse of climate *stabilization*. As the *UN Framework Convention on Climate Change* has stated, beginning with its founding document in 1992: "The ultimate objective of this Convention and any related legal instruments that the Conference of the Parties may adopt is to achieve, in accordance with the relevant provisions of the Convention, *stabilization* of greenhouse gas concentrations in the atmosphere at a level that would prevent dangerous anthropogenic interference with the climate system." Subsequently, the global mean surface temperature has been the primary object of "stabilization targets" to which policymakers make gas concentrations meaningful (NAS 2010). A maximum 2°C increase in temperature since preindustrial times has typically formed what climate policymakers and scientists consider the benchmark for stabilization efforts. Many analysts suggest that strategic targets put forward by scientific assessments are already out of reach, even under scenarios of advanced emission-reduction policies. In *Global Warming of 1.5°C: An IPCC Special Report*, the IPCC (2018) argues that limiting global warming to 1.5°C is necessary to avoid irreversible and significant transformation of the biosphere and "would require rapid, far-reaching and unprecedented changes in all aspects of society." There are minimal signs of such far-reaching changes in the present social and political situation.

Climate crisis and social intransigence about "significant transformation" engender new possibilities. Another project for stabilizing climate, geoengineering, remains outside the purview of revolutionary social transformation and neoliberal climate policies like pricing carbon emissions. Fifteen years ago, Paul Crutzen (who coined the "Anthropocene" concept to represent the geologic epoch marked by human society) argued that a "daunting task lies ahead for scientists and engineers to guide society towards environmentally sustainable management during the era of the Anthropocene." The monumental task that he has continued to promote "may well involve internationally accepted, large-scale geo-engineering projects, for instance to '*optimize*' climate" (Crutzen 2002:23). Climatic instability, in this view, must be met with large-scale stabilization.

Geoengineering involves the use of technologies to deliberately modify Earth's climate. Geoengineering research and proposals basically take two forms. Solar radiation modification interferes with solar radiation, for example, by modifying atmospheric chemistry to enhance the Earth's albedo, or physical capacity to reflect (rather than absorb) solar radiation. Carbon dioxide removal relies on an altogether different approach and requires sequestering carbon by physically removing it from the atmosphere or oceans. James Fleming (2010) demonstrates the basic technocratic worldview that links geoengineering to the climate modification programs of the mid-twentieth century. Following early explorations in geoengineering (like those of Paul Crutzen), which were generally treated with skepticism, many scientists now believe that carbon dioxide removal and/ or solar radiation modification technologies must form at least part of the solution to climatic instability (Crutzen 2006; IPCC 2018; Oxford Geoengineering Programme 2018; Hansson 2024). Among scientific advocates, investigating technical solutions to stabilize climate is often framed as a fail-safe if policy action to sufficiently mitigate carbon emissions or transition from fossil fuels as a primary energy source is politically impossible (Markusson et al. 2014; Schubert 2021). To invoke the language of influential climate scientist Wallace Broecker's 2009 book of the same title, the goal is *"Fixing Climate"* (Broecker and Kunzig 2009).

It remains to be seen what scientific and political struggles, otherwise disconnected, geoengineering might in the future link together to stabilize, or "fix," climate. Acknowledging the contingency of such efforts, a different alternative to climate stabilization emerges. An intervention at once intellectual and political would be to resignify climate entirely and recognize that the knowledge regime of industrial capitalist society is entirely insufficient for present social needs. We may collectively ask: What if a critical climate science were possible—one that conducted analyses of time and space in a way channeled by human needs and values? One that did not did not proceed only to settle a universal reference point, or global "optimum" as a feature of an abstract climate "out there"? Consistent with Max Horkheimer's ([1937] 1975) break with "traditional theory" and his embrace of critical theory, why should anyone assume that science should be oriented toward "stabilizing" climate in order to preserve a ravaging, unsustainable, and oppressive political-economic order? I conclude with some implications that the present analysis might provide for exploring such a position, which, reprising Horkheimer, I call "critical climatology."

Critical climatology can reorient the tools of climate science toward a fundamentally different understanding of the climate-society relationship.

At the outset, a key point is that addressing issues of climatic stability cannot be reduced to climate stabilization by technical means, because the climate-society relationship is centrally marked by both geophysical and social instability. Yet climate sciences have long been separated from knowledge centered on socially stratified relationships to atmospheres.

Recall that during the late eighteenth and early nineteenth centuries, meteorologists, medical geographers, and other scientists did not operate on the basis of a divide between physical and human sciences. However dreadful some of their approaches to human difference may have been, they demonstrate that a divide between "the climate" and "the social" is a historical, rather than natural, one. Since the beginning of the twentieth century, scientists largely have allocated any designation of climatic meaning to geographic zones (chiefly with administrative or economic significance), to global circulation patterns, to mean average temperatures, and the like. Today, in the face of a planetary crisis, climate science and associated stabilization efforts retain the point of view that the world remains representable as global "machinery." The background assumption is that this machinery will work splendidly if only we fix it—make scenarios, set thresholds and targets, economize risk and marketize policy, eschew conservative backwardness, and explore a geoengineered "Plan B"—and human society might return to operating within the global climatic boundary (Rockstrom et al., 2009; Gunderson, Stuart, and Petersen 2018). Some scientists now view themselves as "planetary physicians"—almost akin to the moral climate physicians of the early nineteenth century—yet their diagnosis often proposes a "cure" of geophysical intervention, not social change (see Fleming 2014:341). By contrast, the problem for a critical climatology is not the planet, but rather social relationships, which actively participate in producing the climates that humans and ecosystems inhabit. Rather than operating on the planet with a physician's tools, a critical climatology can enhance the capacity for science to make legible the social-climate relationship. In particular, diverse and radically unequal social relationships to climate are hardly captured by crude medical and mechanical metaphors for the planetary predicament.

Critical climatology can draw some inspiration from the fact that climate science is already moving, albeit unwittingly, toward social analysis. Deborah Coen and Fredrik Jonsson (2022) astutely show how earth system science historically negated the social sciences, including historical analysis, but could successfully incorporate them. Climate research and policy bodies routinely recruit social scientists along with community partners, emphasizing the human dimensions of climate knowledge and "convergence research" that works across disciplinary divides (e.g., as expressed in the design of the NSF Navigating the New Arctic program). Intellectually, social-behavioral trends are undeniably important to concepts of change in the Anthropocene epoch. Thus, the IPCC (2018) has begun to incorporate such concepts as "societal transformation" in the ways it projects plausible future greenhouse gas concentrations in the atmosphere. When climate and other earth sciences integrate such social parameters into their analyses, they implicitly bring in assumptions about human society. As Naomi Oreskes (2015) shows, when climate scientists construct climate-change scenarios, they often include in their underlying models social-behavioral dynamics. Such assumptions need to be made explicit, and they need to be interrogated. Actual work of social scientists, however, is not usually incorporated into climate science. In broader terms, climate knowledge, as it is construed among its producers and users, does not center on the basic structure and trajectory of modern capitalist society. This trajectory involves increasing proportions of greenhouse gasses in the atmosphere but, perhaps more fundamentally, it also involves the instantiation of socially and geographically unequal climates (Roberts and Park 2006). Indeed, chapter 7 demonstrated how preparations for a socially unequal climate are well underway, and on terms that participants present as apolitical.

Critical climatology is reflexive in recognizing that expert knowledge is not inherently removed from power relationships and, indeed, often supports existing distributions of power. Given that many climate change researchers have dealt with political attacks on their science under the climate change countermovement, many are keenly aware of the social context of their work. The March for Science, which in 2017 mobilized scientists and the public to defend scientific values around the United States, reflected this awareness in the form of collective action. However, the point of a critical climatology is not to draw strong boundaries between science and society, but to engage issues of equity and justice in relation to climate action. Sunita Narain (2019), the Indian scholar and active critic of mainstream climate science and policy since the early 1990s, has called the principle of global inequity "the final frontier" and the undeniably "more inconvenient truth." Equity remains controversial, and hence peripheral, to climate governance. Engagement with these issues relies on climate knowledge that centers on accounting for and explaining the reproduction of unequal climates.

When drawn through a political genealogy, history provides some insights that can help advance this project. Consider some historical moments. In industrial, urbanizing England during the early nineteenth century, access to safe air became stratified by social class in ways that it had never been before. Friedrich Engels's ([1845] 1887:109) famous study, *Condition of the Working Class in England*, chronicled the many social and physiological impacts of polluted air and the environments of factory labor, for example, noting that pregnant teenage workers suffered because "the warmth of the factories is the same as that of a tropical climate." In the case of colonial policies and discourses, regarding both "real" tropical climates and territories in general, settlers appropriated land and climates by seizing areas understood to be fertile, profitable, and healthy to white populations. In each of these situations, factory owners and colonizers did not explicitly envision their positions with respect to protected climates as "political" (that is, unless the dominated exposed the politics of inequality through revolt). Generally, meteorological expertise was involved in the process of making unequal climates, either through knowledge that linked historic climates to social and racial difference or by facilitating the rational exploitation of territory for state and capitalistic enterprises. Like Manchester's polluted atmosphere—at once a sign of modern progress to some, and to others, the poison dealt to subordinated people—the climates of the United States have also held, within the understanding and experience of them, relationships of power. And today poor and marginalized communities face climate changes that compound historically created social inequality (Taylor 2014). What, then, is to be stabilized, when the division between natural order and power relations is not so easy to establish? As Taylor (2014) asks in relation to historically based vulnerability to climate change impacts, to what socioclimatic situation must people "adapt"?

For paleoclimatologists and those oriented to determining the historical and future "climatic boundary" of human society, the "relative stability" of the Holocene period (the epoch since the last ice age) represents the "scientific reference point for a desirable planetary state" (Rockstrom et al. 2009; Berger and Loutre 2002). In contrast, critical climatology, by re-embedding an epistemology of climate-in-society, requires recognition that global climate is only from one point of view some stasis to which global society must revert or adapt. Climatic stability, from its not inherently less scientific point of view, is an insufficient optic for climate knowledge to the extent that it does not systematically incorporate analysis of the basic destructive dynamics of fossil-based capitalism.

A critical climatology, in these terms, need not displace the problematic of stabilization per se. We have no reason to believe that society can mitigate climate change without confronting fossil fuel use and replacing it with alternative, renewable energy. In a technical sense, arriving at clear thresholds for global climate change is necessary to inform political deliberation and climate policies to reduce global warming by "stabilizing" the CO_2 level through transparent and accountable emissions reductions. But to borrow the language of the IPCC, climate knowledge can also recognize that measuring climatic stability in technical terms does not determine what constitutes "dangerous" versus not dangerous circumstances (Boykoff, Frame, and Randalls 2010). Climate stabilization, which implicitly rests on a narrative of what constitutes a good world, can obscure as much as it can clarify the social needs that correspond to the stratified spatial and temporal worlds that characterize global society.

A related implication of critical climatology is that the present climate crisis can inspire various degrees of utopian thinking about science and society. Utopian considerations may identify the limits of critical climatology, and among scholars, views on utopian thinking are mixed. Slavoj Zizek (2010) proposes that if we acknowledge in theory that we inhabit the "End Times," then we actively reject such a fate in practice—a contradiction that can inspire radical change. Many activists and writers (see, e.g. Wallace-Wells 2017) have embraced explicit apocalyptic frames to characterize climate crisis. By contrast, Erik Swyngedouw (2010) suggests that amplifying apocalyptic anxiety about climate change excludes meaningful political action and places major social problems like climate change in a "post-political" domain. John Hall (2016) argues that apocalyptic discourse may simply fuel public anxiety, confusion, and powerlessness. The meaning of climate "crisis" and "emergency" within political discourse is perhaps inevitably caught between these views. One way to reconcile such pitched political alternatives is to build visions of science, climate, and society that "escape" the social interests that have long driven the organization and production of climate knowledge and resisted confrontations with capitalism. These interests include the supremacy of the market, economic growth dependent upon state-facilitated ecological devastation, and in broadly cultural assurances, faith that technical innovation will solve environmental problems. Institutionally, such an escape is indeed utopian insofar as, at present, discourse of stabilizing the global climate system remains, at best, linked to these interests.

Yet consider: at one point for nineteenth-century meteorologists, a world of standardized time, simultaneous networks of thousands of observers and instruments, and successful prediction of future weather was also a utopian speculation. Prophecy presented a threat to the basic legitimacy of rational knowledge. To make the utopia of weather prediction into a reality was not simply a matter of working out the technical details. It was about actively reorganizing the relationship between a sprawling array of people and things, about making material and discursive order where that order did not yet exist. Their utopia was about "recovering our Patrimony from Chaos" through the promise that new scientific domains may offer to the flourishing of industrial-capitalist society. Only through this social process was climate reconfigured as a geographically and temporally stable object of measurement, valuation, and government.

Like the nineteenth-century meteorologists' project, a concerted development of a critical climatology can present an avenue for climate knowledge that takes into account, but goes beyond, prevailing climate science. Over the past two centuries industrial society has radically altered the biosphere. The effects will be experienced on timescales that extend for centuries into the future. For the past half century, scientists have produced increasingly accurate projections of local, regional, and global environmental events, some catastrophic, that will be exacerbated by climate change. Most important, the impacts and related consequences are geographically and socially unequally distributed. Given these circumstances, climate knowledge must incorporate social-scientific and geographic insights into the fundamentally uneven ways human societies inhabit space and time. Climate change, according to a critical climatology, is an expression of history as much as it is a determinant of the future. Climate change is thus not only an objective force to which those vulnerable must be made resilient. Knowledge about climate change fundamentally also entails making legible the (re)configuration of society.

Compared to what I have sketched as a possible critical climatology, climate expertise at present is mostly running in the other direction, disavowing a politics of knowledge. Chapter 7 showed that climate security expertise has established as its "nonpolitical" foundation the preservation of a distribution of power that is not driven by the goals of either reducing climate change or alleviating human suffering, except to the extent that they threaten the established order. Rather, the thrust of climate security is to secure some people in some places, reducing others' legibility to government, except to the extent they need to be strategically viewed as threats. Markets provide a complementary, allegedly apolitical, means for making climate change governable through economic, especially neoliberal, legibility. Although further research on market-based climate policy will be needed to connect this central development with the history of climate's economic legibility that this study initiated in part II, a few points warrant consideration.

First, climate policy proposals in the last two decades have taken a dramatic turn toward market solutions, which fundamentally involve making climate change into a range of economically legible units. Thus, as IPCC chairman Rajenda Pachauri stated upon release of the 2013 Climate Change Assessment: "It is only through the market that you can get a large enough and rapid enough response" to stabilize warming at 2°C (quoted in Bawden 2013). The UN REDD+ (Reducing Emissions from Deforestation and Forest Degradation) program in developing countries seeks to govern climate change by constructing counterfactual values of emissions gained from avoided deforestation. However important it may be to finance forest conservation, the financial legibility of forests as carbon sinks open to investment has encountered practical problems, for example, regarding how to protect the rights of those social groups that are managing forests communally (Gauthier 2018). On a different front, take the case of emissions trading, in which measures of the global warming potential of economic activity provide entitlements for industrialized countries to emit greenhouse gases (Felli 2015; Robertson 2012). When a mode of government rests on the ability to commodify, trade, credit, buy, and sell climate change, as these policies do, then actors must ultimately render things—chiefly CO_2 molecules—commensurable with monetary values across time and space. The commensuration process that can identify "global warming potential" or mitigation potential can likewise detach fossil fuels and carbon from their sociopolitical contexts, for example, the unequal social causes of atmospheric emissions and the stratified contexts of energy needs and uses (Lohmann 2016; Wynne 2010; MacKenzie 2009).

Scientific and governmental projects similarly may depend upon making circulations of capital and carbon legible and thus coproduce climate policies that, like climate security, rest on a relatively depoliticized view of the social relationship to climates. For example, governmental emission trading schemes rely on economic assumptions that fail to account for the interested behavior of corporations and their (largely successful) history of lobbying against direct regulation and taxes on GHGs (Spash 2010). Carbon markets instead primarily rely on political opportunities to advance "a utopian faith that marketization can be squared with climate protection," an ideology that persists despite the failure of carbon markets to reduce GHGs (Pearse and Böhm 2014:326; Bond 2012). As another example, mitigation programs and government regulation using dominant economic models that inform the relative costs and benefits of policy actions rest on measures of the "social costs of carbon" (NAS 2016; Nordhaus 2017). Such metrics powerfully represent the relative economic value of preventive action and, in turn, make legible the large economic burden that future generations will face given current rates of GHG emissions. Yet they do not systematically account for social inequality with respect to either the socioeconomic structures causing climate change or the differential social impacts of climate change. In these formulations of climate policy, important social forces and divisions are thus rendered illegible.

The dynamic that pairs the legibility of society within carbon molecules and the simultaneous obfuscation of the social causes and consequences of climate change is hardly surprising. As James C. Scott showed in his study *Seeing Like a State* (1998), legibility projects often obscure as much as they reveal. In the case of proposals to stabilize the future climate, the legibility projects of climate security and (neo-)liberal governmentality monopolize the discursive space where possible alternative visions for climate, science, and society might be considered. Resolving the invented separation of society from climate cannot plausibly be achieved by entrenching society in the industrial-capitalist climates we have inhabited and produced for the last 150 years and more.

Given the obfuscation upon which dominant contemporary climate policy depends, a number of scholars have abandoned the Anthropocene concept, replacing it with alternative accounts of the "Capitolocene" that make legible the class-based stratification that characterizes the causes and impacts of climate change (Moore 2017; Malm 2016). These materialist accounts extend beyond Mike Hulme's (2008) suggestion that resolution to climate crisis involves reorganizing "ideas" and releasing our "fears" expressed in the language of climate crisis. As Taylor (2014:196) concludes, the politics of climate change must involve "urgently working out how to produce ourselves differently." Do we have genealogical resources that can help here?

In this study I have paid special attention to *who* gets to speak about climate and social order and how, based on their epistemic authority. Climate knowledge, I have shown for a series of historical periods, involves struggle, appropriation, and displacement of alternative forms of knowledge and their spokespersons. The project of uncovering the historical production of climate science and its displacement of alternative or subjugated knowledges is unfinished. It remains one task for a reflexive critical climatology. Yet already, by tracing the consistent coproduction of science and government, the present study has provided conceptual tools and historical points of reference for critically evaluating, for the future as in the past, the structure of climate science and the kinds of social order that it helps to anticipate, organize, and govern. For the future, my analysis of climate security is especially important here, and I therefore return to the implications of that analysis.

POLITICIZING CLIMATE SECURITY?
POSSIBLE ALTERNATIVES FOR CLIMATE JUSTICE

The vision made legible by security technologies and pronounced by climate security experts is a contradictory one. Although its proponents seek to depoliticize climate change, especially in the face of those who want to make it a strongly partisan issue, climate security discourse nevertheless presents a normative political vision. In this vision, scarce resources, including energy, food, political representation, and safety from threats to livelihood, must be distributed in accordance with currently existing balances of power and thus unequally. That an alternative distribution of power and resources might exist, now or in the future, is thinkable only as a strategic exercise in alternative scenarios to secure *against*, that is, to render unthinkable. Sociological analysis can help to make visible other possibilities for envisioning climate futures among those currently outside or marginal to the security field—those that view climate change not as a challenge to state security but as an issue of justice. The principle of justice brings into view the normative assumptions undergirding practices of security: security *of what*, and *for whom*?

Those who have advanced a liberal variant of "human security" hope to address the social dislocations associated with climate change while rejecting concern for national security as a sufficient optic for security practices. Serious efforts are underway within this framework, for example, to institutionalize humanitarian protections measures for persons dislocated by climate change (Nansen Initiative 2015). Christian Webersik (2010:127) explains, in line with many NGOs and the position of the UN, that "climate change impacts are primarily a human security issue." The goal, Webersik continues, "is *not* to 'securitize' nontraditional threats such as human rights, transnational crime, and the environment to gain the attention of policy makers and thereby mobilize resources, but to better understand the implications of change for all human beings (with less emphasis on the state)." Past work to transform USAID and development aid into "climate resilient" forms serving climate adaptation objectives proceeded through a similar logic, exposing possible lines of contention between development and military sectors of the U.S. bureaucratic field (see USAID 2012:1, 11). Some scholars have argued that the policy frameworks of sustainable development and human security do not present achievable policy goals but instead reflect "the depoliticization of public debate in liberal democracies, whereby politics have been reduced to the search for technocratic solutions" rather than "a genuinely antagonistic struggle between alternative visions" (Kallis, Demaria, and D'Alisa 2015;

Pelling 2011). Yet as the climate security field matures, antagonistic lines of opposition are indeed forming within it. By 2023, climate security was formally incorporated into UNFCCC negotiations, notably at COP 28. International development organizations, for example the UN Development Programme, have called for "global climate security" through financing (built into multilateral climate policies) for conflict prevention in climate-impacted regions. The developmental vision of equity in climate security highlights how militaristic visions of national security perpetuate climate inequity rather than promote human security (UNDP 2021).

Others may continue this critical but proactive engagement with climate security. Experts working outside national military contexts may readily draw upon the nexus of transnational climate science, international politics, and the climate movement to bring out the political terms and implications of climate security more directly. Kelly Moore's (2007) historical analysis of "disrupting science" shows that physical scientists have sometimes become organized political actors, specifically at moments during which at the same time their expertise is perceived to be under threat of misappropriation or rejection, and resources exist for successfully articulating a novel vision of science and society. Her account of scientists' organized resistance to nuclear arms during the Cold War documents one case in which scientists and social movements successfully developed a challenge to prevailing visions of national security and to security technologies. This kind of approach could also gain traction among climate scientists and social movements today that challenge prevailing logics of governing climate. In 2019 the UK and other governments at various political scales declared a "climate emergency" under pressure from social movements and scientific warnings. It remains to be seen what these statements have meant for policy (Brown 2019). The climate justice movement, informed by democratic principles, has organized around the idea that governments, the capitalist system, and mainstream environmentalism cannot be relied upon to support a survivable future (Schlosberg and Collins 2014). The discourse of climate justice emerged through social movements in the Global South, especially around the time of COP 15 in 2009, held in Copenhagen, Denmark, which failed to produce any international legally binding emissions reduction target. In one of many demonstrations, protesters chained themselves to train tracks, displaying a banner that read, "Greed wrecked Copenhagen, now it's up to us all!" (Rising Tide North America n.d.:9). The climate movement became a new form of contentious politics that increasingly saw itself as operating outside mainstream environmentalism and its expression in mainstream policy. As Rising Tide put it, "The climate movement is dead, long live the climate movement!"

The climate justice movement poses a fundamental challenge: "Climate change demands that we ask what kind of world we want to live in." Rising Tide responds: "No effort to create a livable future will succeed without the empowerment of marginalized communities and the dismantling of the systems of oppression that keep us divided" (Rising Tide North America 2018). Such discourse takes the sentiment of emergency to the radical conclusion of social revolution.

Some climate justice activists point to global capitalism and state complicity to advance a vision of local autonomous and noncapitalist social relations to confront the root causes of climate change. Others thread alternative visions into climate action through historical challenges against colonialism. Thus, Canadian decolonization activist Bev Sellers narrates an imagined alternative historical pathway and, relating it to the ongoing experience of environmental and climate injustices, concludes: "The fantasy I have of what could have been did not happen" (Sellers 2017:7). Whyte (2017) argues that Native Americans, who have faced historically forced relocation to different climates in North America, experience climate change as a kind of "colonial déjà vu." By inhabiting what Whyte calls their "ancestors' dystopia," some Native Americans' orientations to climate justice conjoin climate action with decolonization. Indigenous Climate Action, as a transnational expression of this position, aims to "to develop climate policies, mitigation strategies, and alternative climate solutions that are driven by Indigenous worldviews, rights frameworks, and knowledge systems" and form part of the decolonial struggle.[1] In a similar way for women, the ecofeminist organization the Women's Earth and Climate Action Network (WECAN) seeks to confront the oppressive basis on which the present social order is generating a future that "the women of WECAN, and our allies worldwide, cannot and will not tolerate." WECAN's (2016) manifesto concludes, "In order to live in harmony with the Earth and safeguard our world for present and future generations, we must take back power and revolutionize the roots of modern society." Letze Generation (The Last Generation) in Germany, youth activism, and the Fridays for Future protests increasingly connect intergenerational justice claims to solidarity claims with victims of global climate injustice (Thew, Middlemiss, and Paavola 2020). And recently, intellectuals and activists have called for escalating tactics, including strategic destruction of fossil fuel infrastructure (Malm 2021). Perhaps this is among the very kinds of political confrontation that mainstream climate security experts seek to stave off.

In the present study, it is not possible to claim what *must* be done from the standpoint of climate security or climate justice. In reality, climate security experts and climate justice activists are worlds apart. However, perhaps such a separation is a point worth reflecting on. Despite having media-savvy representatives, climate security experts are not engaged in analyzing or discussing otherwise high-profile climate justice activism, for instance, the protests at Standing Rock, North Dakota, in 2016 against the Dakota Access Pipeline. Climate security organizations remained remarkably silent on popular protest actions regarding climate change in 2018 and 2019. To so engage would risk entering politics in a way that many experts understand as anathema. By avoiding contentious politics and pulling the discourse of climate crisis into a depoliticized view of national security, the climate security field will likely gain adherents. For its part, the climate justice movement appears unlikely at present to provide a basis for transformation in government, and its agenda is more likely to be incorporated into reform environmentalism through equity mechanisms in sustainable development programs.

Despite the challenges that the climate justice movement faces, it may present principles of political order that at some future point in time can expose the narrowness of the political vision that undergirds climate security. New political forces could exploit possible relationships and divisions within climate expertise. On this front, it will be necessary to consider the following. How will social scientists invested in the study of climate change and migration relate to those who are invested in anti-immigrant policies? How will regional climate modelers engage the socially inequitable access to and deployment of "climate services" that can inform resource management governance? Will heterodox economic models (for example, in the field of ecological economics) challenge the political-economic assumptions of those that presume economic growth to be the only path toward sustainable social welfare? (Anderson and M'Gonigle 2012).

Within the epistemological and institutional space opened by climate justice, then, it might become clear to broader political constituencies that climate security, like other supposedly apolitical approaches to climate change, places limits on how to collectively envision alternative futures for society. Within the space of prevailing orientations to climate security, human suffering, complex phenomena, and adaptation to uncertain circumstances are reduced to threats to be addressed by a state oriented generally to the sustainability of present economic interests. To emphasize this situation, Simon Dalby (2013) criticizes an "imperial optic" of climate security and related geopolitical formulations of climate impacts. for a recognition of unequal global interconnectedness, he (2013:39, 42) writes, is "elided in favor of simple specifications of 'here' and 'there' and usually virtuous proximity and threatening or morally flawed others elsewhere." Other research has found that crisis narratives embedded in security practices can deepen vulnerability of marginal communities to both environmental stress and unequal power relations (Chandler and Hynek 2011; Baldwin and Bettini 2017; Davis 2016).[2]

Similar intellectual movements of reducing social life to climate mark the historical development of climate knowledge as it has been traced in this study. Climate change during the eighteenth and nineteenth centuries informed social, especially racial, stratification. Understandings of climate change at the time made legible the progress of (or limits on) civilization, savagery, and barbarity. In turn, when climate was "stabilized" by science, climatologists understood climate to determine regional and national economic futures. And today, concern over climate apocalypse strongly reduces the present to the destabilized climatic order out of which future society will emerge. The reduction of social complexity to climate has in various historical and contemporary contexts served political purposes, especially the tendency to naturalize one group's power as inevitable in the face of challengers (Hulme 2011; Livingstone 2015; Mahony and Caglioti 2017; Baldwin and Bettini 2017; Felli 2021). The narrative of climate crisis in these terms is penned in the ink of symbolic power, not of nature alone.

One somewhat ironic problem for developing an alternative to reducing social futures to climate change, however, is that climate security expertise appears safe and apolitical even to those mobilizing for action on climate change. Organizations associated with the climate justice movement have yet to formulate a critique of climate security expertise, security technologies, or related developments in climate science. Rather, those whose political organizing relies on visions of "emergency" have drawn upon the dramatic apocalyptic imagery of climate war. Understanding contemporary resource conflicts as foreshadowing a dire future matches the aims of both radical activists and security experts. Thus, some social movements utilize security expertise to bolster their claims. For example, the UK-based Environmental Justice Foundation, in its short film *Beyond Borders* (EJF 2017), aims to sensitize its audiences to the problem of climate refugees and the necessity of a legal instrument for their protection. To make the case they have relied almost entirely on retired U.S. military officers affiliated with a U.S.-based climate security think tank. Synthesis of climate security and environmental justice has thus far prevailed over critique.

Such is the case despite frequently stated rejections of militarism, evident for example in the foundational *Bali Principles of Climate Justice* (International Climate Justice Network 2002), which declared that "Climate Justice opposes military action, occupation, repression and exploitation of lands, water, oceans, peoples and cultures, and other life forms." Already in 1987, the UN Brundtland Commission (1987) had argued, "There are, of course, no military solutions to 'environmental insecurity,'" a statement that runs counter to climate security expertise at present. It is also clear that militarism, forming what environmental sociologists have analyzed as the "treadmill of destruction," presents a fundamental cause of environmental damage, contributes substantially to global warming, and exacerbates climate insecurity (Clark and Jorgenson 2012; UNDP 2021). The U.S. military is expectedly the clear positive outlier. Efforts to improve the energy efficiency of militaries have not eroded the basic logic by which expansionist militarism, during war and peace, has increased the environmental impacts of militaries and with disproportionate impacts on nations in the Global South and on vulnerable populations (Clark, Jorgenson, and Kentor 2010; Gould 2007; Hooks and Smith 2004). The lack of critique of climate security in this light is surprising insofar as the burgeoning field of climate security research and its level of governmental attention rest on the deepening of American military preparedness as a necessary condition for confronting the climate threat. Thus, like the case of "climate stabilization" and the difficulties that would face efforts to transform climate knowledge in a way that considers the instability of social relations to climate, movements for climate justice face difficulties in developing social visions of climate change substantiated through alternative forms of climate expertise.

As the present study shows, however, it is possible to bring to light the contingency of one or another configuration of science and government. Doing so may help to shape alternatives. Like movements to organize the "rights of Mother Earth," dominant political categories (in that case, property rights) can themselves be appropriated within a discourse of environmental justice.[3] A critical engagement of climate justice with the politics of climate security expertise is one way forward, and it may be supported through what I have sketched as critical climatology. The climate justice movement is ostensibly counter to the climate security field. Although it is practically orthogonal to it politically, it opens up a critical foundation for an alternative configuration of science, government, and social order.

The creation of alternatives is a fundamentally historical project. This is not because history strongly determines the future. Precisely the opposite: it is because transformation, at times featuring struggles for particular modes of knowing (in the face of alternatives), marks the history of climate knowledge. As Pierre Bourdieu (1986:241) maintains, "the social world is accumulated history." In this book, I have pulled out some of the threads of "accumulated history" that may be useful for uncovering past transformations and using them to clarify some of the spaces of political possibilities today that otherwise might be less visible.

So what comes next? Climate researchers of all kinds clearly need to bridge the geophysical and social sciences and impacted communities, an effort already underway, for example through the American Geophysical Union's Thriving Earth Exchange. Academic and educational institutions likewise must cultivate spaces for research, teaching, and learning across disciplines and thus "remake" climate as a category of knowledge. From the perspective of a critical climatology, the hierarchy of climate science, at present dominated by physical sciences under the inertia of their institutional development, is no longer functional. Public audiences, moreover, need to consider their own assumptions about the climate future and find ways of rejecting the polarizing, warring voices of dismissive skepticism and fatalistic forms of apocalypticism. The former only provides a baseless and unsustainable hope in the present social relationship to climate. The latter may amount to an equally baseless hopelessness. Social and political movements provide much-needed spaces in which deliberation, communication, and contestation can occur without the weight of a ruined planet on one's own shoulders. Historical perspective affords a new view that brings into relief the openness of how scientists, governments, and publics may practice and interpret their activities with reference to climate and society. To be sure, humans cannot control climate. Indeed, although this lesson should have been learned through historical failures, we may need to revisit this temptation when it arises. But neither can climate control the trajectory of society. The legibility of paths forward requires knowledge and social practices that fully embrace a "climate" that places scientific facts *and* social values at the center. Only then can we deliberate and remake the social relationship to climate and thus place hope in a more desirable and sustainable relationship to the environments we must collectively inhabit.

Notes

1. With government, social theorist Michel Foucault (1991:95) writes, "it is a question not of imposing law on men, but of disposing things." To govern means to order "men [*sic*] in their relations, their links, their imbrication with those other things which are wealth, resources, means of subsistence, the territory with its specific qualities, climate, irrigation, fertility, etc." (Foucault 1991:93). Governmentality, to use Foucault's term, is a unique kind of power developed through knowledge and the formation of social categories, populations, and territories in such a way that they become amenable to security, management, and discipline (Foucault 2004, 2008, 1980). Foucault elaborated on the power-knowledge relationship in these terms by investigating disciplinary knowledges and institutions. Foucault's later (1991, 2008) historical analyses of "governmentality" pursue concerns with power and knowledge through a specific theory of government, and these studies thus serve as a convenient point of conceptual departure here. Foucault and his followers have paid scant attention to climate (see Foucault 1991:93–95, quoted above; 2004:21–22; 1980:175).

2. Jasanoff (1996:397) argues that predominant accounts of scientific controversies had often fallen short of integrating into analysis the wider interplay between "mutually supporting forms of knowledge and forms of life" that may often unfold simultaneously as struggles in science and in politics. This scholarship recognizes the possible limits to focusing on the microcontext of science that

roots the achievement of scientific facts in terms of "epistemic cultures," tacit knowledge, and practices, especially in laboratory settings (Knorr-Cetina 2003; Collins 1974; Pickering 1993; Latour and Woolgar 1986).

3. As scholars have shown, the general power of science often hinges on *inscription devices* that grant scientists and technologists the capacity to get parts of nature and material objects to "write" in such a way that they can be captured by scientists or made to travel, for instance, as data in support of a claim or theory (Latour 1983, 1988).

4. As sociologist Emile Durkheim has established, both the state and science are at their core producers of collective representations—of society and of nature, respectively.

5. Political ecologists and scholars from developing countries have addressed parallel issues surrounding climate policy. Bumpus and Liverman (2010) and Agarwal and Narain (1991), for instance, question climate-change mitigation programs as "carbon colonialism" when afforestation projects (to offset carbon emissions) infringe upon other, nonmarketized uses of forests. For other work in this direction, see the works collected in Beattie, O'Gorman, and Henry (2014), Cushman (2013), and Whyte (2017).

6. Major recent interdisciplinary reviews of historical studies of climate knowledge (Mahony and Endfield 2018; Mahony and Caglioti 2017) cite minimal sociological work on the topic. Canonical scholarship building into what Adams, Clemens, and Orloff (2005) have called the "three waves of historical sociology" did not substantially engage either historical climate change or the issue of global warming.

7. One ambitious example of such an approach is Michael Mann's (1993) account of "the sources of social power," which builds a conceptual apparatus that allows for case-comparative historical investigation of power relations across centuries and continents.

8. Critical studies of climate governance have thus far focused successfully on local and global scales in recent and historical contexts (Fleming, Jankovic, and Coen 2006; Jasanoff and Martello 2004); given the present situation, they might be enhanced by investigating how climate knowledge is deployed through state-making to govern climate change impacts. Genealogical analysis can thus contribute to what are at present quite separate efforts to critically analyze climate adaptation policies (Taylor 2014; Buxton and Hayes 2015), to reassess the basic category of "climate" (Hulme 2008, 2017), and to critique resurgent forms of climatic "reductionism" that reduce complex social outcomes to climatic causes (Livingstone 2015).

9. Historians of science concerned with geoengineering (Fleming 2010; Harper 2017) focus primarily on the technologies proposed and the individual or political hubris associated with technocratic mastery of nature. Some climate and social scientists have argued that presenting climate change politics as

a global stabilization effort obscures the politics of climate (Victor and Kennel 2014; Geden and Beck 2014); fails to adhere to time and spatial scales that connect to human society (Mahony and Hulme 2016; Brace and Geoghegan 2010; Livingstone 2012; Norgaard 2011); and mischaracterizes the uneven geography of climate-change causes, consequences, and effects (Roberts and Park 2006; Taylor 2014).

CHAPTER 1. GOVERNING CLIMATE IN EARLY AMERICA, 1770–1840

1. The significance of popular knowledge and almanacs is raised again in chapter 4, in the context of official efforts at producing weather forecasts.

2. Aristotle (n.d.).

3. Alexander von Humboldt's (1817) representation of temperature across space in the form of isothermal lines, based on instrument readings, not latitudinal geometrics, changed how meteorologists addressed the latitude-climate issue. However, the relationships between civilization, "the powers of man," and climate were often transposed from a latitudinal determinism to an isothermal one (Forry 1842a:21). Samuel Forry (1842a:32) summarized this logic: "The former division of the surface of the earth into five zones has been superseded by a more precise arrangement: places having the same mean annual temperature are connected by Isothermal lines, and the spaces between them are called *isothermal* zones." He went on to discuss isotherms as socially determinative zones.

4. English philosopher James Dunbar (1781:222) put it this way: "The genius of mankind, far from being equal, must have been as various as the situations in which they are placed." Dunbar then provided a grand differentiation of nations according to climate, relying on an implicit Aristotelian framework that categorized climates through three categories—hot, temperate, and cold. On the one hand, the "original affluence" of hot tropical climates "will be cultivated slowly, and with inferior ardor," whereas those countries "more pernicious in nature" would have to make up for natural deficiencies with "the resources of industry and invention" (1781:223). English physician William Falconer (1781) likewise constructed historical comparisons to explain the rise of European civilization as contingent on temperate—as opposed to cold or hot—climatic conditions. Comparing ancient Rome and Russia, for instance, Falconer (1781:22) argued the Romans became "more enterprising," and hence superior in battle, because of climate.

5. Golinski (2010:74, 79) presents a dichotomy between climate- and race-based arguments concerning "civilization." However, the relationship between race, climate, and civilization has some continuity from natural history to twentieth-century human sciences. The work of the early twentieth-century

geographer Ellsworth Huntington perhaps most explicitly synthesizes natural and social history through a climatic explanation of racial classifications. "It matters little whether we are dealing with the red race, the black or the white," he concludes, the physical environment and climate remain the basic causes of "man's progress" in mental capacity and in political and social organization (Huntington 1919:172). He calls his approach a break from physical geography, introducing a "new" (human) geography chiefly through his "climatic hypothesis of civilization," which holds that the race-climate relationship explains the level and "distribution" of civilization across time and space (Huntington 1915:v, 217, 148).

6. Improvement was not the only pattern that observers defended; however, it was the most commonly held or cited view and the most long-standing controversy concerning American climate change. Others, for example, argued that the American climate was growing more capricious and extreme (see Wilson 1815).

7. APS members and physicians took up the issue of climate in an uneven manner, as some continued to assume climate change (e.g., Antill 1771); others cited authorities including Benjamin Rush, Lionel Chalmers, and Hugh Williamson; and still others proceeded by directly testing ideas through experiment and data collection (e.g., Currie 1792).

8. In the United States, federal forest policy, culminating in the establishment of the U.S. Forest Service, did not emerge until the early twentieth century. In chapter 3 I address the contradiction between the desiccation theory of climate change and "climatic stability" in the context of U.S. climatology in the late nineteenth century.

9. Medical topographers frequently invoked Hippocrates as an authoritative source. For example, Stephen Williams (1836), a Massachusetts professor of medicine, responded to a call in the *Boston Medical and Surgical Journal* for a "Medical Topography of Massachusetts." He framed his account of diseases in the west of the state first by citing "the immortal Hippocrates" at length to answer the question: "Who is the physician that is an honor to his profession?" Williams framed his own work with respect to Sydenham as the "English Hippocrates," and that of the Philadelphia physician Benjamin Rush as the "American Hippocrates."

10. Wilson (1815:5–6, 113–26) largely disagreed with the earlier physicians who had argued cultivation tempered American extremes in temperature. He provided meteorological data to argue that a "new era" with a more capricious New England climate commenced in 1804, and he used this change to explain the rise of epidemic fever.

11. "Dephlogisticated air" refers here to what Joseph Priestly formally discovered as oxygen. "Phlogiston," considered the substance burned in combustion, was often measured as an element in the purity, and hence healthiness, of air.

12. Despite both theories emphasizing heat and moisture, "exhalations" in this context refers to *miasmata*, discussed later, not the mineral exhalations of Aristotelean meteorology addressed previously.

13. David Ramsay (1790:19), like Chalmers a South Carolina physician, provided a similar evaluation of local miasma: "Our summer and autumnal fevers chiefly arise from the separate or combined influence of heat, moisture, and marsh miasmata." Chalmers (1776:11) ultimately prescribed that combating disease in the Carolinian climate would have to involve a policy of land and marsh clearance: "Till the land is more cleared," he wrote, "our atmosphere cannot be wholly renewed, even by a hurricane." If authorities could formalize these practices, Chalmers held, then they could mitigate the climatic effects on disease.

14. In addition to containing the effects of miasma by regulating soldiers' bodies in relation to the tropical environment, other physicians recommended pharmaceutical containment. For instance, John Hunter (1788:108), like Jackson a Scottish military surgeon, recommended that soldiers use cream of tartar and rhubarb to ward off fevers, "if experience has shown that they agree with a particular constitution." Therefore, physicians did not treat the practical requirements to understand climates and bodies as separable problems.

15. Future actors, including meteorologists and those involved in the sanitary movement, later succeeded in incorporating vital statistics into formal government policy. The city of Boston and the state of Massachusetts were particularly important in institutionalizing the collection and use of vital statistics. See, for example, Shattuck (1841) and his later *Report of the Sanitary Commission of Massachusetts* (1850).

16. A direct link between mental illness and medical meteorology is tenuous, even though some physicians outlined the "atmospheric origins" of insanity and nervous illnesses (e.g., Forster [1817] 1829). Nonetheless, moral evaluations of the climate-body relationship match Foucault's [1965] 1988 theory that the categories of madness and mental illness demonstrate, over time, the discursive formulation of civilization with the containment of its opposites: unreason, madness, and folly.

17. These "sins" appear related to Barton's (1837) general consideration of diet and digestion. Cold climates, he and others argued, demanded a fuller diet, while one must consume less meat, salt, and alcohol in hot climates because "the calorific process" of such a diet was excessive, promoted disease, and retarded acclimation.

18. Reviewed and reproduced also in the *New York Journal of Medicine* 2 (1849):237–46.

19. Scholars' emphasis on periods partly reflects the conventions of historiography of science and the relative distance between historians of pre-twentieth-century meteorology and scholars of contemporary climate science and politics. In his analysis of "moral climatologies" of racial difference, however, geographer David Livingstone (1991, 2002) shows how climate has continuously, if dynamically, entered moral evaluations of social order, thus suggesting the value of a *longue durée* perspective.

CHAPTER 2. METEOROLOGICAL FRONTIERS

1. The "Meteorological Register" (in James 1823) terminated in May 1820, three months prior to this desertion. However, James (1823:xliiv) also claimed the recordings from Thomas Says's journals were "too voluminous" to compile in full.

2. As reproduced in Benson (1988:371). See also James (1823:486–87) for an account of this incident.

3. Governing climate in these terms remains roughly consistent with analyses that decenter a theory of "the state" as a necessarily monolithic center of power in society (Foucault 1991; Mitchell 1991; Gordon 1991; Rose, O'Malley, and Valverde 2006). For these scholars, analysis must follow power relations as they "extend beyond the limits of the state" (Foucault 1980:122). Thus, Rose and Miller (1992:177) suggest, "[The] question is no longer one of accounting for government in terms of 'the power of the State', but of ascertaining how, and to what extent, the state is articulated into the activity of government." This approach to government helps raise the basic question: How did the process of state formation "articulate" with the ways of knowing and governing climate already addressed?

4. Here Bourdieu's approach to political power is closest to governmentality scholars inspired by Foucault, for whom "government defines a discursive field in which exercising power is 'rationalized'" (Lemke 2001:191).

5. In figure 1, note the frontier forts from which data was collected. Other observation networks, especially in New York, and by the 1840s through the Smithsonian Institution, reflect processes similar to those within the Army Medical Department.

6. Government policy only weakly structured practices aimed at "civilizing" Indigenous tribes at this time. Most Native policy consisted of war- and treaty-making in order to establish and regulate trade, property rights, and land use (Banner 2007; Frymer 2017). Yet the War Department introduced a Civilization Fund in 1818, with appropriations for civilians to introduce agricultural and educational reforms within western borderlands. Later controversies surrounding removal of Natives west of the Mississippi River also hinged on a discourse of how to measure "civilization" among tribes. The mission, as James Madison declared in his 1809 inaugural address, was clear: "carry on the benevolent plans," meaning "the conversion of our aboriginal neighbors from the degradation and wretchedness of savage life to a participation of the improvements of which the human mind and manners are susceptible in a civilized state" (Hunt 1908:49).

7. Jefferson served as president of the APS from 1791 to 1793 and was a member from 1780 until 1826. Through Jefferson, the APS was central to Captain Lewis's training (Bedini 1990). Benjamin Rush was responsible for providing medical-scientific training. Because Rush was a central figure in advancing climate-change theory among physicians, the issues of climate change and disease were likely a common understanding among expedition members.

8. Humboldt's approach to geography especially influenced frontier cartographies. On the import of Humboldtian cartography on medical meteorology, see Rupke and Wonders (2000), and on expeditions, see Ponko (1997). Meteorological records from the Fremont Expedition, addressed later, show how Fremont implemented his prior meteorological training with the Humboldtian cartographer Joseph Nicollet (Frémont 1845:673; Nicollet 1839).

9. Analysis of Pike's expedition is drawn from Pike's 1811 account, reproduced in Maguire's (1889) edition and appended with letters between Pike, Wilkinson, and Spanish officials, and from Pike's papers, first recovered by Bolton (1908).

10. Pike to Wilkinson, July 5, 1807 (in Maguire 1889:390).

11. For a typical example of influential, practical emigrant information written for prospective settlers in the U.S. Far West, see Disturnell (1849).

12. Forry situated his account of race, climate, and civilization with reference to anthropologist James Pritchard, whom he defended, and Samuel Morton, a leading polygenist. He upheld the cranial and physical superiority of Caucasians. Sources of authority included Morton's (1839) craniometric studies and Francis Peron's ([1809] 2012) dynamometric studies of "savages'" inferior strength, based on Peron's work during imperial French expeditions to Tasmania and elsewhere in Australasia. Yet he rejected polygenism with climatological arguments, especially concerning the aboriginal "American" racial type. Analysis of climate helped Forry provide a materialist ground for racial difference: "It thus appears that mind dwells in a material tabernacle, and is acted upon by material causes" (1842b:130), a position from which he justified the civilizing obligation of Anglo-Saxons.

13. Forry (1842a:128) claimed *The Climate of the United States* provided "some general laws towards the basis of a system of medical geography." It was reviewed favorably in similar terms (Caldwell 1843; see also Forry 1848:299–301). Forry's posthumously published text *Vital Statistics* extended the governmental implications of medicine. "The object of the science of medicine," Forry (1848:290) reasoned, "is not alone the cure of diseases: but it has, as will be seen, the most intimate relations with the social organization. . . . [Medicine] spreads the wings of its solicitude over all society."

14. Decades later Köppen began producing work in paleoclimatology, especially informed by Milutin Milanković's astronomical theory of ice ages (Köppen and Wegener [1924] 2015). Milankovitch (1920), a Serbian engineer and mathematician, argued that planetary orbital oscillations explained climate change over geologic timescales. Köppen, Wegener, and Milankovic's works on climate change were primarily oriented to debates among geologists, for whom climate change was centered less around human history than around geologic time.

15. Ratzel is best known for his essentially biopolitical concept of Lebensraum, meaning "living space." In particular, he argued that environmentally determined racial groups formed naturally into a hierarchy of space-seeking nations with essential characteristics. In Malthusian and Darwinist terms,

then, races inevitably culminate in a geopolitical "struggle of nations" [*Völker-kampf*]. For example, Ratzel ([1901] 2018:71) claimed, "A small Indian tribe in the South American virgin forest has needs and expectations regarding space that are very different from those of a European for whom the well-being of his people can only lie in grasping the whole world [*Weltumfassung*]." A nationalist and colonialist in his own time, Ratzel's political geography directly influenced post–World War I German Geopolitik and Hitler's Nazism (Klinke and Bassin 2018).

CHAPTER 3. CLIMATE DOES NOT CHANGE

1. William Meyer (2014) likewise argues in his cultural history of weather that pervasive anxiety about climate change in the early United States does not mean that climate as a geophysical system had universally changed. Rather, social relations to weather changed, for example, through new cultural anxieties about weather impacts on standards of living or new land-use and settlement patterns that made sectors of society newly exposed to vulnerabilities like flooding or tornadoes.

2. By the 1870s, many scientists sought to explain earth formation, ice ages, the earth's heat budget, and the cycle of atmospheric gases with reference to the terrestrial. Scientists in these fields thus precipitated new controversies and overturned religious and scientific orthodoxy concerning Earth's history, especially assumptions that the earth was divinely created merely thousands of years earlier. Geological theory reconciled the relative stability of human time with revelation of past, periodic cycles of ice ages, which had become by the end of the nineteenth century a matter of scientific consensus (Kruger 2013). Such formulations of climate change primarily served to disconnect human history from climate rather than according it centrality as in prior periods. An important exception in the geological literature is the carbon dioxide/combustion theory of climate change. Scientists working in several fields expressed interest in carbon theories of climate change by the turn of the twentieth century (Fleming 2007; Crawford 1996). Yet the scientists often credited as the "founders" of theorizing that global warming was a result of burning coal, from Svante Arrhenius (1896, 1908) to engineer Guy Callendar (1938), tended to view climate warming as an economic gain. This view marks a departure from the anxiety about climate change in earlier work on carbonic acid and human modification of climate. For example, Charles Fourier ([1822] 1847, cited in Locher and Fressoz 2012:586) outlined in 1822 how "climatic disorder" resulting from abuses of forests and the atmosphere "is a vice inherent to civilized cultures that disrupts everything due to the battle between individual and the collective interest." Science at the turn of the twentieth century reflected less on the negative impacts of

industrial society on the environment (Driver 2001). Nils Eckholm (1901), for example, presents plans for deliberate climate control, arguably initiating consistent efforts to modify climates through strategic intervention. Important work on how human-modified climate related to the political economy of industrial capitalism includes that of John Tyndall (1861), taken up especially by Arrhenius and geologist T. C. Chamberlin in their work on carbon and ice ages. On Chamberlin, see Fleming (2000a). As Fleming (1998b, 2007) describes, these scholars' work on the carbon cycle and climate was largely abandoned, however, and only taken up again by Callendar in the 1930s.

3. On the role of assimilationist land policies within the larger process of settler colonialism, see Wolfe (2006).

4. More recently, climatologists have recognized (and computationally corrected for) the fact that administrative geography unfortunately does not represent "climatological homogeneity" (see Guttman and Quayle 1996).

5. Meteorologists had long been developing scientific laws, but regarding laws of climate, such statements generally took the form of declarations of what might be possible rather than a list of enumerated laws (e.g., Blodget 1857; see also figure 13 in chapter 6). But the quest for laws had long been the goal for atmospheric scientists. As one example, the preeminent meteorologist James Espy, in his 1841 text *The Philosophy of Storms*, framed his work on the convective nature of storms as indicating " the light of truth" in which atmospheric motion was "not accidental, but subject to laws as fixed as those which govern the planetary motions" (Espy 1841:vi).

CHAPTER 4. ECONOMIC RATIONALIZATION OF WEATHER

1. On Abbe's and Lapham's economic justifications for their efforts, see Miller (1931), Craft (1999), and Fleming (1990).

2. On volunteers, see Marvin (1896) and U.S. Weather Bureau (1897); see USDA (1892:80–83) for a typical register of voluntary observers.

3. For examples of difficulties with self-recording instrumentation, see U.S. Signal Service (1890:655). See also Waldo (1901) and U.S. Signal Service (1884:29–33). On the broader context of early computing technology, including that used to reduce climatological data after the 1880s, see Ager (2003) and Nebeker (1995).

4. T. B. Maury (not to be confused with Matthew Maury, addressed previously), was, according to Cleveland Abbe, a meteorologist employed by the Signal Service as a civilian professor to train the Service's corps of weather officers (see Abbe 1991).

5. Such boundary-making had long marked meteorology as a scientific enterprise (see chapter 1). Espy (1841:vi) defended his meteorological work as holding the capacity to "penetrate mysteries heretofore thought inscrutable." Penetration

of deductive philosophy, Espy claimed, would entail a "true system of meteorology" that "will be the death of superstition on this subject" (1841:vii).

6. Ironically, the article, which also appeared in the *Farmers Journal*, was heavily plagiarized by F. J. Walz, who published "Fake Weather Forecasts" in *Popular Science Monthly* in 1905. Walz, a Weather Bureau District forecaster based in Louisville, Kentucky, sought to persuade society to relinquish false "weather prophets," making direct appeals to public media: "Will not the newspapers, the great enlighteners and disseminators of truth and knowledge in the present age, help these investigators by discouraging and discountenancing the publication of weather predictions founded upon such baseless theories[?]" (Walz 1905:503, 513).

7. For a typical tabular representation of weather featuring "excess" and "deficiency" of normalized values, see *Monthly Weather Review*, April 1889, 142).

8. For an example, see *Monthly Weather Review*, March 1919, 193.

9. Evaluative knowledge of normal weather events fits Ian Hacking's (1990:160) account of statistical normalization. Hacking (1990:ix, 160) shows that "normal" only designated "usual" or "typical" beginning in the nineteenth century and was used as an objective and neutral "bridge between 'is' and 'ought.'" By historical comparison, the formal use of "climate normals" within the Weather Bureau has at times led climatologists to emphasize the statistical (rather than experiential and evaluative) basis of what climatologists label "normal" (see, e.g., Landsberg 1955). A comparison to recent years that illustrates the descriptive-evaluative context of climate "normals" established historically during the 1870–1920 period regards recent framings of "climate events" as indicating a "new normal" (given climate change). Examples include Gill (2017), Watts (2017), Schapiro (2016), and World Bank (2014).

10. Caliskan and Callon (2009:370) write, "The study of economization involves investigating the processes through which activities, behaviors and spheres or fields are established as being economic." More generally, commensuration is a social process, as Espeland and Stevens (1998:315) define it: "the expression or measurement of characteristics normally represented by different units according to a common metric." In capitalist society, then, as Polanyi and others have shown, the process of economizing nature involves the social process of valuating aspects of nature in terms commensurable with money (see also Bigger and Robertson 2017; Fourcade 2011).

CHAPTER 5. THE CLIMATE STATE AND THE ORIGINS OF A CLIMATE SCIENCE FIELD, 1930–1980S

1. The concept of a climate state refers to a sociopolitical formation, but borrows the term from climatologists, for whom "climate state" means the "standard

normal," typically represented by statistical averages derived from a running thirty-year time series. Global warming has upset climatologists' understanding of the "climate state," leading to debate over whether a climate "state change" requires a new approach (Arguez and Vose 2011:701; IPCC 2014:1451).

2. For example, Bourdieu (1996a) emphasizes that the field of artistic production emerged through the opposition of the cultural ("art for art's sake") and economic ("art for money"). Bourdieu (1975, 1991, 2004) theorizes the scientific field as a social struggle over objectivity, emphasizing that this field, when autonomous, can transcend its field-specific competitive "game" to advance the "progress of reason" (for a critique, see Mialet 2003). Bourdieu acknowledges that the power scientists have in society and the extent to which scientific controversies can be waged more narrowly among "competitor-peers" without direct political and economic pressure is clearly historically variable (Bourdieu 1975:23; 1991).

3. For government evaluations of weather and climate modification research in this period, see U.S. Advisory Committee on Weather Control (1957), National Science Foundation (1965), and U.S. Congress, Senate (1965–1966).

4. Daniel Hirschman (2016, ch. 4) argues that some economists succeeded from the 1930s through the 1950s in constructing "the economy" as a sociotechnical object to be managed by the state. In Timothy Mitchell's (1998:90) terms, "the economy" became "a new language in which the nation-state could speak for itself and imagine its existence as something natural [and] subject to political management." By the 1960s, economists could take the economy for granted and construct a world in which it was taken for granted that government required their expertise (Hirschman and Berman 2014; Mitchell 2005; Mudge 2018).

5. I address Charney and Smagorinsky later. Bert Bolin became the first director of the Intergovernmental Panel on Climate Change in 1988.

6. For data on these science policy perspectives, see Waterman (1954), Kaplan (1954), and Berkner (1954). For Commerce and Defense Department perspectives, see Weeks (1954) and Quarles (1954).

7. Primary data on NCAR was collected from the UCAR Digital Archives, available at archives.ucar.edu.

8. Important to the success of GFDL, Smagorinsky's career paralleled those of other first-wave climate scientists. Smagorinsky enrolled in the U.S. Army Air Force meteorology program at NYU during World War II. In 1950 Jule Charney, whom Smagorinsky met at the MIT, recruited him to the NWP Project at Princeton University. Smagorinsky's modeling work and recognition by von Neumann prepared him for later opportunities through the GFDL.

9. Lahsen (2008) has shown that climate science inhabited a unique cultural space that featured a larger division in the scientific field, between the previously dominant Cold War physics and the rise of post-1970s environmental sciences.

10. Within two years of the establishment of the EPA in 1970, Congress established the Congressional Office of Technology Assessment to effectively oversee

research activities. At the same time, federal funding for physics continued to decline.

11. See, as examples, the statements and publications by the Cornwall Alliance (2024) and the CO2 Coalition (2024).

12. Regarding a prevailing neoliberal orientation to environmental policy, beginning in the 1970s, and the related concerted political movement against climate science and environmentalism, see Brulle (2000), Parr (2013), and Ciplet and Roberts (2017).

13. In Latour's (1988:56) terms, scientists came to generate "new source[s] of power with which to conquer the state."

CHAPTER 6. GOVERNING CLIMATE FUTURES

1. Other starting points for understanding climate security are possible, especially normative theories of security and justice. For example, within a national security framework, analysis may help determine factors that stabilize or preserve state strength. In turn, those theorizing "international security" investigate the basic, realist corollary to national security by drawing implications for a peaceful interstate system. Alternatively, "human security," linked to human rights discourse, renders human individuals and their livelihoods as the referent objects of security. "Environmental security," although not used consistently by scholars and government officials, generally formulates human security as beset by changing environmental conditions, including climate change (MacDonald 2013). A proliferation of modifiers even within the field organized around "climate security" indicates a lack of conceptual clarity (cf. Morgan and Orloff 2017:2). Sharon Burke (2009), now president of the Center for Climate and Security, has called for "natural security," others have called for "planetary security" or "sustainable security," and the list continues.

2. In *Security, Territory, Population*, Michel Foucault defines the "mechanism of security" as involving "management of an open series [of events] that can only be controlled by an estimate of probabilities" (Foucault 2004:20). Foucault argues the "space of security" involved a "milieu" (as opposed to a territory), defined by "what is needed to account for action at a distance of one body on another" (2004:21). For example, using the case of smallpox vaccination, nineteenth-century efforts to regulate contagious disease formed a security apparatus insofar as strategies to contain the disease were conducted not through direct confrontation with the disease but by acting upon risk factors. Importantly, a governmentality specific to security presupposes the "rationalization of chance and probability," especially through technologies that facilitate "normalization" (of people in society) and problematization via assessment of "risk," for example population statistics or, in this case, the strategic projection

of security risks associated with climate change (Foucault 2004:59, 63; see also Parenti 2015; Tuathail 1996).

3. For the complete Lieberman-Warner bill, see Climate Security Act (2008).

4. The figure of 150 million cross-border refugees refers to a 2009 headline in the *Guardian* newspaper that highlighted a report by the Environmental Justice Foundation (Vidal 2009). Larger figures circulate widely—for example, "half a billion" (Titley 2017),"750 million" (Miller 2017:22), and "2 billion refugees by 2100" (Jamail 2018). Many figures are reproduced with little understanding of underlying evidence and without qualification regarding types of human mobility.

5. Between 1993 and 2008, the budget of the U.S. Immigration and Naturalization Service (since 2003, the U.S. Citizenship and Immigration Service) tripled. The number of Border Control agents in 2016 (21,370 agents) was nearly double that employed in 1993, and ICE agents tasked with internal Enforcement and Removal Operations nearly tripled from 2,710 in 2003 to 7,995 in 2016 (American Immigration Council 2021).

6. Myers and the Climate Institute's work on "climate refugees" illustrates this process. Their work generated wider impact compared to the work of the migration scholars who criticized Myers's findings as faulty or irresponsibly overblown. For example, Myers's estimates of climate refugees were utilized in the 2009 UN Secretary-General report on the security implications of climate change. The widely cited Stern Review commissioned by the UK Parliament and titled *The Economics of Climate Change* also used Myers's findings as a basis for calculating the financial costs of climate change (Stern 2007:77, 111). Scientists (e.g., Ehrlich and Ehrlich 2013:4), popular progressive writers (e.g. Miller 2017:21–22), scenario-based futurists (Randers 2012:181–185), environmental NGOs (e.g., EJF 2011:2), and media reports persistently drew upon Myers's estimates of environmental refugees.

7. The phrase "chronic security anarchy" was used in 2013 by early climate-conflict scholar Thomas Homer-Dixon, in a press review of climate-conflict research: "The world will be a very violent place by mid-century if climate change continues as projected. Climate change will increase the likelihood that large zones of the world will see central institutions disintegrate and evolve into a form of chronic security anarchy" (Freedman 2013).

8. In a 2007 editorial, UN Secretary-General Ban Ki-Moon linked the climatic causes of the Darfur crisis to the contemporary importance of ongoing climate negotiations in 2007 (Ki-Moon 2007).

9. See Giorgi and Mearns (1999, 2001) for early reviews of regional climate modeling efforts.

10. For extensive annual compilations of climate "event attribution" studies, see the American Meteorological Society's annual special report, "Explaining Extreme Events from a Climate Perspective," published since 2011 (AMS 2011–2016).

11. The directorate of Space, Security and Migration (ISPRA) within the European Commission's Joint Research Center parallels the case of the GMES by also bridging environmental and social monitoring.

12. As Paul Norwood and Benjamin Jensen (2016) describe, anticipation through simulated rehearsal is also a feature of defense training via advances in war-gaming technologies. On early climate change war-game development, see also Burke and Parthemore (2009).

CHAPTER 7. FUTURE STRUGGLES

1. The metaphor of organizational ecologies diverges slightly from Pierre Bourdieu's field theory (Liu and Emirbayer 2016); in the present case it denotes how the struggle over competing forms of capital (to use the terms of field theory) is not complete, and power relations unfold alongside a more functional division of activities.

2. Of note, political polarization on climate change has increased since that time, both among the American public and within partisan maneuvering. For example, in the 2008 presidential race, the Republican candidate, Senator John McCain, ran on a campaign that strongly emphasized his credentials (over Barack Obama's) to confront climate change. Growing popular support for climate policy notwithstanding, the fractures within the Republican Party have made the United States, occupied strongly by fossil fuel and anti–climate policy interests, an anomaly among states with right-wing parties in power (Davenport and Lipton 2017).

3. Video interview at Mathiesen (2015).

4. John Podesta, who was also former president Bill Clinton's chief of staff and longtime political strategist, founded the CAP in 2007 as an "independent nonpartisan policy institute" with a mission to "improve the lives of all Americans, through bold, progressive ideas." Broadly, the organization's staff and board include people with close ties to the Democratic Party (compared, for example, to the CNA MAB), yet CAP's work on climate security, like for other organizations, avoids reference to political party platforms (see Podesta and Ogden 2007).

5. Cato Institute senior fellow and libertarian critic Pat Michaels likewise has rebuffed the political undertone of climate security expertise, claiming in a 2008 televised interview on *Russia Today*: "When you can't sell an issue to the American public very well, you try and sell it as a military issue" (Michaels 2008).

6. Interviews I have conducted with climate change researchers indicate that researchers are often hesitant to identify as "climate scientists" unless they are longtime climate modelers who have participated in developing GCMs or related products (Baker et al. 2020). The term is often a sign of prestige rather than an objective credential, but it is also regularly adopted by peripheral actors in

climate science, especially climate "skeptics" (see, e.g., science reporting of the CO2 Coalition, the Friends of Science, and other contrarian groups that reject the consensus on global warming but regularly adopt the term "climate science").

7. Data on Bowman's corporate positions were retrieved from British Petroleum (bp.com/en/global/corporate/who-we-are/board-and-executive-management/the-board/admiral-frank-bowman.html), and Reuters (https://www.reuters.com/finance/stocks/officer-profile/MSD/1190217). These sources are no longer available.

8. Adaptation and capacity-building are also inflected in national security strategy, however, perhaps demonstrating what David Chandler and Nik Hynek (2011) critique as a general co-optation of "human security" discourse. The NIC (2008:46–47), for example, promotes efforts at "enhancing capacity" and increasing climate "resiliency" because such measures "reduce future risk to national security." The U.S. Interagency Climate Change Adaptation Task Force (2011:22) likewise identifies "climate-resilient development strategies" as those that simultaneously "address the impacts of climate change that exacerbate conflict and social, economic, and political instability abroad." Kent Butts and Joshua Busby similarly link anti-terrorism strategies and climate-resilient development, given "USAID's substantial capacity to address the terrorist insurgency" by promoting stability in "strategically significant places" (Butts 2008:146; see also Busby 2008a, 2008b).

9. In a public address at the American Geophysical Union's annual meeting in December 2019, Goodman suggested that each regional U.S. military command around the globe should house a climate scientist (personal observation).

10. Statements by Bernie Sanders and his campaign were retrieved from CNN (2015) and Sanders's 2016 campaign website, since removed.

11. For inventories up to 2020, see Sabin Center (2020).

12. For a draft of the rejected testimony with critical commentary, see Schoonover (2019).

13. Concern for such a situation may be warranted, especially in developing countries where populations may face "carbon colonialism" and where authoritarian environmental policies allegedly are guided by market principles (Goldman 2004; Bumpus and Liverman 2010).

CONCLUSION

1. Indigenous Climate Action (n.d.).

2. Carol Farbotko and Heather Lazrus (2012), for example, show how portrayals of citizens of the island nation of Tuvalu as among the "first climate refugees" replaces situated experiences of mobility and environmental change with an imposed narrative in which climate change will determine the future of Tuvaluan

society. The narrative, they conclude, represents a "political appropriation of the space of an already marginalized population" (Farbotko and Lazrus 2012:385).

3. The term quoted refers to the World People's Conference on Climate Change and the Rights of Mother Earth, held in Cochabamba, Bolivia, April 22, 2010 as an alternative climate justice forum to supplement policy negotiations in Copenhagen in 2009. For the "Peoples Agreement," see World People's Conference on Climate Change (2010). For a broader discussion of the relationship of environmental justice to climate justice discourse, see Schlosberg and Collins (2014).

References

Abbe, Cleveland. 1889. "Is Our Climate Changing?" *New York Times*, February 3, 4.

Abbe, Cleveland. 1893. "The Meteorological Work of the U.S. Signal Service, 1870 to 1891." *Report of the Chicago Meteorological Congress* (Part II):1–53.

Abbe, Cleveland. 1991. "Personal View of Professor Cleveland Abe." Pp. 14–16 in *The Beginning of the National Weather Service: The Signal Service Years (1870–1891) as Viewed by Early Weather Pioneers*, edited by G. K. Grice. Washington, DC: U.S. National Weather Service.

Abbe, Cleveland, E. B. Elliott, H. A. Newton, and C. S. Peirce. [1879] 1880. "Report on Standard Time." *Proceedings of the American Metrological Society* 2: 17–44.

Abbott, Andrew. 2005. "Linked Ecologies: States and Universities as Environments for Professions." *Sociological Theory* 23(3):245–74.

Ackerknecht, Erwin H. 1948. "Anticontagionism between 1821 and 1867." *Bulletin of the History of Medicine* 22:562–93.

Adams, Julia, Elisabeth S. Clemens, and Ann S. Orloff, eds. 2005. *Remaking Modernity: Politics, History, and Sociology*. Durham, NC: Duke University Press.

Advisory Committee on Weather Services. 1953. *Weather Is the Nation's Business: Report to the Department of Commerce*. Washington, DC: Government Printing Office.

Agarwal, Anil, and Sunita Narain. 1991. *Global Warming in an Unequal World: A Case of Environmental Colonialism*. New Delhi, India: Center for Science and Environment.

Ager, Jon. 2003. *The Government Machine: A Revolutionary History of the Computer*. Cambridge, MA: MIT Press.

Albert, Mathieu, and Daniel L. Kleinman. 2011. "Bringing Pierre Bourdieu to Science and Technology Studies." *Minerva* 49(3):263–73.

Alberts, Paul. 2013. "Foucault, Nature, and the Environment." Pp. 544–61 in *A Companion to Foucault*, edited by C. Falzon, T. O'Leary, and J. Sawicki. Malden, MA: Blackwell.

Allen, Douglas R. 2001. "The Genesis of Meteorology at the University of Chicago." *Bulletin of the American Meteorological Society* 82(9):1905–9.

Alperen, Martin J. 2017. *Foundations of Homeland Security: Law and Policy*. Hoboken, NJ: Wiley.

Alter, J. Cecil. 1915. "Weather Bureau Exhibit at San Francisco, 1915." *Monthly Weather Review* 43(9):452–55.

Amer, Muhammad, Tugrul U. Daim, and Antonie Jetter. 2013. "A Review of Scenario Planning." *Futures* 46:23–40.

American Climatological Association. 1892. "Introduction." *The Climatologist* 2(3):1.

American Immigration Council. 2021. "The Cost of Immigration Enforcement and Border Security." Accessed May 20, 2024. www.americanimmigration council.org/research/the-cost-of-immigration-enforcement-and-border -security.

American Meteorological Society (AMS). 2011–2016. "Explaining Extreme Events from a Climate Perspective." Special supplement to *Bulletin of the American Meteorological Society* 93–98. Accessed April 1, 2018. ametsoc.org /ams/index.cfm/publications/bulletin-of-the-american-meteorological -society-bams/explaining-extreme-events-from-a-climate-perspective/.

American Security Project (ASP). 2012. *Climate Security Report*. Accessed June 19, 2014. www.americansecurityproject.org/climate-security-report/.

American Security Project (ASP). 2014. *The Global Security Defense Index on Climate Change*. Accessed January 7, 2016. https://www.americansecurity project.org/climate-energy-and-security/climate-change/gsdicc/.

Amster, Ellen J. 2013. *Medicine and the Saints: Science, Islam, and the Colonial Encounter in Morocco, 1877–1956*. Austin: University of Texas Press.

Anderson, Benedict. 1991. *Imagined Communities: Reflections on the Origin and Spread of Nationalism*. New York: Verso.

Anderson, Blake, and Michael M'Gonigle. 2012. "Does Ecological Economics Have a Future?" *Ecological Economics* 84:37–48.

Anderson, Katharine. 1999. "The Weather Prophets: Science and Reputation in Victorian Meteorology" *History of Science* 37:180–203.

Anderson, Katharine. 2005. *Predicting the Weather: Victorians and the Science of Meteorology*. Chicago: University of Chicago Press.

Anderson, Warwick. 2006a. *The Cultivation of Whiteness: Science, Health and Racial Destiny in Australia*. Durham, NC: Duke University Press.

Anderson, Warwick. 2006b. *Colonial Pathologies: American Tropical Medicine, Race, and Hygiene in the Philippines*. Durham, NC: Duke University Press.

Andersson, Jenny. 2018. *The Future of the World: Futurology, Futurists, and the Struggle for the Post Cold War Imagination*. London: Oxford University Press.

Andreas, Peter, and Timothy Snyder, eds. 2000. *The Wall around the West: State Borders and Immigration Controls in North America and Europe*. New York: Rowman & Littlefield.

Anstey, Peter. 2011. "The Creation of the English Hippocrates." *Medical History* 55(4):457–78. https://doi.org/10.1017/S0025727300004944.

Antill, Edward. 1771. "An Essay on the Cultivation of the Vine, and the Making and Preserving of Wine, Suited to the Different Climates in North-America." *Transactions of the American Philosophical Society* 1:117–98.

Antonio, Robert, and Brett Clark. 2015. "The Climate Change Divide in Social Theory." Pp. 333–68 in *Climate Change and Society: Sociological Perspectives*, edited by R. Dunlap and R. Brulle. New York: Oxford University Press.

Antonio, Robert J., and Robert J. Brulle. 2011. "The Unbearable Lightness of Politics: Climate Change Denial and Political Polarization." *Sociological Quarterly* 52(2):195–202.

Archibald, Douglas T. 1897. *The Story of the Earth's Atmosphere*. New York: Appleton.

Arguez, Anthony, and Russell S. Vose. 2011. "The Definition of the Standard WMO Climate Normal: The Key to Deriving Alternative Climate Normals." *Bulletin of the American Meteorological Society* 92(6):699–704.

Aristotle. n.d. *Meteorologica*. Book II, Part 5. Accessed June 9, 2020. classics .mit.edu/Aristotle/meteorology.2.ii.html.

Arnold, Daniel. 1996. *Warm Climates and Western Medicine: Emergence of Tropical Medicine, 1500–1900*. London: Rodopi.

Arrhenius, Svante. 1896. "On the Influence of Carbonic Acid in the Air upon the Temperature of the Ground." *Philosophical Magazine and Journal of Science* 41:237–76.

Arrhenius, Svante. 1908. *Worlds in the Making: The Evolution of the Universe*. Translated by H. Borns. New York: Harper.

Asaka, Ikuko. 2017. *Tropical Freedom: Climate, Settler Colonialism, and Black Exclusion in the Age of Emancipation*. Durham, NC: Duke University Press.

Aschbacher, Josef, and Maria P. Milagro-Pérez. 2012. "The European Earth Monitoring (GMES) Programme: Status and Perspectives." *Remote Sensing of Environment* 120:3–8.

Ash, Eric. 2010. "Expertise and the Early Modern State." *Osiris* 25(1):1–24.

Ashutosh, Ishan. 2018. "Mapping Race and Environment: Geography's Entanglements with Aryanism." *Journal of Historical Geography* 62:15–23.

Baker, Zeke. 2017. "Climate State: Science-State Struggles and the Formation of Climate Science in the US from the 1930s to 1960s." *Social Studies of Science* 47(6):861–87.

Baker, Zeke. 2018. "Meteorological Frontiers: Climate Knowledge, the West, and US Statecraft, 1800–50." *Social Science History* 42(4): 731–61.

Baker, Zeke, Julia Ekstrom, and Louise Bedsworth. 2018. "Climate Information? Embedding Climate Futures within Temporalities of California Water Management." *Environmental Sociology* 4(4):419–33.

Baker, Zeke, Julia A. Ekstrom, Kelsey D. Meagher, Benjamin L. Preston, and Louise Bedsworth. 2020. "The Social Structure of Climate Change Research and Practitioner Engagement: Evidence from California." *Global Environmental Change* 63:102074.

Baldwin, Andrew, and Giovanni Bettini, eds. 2017. *Life Adrift: Climate Change, Migration, Critique.* New York: Rowman & Littlefield.

Banner, Stuart. 2007. *How the Indians Lost Their Land: Law and Power on the Frontier.* Cambridge, MA: Harvard University Press.

Bardsley, Douglas, and Graeme Hugo. 2010. "Migration and Climate Change: Examining Thresholds of Change to Guide Effective Adaptation Decision-Making." *Population and Environment* 32:238–62.

Barnes, Hannah. 2013. "How Many Climate Migrants Will There Be?" *BBC News Magazine.* September 2. Accessed January 18, 2018. www.bbc.com /news/magazine-23899195.

Barnett, Jon. 2001. *The Meaning of Environmental Security: Ecological Politics and Policy in the New Security Era.* London: Zed Books.

Barnett, Jon. 2003. "Security and Climate Change." *Global Environmental Change* 13:7–17.

Barry Roger G. 2015. "The Shaping of Climate Science: Half a Century in Personal Perspective." *History of Geo- and Space Sciences* 6:87–105.

Bartky, Ian R. 1989. "The Adoption of Standard Time." *Technology and Culture* 30(1):25–56.

Bartky, Ian R. 2000. *Selling the True Time: Nineteenth-Century Timekeeping in America.* Stanford, CA: Stanford University Press.

Barton, Edward H. 1837. *Introductory Lecture on Acclimation, Delivered at the Opening of the Third Session of the Medical College of Louisiana.* New Orleans: Commercial Bulletin Printing.

Barton, Edward H. 1852. "Report on the Epidemics of Louisiana, Mississippi, Arkansas and Texas." *Transactions of the American Medical Association* 5:571–662.

Barton, William. 1791 [1793]. "Observations on the Probabilities of the Duration of Human Life and the Progress of Population in the United States of

America." *Transactions of the American Philosophical Society* 3:25–63, 134–38.

Bashford, Alison. 2004. *Imperial Hygiene: A Critical History of Colonialism, Nationalism and Public Health.* New York: Palgrave.

Bates, Charles, and John Fuller. 1986. *America's Weather Warriors: 1814–1985.* College Station: Texas A&M University Press.

Bawden, Tom. 2013. "Plan To Use Financial Markets to Halt Climate Change Is 'Doomed.'" *The Independent*, October 1. Accessed May 1, 2024. https://www .independent.co.uk/climate-change/news/plan-to-use-financial-markets-to -halt-climate-change-is-doomed-8852011.html.

Beattie, James, Emily O'Gorman, and Matthew Henry, eds. 2014. *Climate, Science, and Colonization: Histories from Australia and New Zealand.* New York: Palgrave.

Bedini, Silvio A. 1990. *Thomas Jefferson: Statesman of Science.* New York: Macmillan.

Benestad, Rasmus E. 2005. "Climate Change Scenarios for Northern Europe from Multi-model IPCC AR4 Climate Simulations." *Geophysical Research Letters* 32(17).

Benjamin, Walter. [1940] 2005. "On the Concept of History." Accessed October 2, 2018. www.marxists.org/reference/archive/benjamin/1940 /history.htm.

Benson, Michael, ed. 1988. *From Pittsburgh to the Rocky Mountains: Major Stephen Long's Expedition, 1819–1820.* Golden, CO: Fulcrum.

Berger, André, and Marie-France Loutre. 2002. "Climate: An Exceptionally Long Interglacial Ahead?" *Science* 297(5585):1287–88.

Berkner, Lloyd V. 1954. "International Scientific Action: The International Geophysical Year 1957–58." *Science* 119(3096):569–75.

Bernstein, Peter L. 1998. *Against the Gods: The Remarkable Story of Risk.* New York: Wiley.

Bettini, Giovanni. 2014. "Climate Migration as an Adaption Strategy: Desecuritizing Climate-Induced Migration or Making the Unruly Governable?" *Critical Studies on Security* 2(2):180–95.

Bigelow, Frank H. 1900. "Work of the Meteorologist for the Benefit of Agriculture, Commerce, and Navigation." Pp. 71–92 in *Yearbook of the U.S. Department of Agriculture, 1899.* Washington, DC: Government Printing Office.

Bigger, Patrick, and Morgan Robertson. 2017. "Value Is Simple: Valuation Is Complex." *Capitalism Nature Socialism* 28(1):68–77.

Biggs, Michael. 1999. "Putting the State on the Map: Cartography, Territory, and European State Formation." *Comparative Studies in Society and History* 41(2):374–405.

Block, Ben. 2008. "A Look Back at James Hansen's Seminal Testimony on Climate, Part One." Worldwatch Institute/Grist. Accessed January 20, 2019. https://grist.org/article/a-climate-hero-the-early-years/.

Block, Fred L. 2011. "Innovation and the Invisible Hand of Government."
Pp. 1–26 in *State of Innovation: The US Government's Role in Technology Development*, edited by Fred L. Block and Matthew R. Keller. New York: Paradigm.

Blodget, Lorin. 1857. *Climatology of the United States [. . .]*. Philadelphia: Lippincott.

Bloor, David. 1974. *Knowledge and Social Imagery*. Henley, UK: Routledge and Kegan Paul.

Bolin, Bert, ed. 1959. *The Atmosphere and the Sea in Motion*. New York: Rockefeller Institute Press.

Bolin, Bert. 1999. "Carl-Gustav Rossby: The Stockholm Period, 1947–1957." *Tellus* 51A–B(1):4–12.

Bolin, Bert. 2007. *A History of the Science and Politics of Climate Change: The Role of the Intergovernmental Panel on Climate Change*. New York: Cambridge University Press.

Bolton, Herbert, ed. 1908. "Papers of Zebulon M. Pike, 1806–1807." *American Historical Review* 13(4):798–827.

Bond, Patrick. 2012. "Emissions Trading, New Enclosures and Eco-Social Contestation." *Antipode* 44(3):684–701.

Bordo, Michael D., and Anna J. Schwartz, eds. 1984. *A Retrospective on the Classical Gold Standard, 1821–1931*. Chicago: University of Chicago Press.

Boulding, Elise. 1983. "Setting New Research Agendas: A Social Scientist's View." Pp. 3–8 in *Social Science Research and Climate Change*, edited by R. S. Chen, E. Boulding, and S. Schneider. Dordrecht: Springer.

Bourdieu, Pierre. 1975. "The Specificity of the Scientific Field and the Social Conditions of the Progress of Reason." *Social Science Information* 14(6):19–47.

Bourdieu, Pierre. 1986. "The Forms of Capital." Pp. 241–56 in *Handbook of Theory and Research for the Sociology of Education*, edited by John Richardson. Westport, CT: Greenwood.

Bourdieu, Pierre. 1991. "The Peculiar History of Scientific Reason." *Sociological Forum* 6(1):3–26.

Bourdieu, Pierre. 1993. *The Field of Cultural Production: Essays on Art and Literature*. Cambridge, UK: Polity.

Bourdieu, Pierre. 1994. "Rethinking the State: The Genesis and Structure of the Bureaucratic Field." Translated by L. Wacquant and S. Farage. *Sociological Theory* 12(1):1–18.

Bourdieu, Pierre. 1996a. *Rules of Art: Genesis and Structure of the Literary Field*. Translated by Susan Emanuel. Stanford, CA: Stanford University Press.

Bourdieu, Pierre. 1996b. *The State Nobility: Elite Schools in the Field of Power*. Stanford, CA: Stanford University Press.

Bourdieu, Pierre. 2004. *Science of Science and Reflexivity*. Translated by R. Nice. Chicago: University of Chicago Press.

Bourdieu, Pierre. 2015. *On the State: Lectures at the College de France, 1989–1992*. Edited by P. Champagne, R. Lenoir, F. Poupeau, and M. Riviere. Translated by D. Fernbach. Cambridge, MA: Polity Press.

Bourdieu, Pierre, and Loic J. D. Wacquant. 1992. *An Invitation to Reflexive Sociology*. Chicago: University of Chicago Press.

Boykoff, Maxwell T., David Frame, and Sam Randalls. 2010. "Discursive Stability Meets Climate Instability: A Critical Exploration of the Concept of 'Climate Stabilization' in Contemporary Climate Policy." *Global Environmental Change* 20(1):53–64.

Brace, Catherine, and Hilary Geoghegan. 2010. "Human Geographies of Climate Change: Landscape, Temporality, and Lay Knowledges." *Progress in Human Geography* 35(3):284–302.

Bradley, Raymond S., and Philip D. Jonest. 1993. "'Little Ice Age' Summer Temperature Variations: Their Nature and Relevance to Recent Global Warming Trends." *Holocene* 3(4):367–76.

Bramwell, Lincoln. 2012. "The 1911 Weeks Act: The Legislation That Nationalized the U.S. Forest Service." *Journal of Energy and Natural Resource Law* 30(3):325–36.

Brandt, Patrick T., John R. Freeman, and Philip A. Schrodt. 2014. "Evaluating Forecasts of Political Conflict Dynamics." *International Journal of Forecasting* 30:944–62.

Braun, Bruce, and Sarah J. Whatmore, eds. 2010. *Political Matter: Technoscience, Democracy, and Public Life*. Minneapolis: Minnesota University Press.

Broecker, Wallace S., and Robert Kunzig. 2009. *Fixing Climate: What Past Climate Changes Reveal about the Current Threat—and How to Counter It*. New York: Hill and Wang.

Brooks, Charles F. 1932. "Robert DeCourcy Ward, Climatologist." *Annals of the Association of American Geographers* 22(1):33–43.

Brown, Lindsay. 2019. "Climate Change: What Is a Climate Emergency?" BBC, May 3. Accessed May 14, 2019. https://www.bbc.com/news/newsbeat -47570654.

Browne, Des, and Michael Shank. 2015. "A Clear and Present Danger to Planet Earth: Climate Change." *National Interest*, January 26. Accessed February 8, 2018. http://nationalinterest.org/feature/clear-present-danger-planet-earth -climate-change-12107.

Brozović, Danilo. 2023. "Societal Collapse: A Literature Review." *Futures* 145:103075. https://doi.org/10.1016/j.futures.2022.103075.

Brubaker, Rogers, and Frederick Cooper. 2000. "Beyond 'Identity.'" *Theory and Society* 29(1):1–47.

Brulle, Robert J. 2000. *Agency, Democracy, and Nature: The US Environmental Movement from a Critical Theory Perspective*. Cambridge, MA: MIT Press.

Brulle, Robert J. 2014. "Institutionalizing Delay: Foundation Funding and the Creation of U.S. Climate Change Counter-movement Organizations." *Climatic Change* 122(4):681–94.

Brutschin, Elina, and Samuel R. Schubert. 2016. "Icy Waters, Hot Tempers, and High Stakes: Geopolitics and Geoeconomics of the Arctic." *Energy Research & Social Science* 16:147–59. https://doi.org/10.1016/j.erss.2016.03.020.

Bryson Reid A. 1997. "The Paradigm of Climatology." *Bulletin of the American Meteorological Society* 78(3):449–51.

Budyko, Mikael I. 1969. "The Effect of Solar Radiation Variations on the Climate of the Earth." *Tellus* 21:611–19.

Buff, Rachel I., ed. 2008. *Immigrant Rights in the Shadows of Citizenship*. New York: New York University Press.

Buhaug, H., J. Nordkvelle, T. Bernauer, T. Böhmelt, M. Brzoska, J. W. Busby, A. Ciccone, et al. 2014. "One Effect to Rule Them All? A Comment on Climate and Conflict." *Climatic Change* 127:391–97.

Bumpus, Adam. G., and Diana M. Liverman. 2010. "Carbon Colonialism? Offsets, Greenhouse Gas Reductions, and Sustainable Development." Pp. 203–24 in *Global Political Ecology*, edited by R. Peet, P. Robbins, and M. Watts. London: Routledge.

Burke, M. B., E. Miguel, S. Satyanath, J. A. Dykema, and D. B. Lobell. 2009. "Warming Increases the Risk of Civil War in Africa." *Proceedings of the National Academy of Sciences* 106:20670–74.

Burke, Sharon. 2009. *Natural Security*. Washington, DC: Center for a New American Security.

Burke, Sharon, and Christine Parthemore. 2009. *Climate Change War Game: Major Findings and Background*. June 6. Washington, DC: Center for a New American Security.

Burkett, Paul, and John Bellamy Foster. 2006. "Metabolism, Energy, and Entropy in Marx's Critique of Political Economy." *Theory and Society* 3:109–56.

Busby, Joshua W. 2007. *Climate Change and National Security: An Agenda for Action*. New York: Council on Foreign Relations Press.

Busby, Joshua. 2008a. "Under What Conditions Could Climate Change Pose a Threat to U.S. National Security?" Pp. 142–55 in *Global Climate Change: National Security Implications*, edited by Carolyn Pumphrey. Carlisle, PA: Strategic Studies Institute.

Busby, Joshua W. 2008b. "Who Cares about the Weather? Climate Change and US National Security." *Security Studies* 17:468–504.

Bush, Vannevar. 1945. *Science, the Endless Frontier: Report to the President on a Program for Postwar Scientific Research*. Washington, DC: Government Printing Office.

Butts, Kent H. 1993. "Environmental Security: What Is DOD's Role?" Strategic Studies Institute, May 28. Carlisle, PA: US Army War College.

Butts, Kent H. 2008. "Climate Change: Complicating the Struggle against Extremist Ideology." Pp. 126–41 in *Global Climate Change: National Security Implications*, edited by Carolyn Pumphrey. Carlisle, PA: Strategic Studies Institute.

Buxton, Nick, and Ben Hayes. 2015. *The Secure and the Dispossessed: How the Military and Corporations Are Shaping a Climate-Changed World*. London: Pluto Press.

Buys-Ballot, Christophorus H. D. 1872. *Suggestions on a Uniform System of Meteorological Observations*. Utrecht: Manssen.

Buzan, Barry, Ole Wæver, and Jaap de Wilde, eds. 1998. *Security: A New Framework for Analysis*. Boulder, CO: Lynne Rienner.

Byers, Horace R. 1970. "Recollections of the War Years." *Bulletin of the American Meteorological Society* 51(3):214–17.

Byers, Horace R. 1960. *Carl-Gustaf Arvid Rossby, 1898–1957: A Biographical Memoir*. National Academy of Sciences Biographical Memoirs. Washington, DC: National Academies Press.

Byers, Horace R., Joseph Kaplan, and Edward Minser. 1946. "The Teaching of Meteorology in Colleges and Universities." *Bulletin of the American Meteorological Society* 27(2):95–98.

Caldwell, Charles. 1843. "Review: Samuel Forry's 'The Climate of the United States.'" *Western Journal of Medicine and Surgery* 7(2):142–53.

Caliskan, Koray, and Michel Callon. 2009. "Economization, Part 1: Shifting Attention from the Economy towards Processes of Economization." *Economy and Society* 38(3):369–98.

Callendar, Guy Stewart. 1938. "The Artificial Production of Carbon Dioxide and Its Influence on Temperature." *Quarterly Journal of the Royal Meteorological Society* 64(275):223–40.

Callison, Candis. 2024. "Rethinking Our Histories and Relations with Climate Change." Pp. 19–26 in *Climate, Science and Society: A Primer*, edited by Z. Baker, T. Law, M. Vardy, and S. Zehr. London: Routledge.

Callon, Michel. 1986. "Some Elements of a Sociology of Translation: Domestication of the Scallops and the Fishermen of St. Brieuc Bay." Pp. 196–233 in *Power, Action, and Belief: A New Sociology of Knowledge?*, edited by J. Law. London: Routledge.

Calvert, E. B. 1931. *The Weather Bureau*. U.S. Department of Agriculture Miscellaneous Publication No. 114. Washington, DC: Government Printing Office.

Calvert, Jane. 2006. "What's Special about Basic Research?" *Science, Technology and Human Values* 31(2):199–220.

Camic, Charles. 2011. "Bourdieu's Cleft Sociology of Science." *Minerva* 49(3):275–93.

Campbell, Hardy W. 1902. *Soil Culture Manual*. Holdrege, NE: Campbell.

Campbell, Hardy W. 1916. *Progressive Agriculture: Tillage, Not Weather, Controls Yield*. Lincoln, NE: Woodruff Bank Note Company.

Campbell, Kurt M., ed. 2009. *Climatic Cataclysm: The Foreign Policy and National Security Implications of Climate Change*. Washington, DC: Brookings Institution Press.

Campbell, Kurt M., Jay Gulledge, J. R. McNeill, John Podesta, Peter Ogden, Leon Fuerth, R. James Woolsey, Alexander T. J. Lennon, Julianne Smith, Richard Weitz, and Derek Mix. 2007. *The Age of Consequences: The Foreign Policy and National Security Implications of Global Climate Change*. Washington, DC: Center for Strategic and International Studies.

Campbell, Kurt M., and Christine Parthemore. 2007. "National Security and Climate Change in Perspective." Pp. 1–25 in *The Age of Consequences: The Foreign Policy and National Security Implications of Global Climate Change*. Washington, DC: Center for Strategic and International Studies.

Carmichael, Jason T., Robert J. Brulle, and Joanna K. Huxster. 2017. "The Great Divide: Understanding the Role of Media and Other Drivers of the Partisan Divide in Public Concern over Climate Change in the USA, 2001–2014." *Climatic Change* 141(4):599–612.

Carroll, Patrick. 2006. *Science, Culture, and Modern State Formation*. Oxford: Oxford University Press.

Carroll, Patrick. 2012. "Water and Technoscientific State Formation." *Social Studies of Science* 42(4):489–516.

Cassedy, James H. 1969. "Meteorology and Medicine in Colonial America: Beginnings of an Experimental Approach." *Journal of the History of Medicine and the Allied Sciences* 24:193–204.

Cassedy, James H. 1984. *American Medicine and Statistical Thinking, 1800–1860*. Cambridge, MA: Harvard University Press.

Cassedy, James H. 1986. *Medicine and American Growth, 1800–1860*. Madison: University of Wisconsin Press.

Castles, Stephen, and Mark J. Miller. 2009. *The Age of Migration: International Population Movements in the Modern World*, 4th ed. Basingstoke: Palgrave Macmillan.

Castree, Noel. 2001. "Socializing Nature: Theory, Practice, and Politics." Pp. 1–21 in *Social Nature: Theory, Practice and Politics* edited by N. Castree and B. Braun. New York: Wiley Publishing.

Castree, Noel, and Bruce Braun, eds. 2001. *Social Nature: Theory, Practice and Politics*. New York: Wiley.

Center for a New American Security (CNAS). 2009. "Press Release." Accessed August 2, 2013. cnas.org/press/press-release/cnas-vice-president-of -natural-security-sharon-burke-testifies-on-climate-change-and-national -security.

Center for Climate and Security (CCS). 2016. "Climate and Security Consensus Project." Accessed February 3, 2017. climateandsecurity.files.wordpress.com /2016/09/climate-and-security-consensus-project-statement-2016_09.pdf.

Central Intelligence Agency (CIA). 1974. *A Study of Climatological Research as It Pertains to Intelligence Problems*. Accessed January 6, 2018. climatemonitor.it/wp-content/uploads/2009/12/1974.pdf.

Central Intelligence Agency (CIA). 1976. *USSR: The Impact of Recent Climate Change on Grain Production*. October. Partly declassified report accessed January 6, 2018. cia.gov/library/readingroom/docs/CIA-RDP08S01350R000 602140002-3.pdf.

Central Intelligence Agency (CIA). 2009. "CIA Opens Center on Climate Change and National Security." September 25. Accessed May 12, 2019. https://www .cia.gov/news-information/press-releases-statements/center-on-climate -change-and-national-security.html.

Chalmers, Lionel. 1776. *Account of the Weather and Diseases of South-Carolina*. London: Dilly.

Chandler, David, and Hynek, Nik, eds. 2011. *Critical Perspectives on Human Security: Rethinking Emancipation and Power in International Relations*. London: Routledge.

Charney Jule G., R. Fljortoft, and John von Neumann. 1950. "Numerical Integration of the Barotropic Vorticity Equation." *Tellus* 2(4):237–54.

Charney, Jule G., et al., eds. 1979. *Carbon Dioxide and Climate: A Scientific Assessment*. Washington, DC: National Academy of Science.

Chatterton, Paul, David Featherstone, and Paul Routledge. 2013. "Articulating Climate Justice in Copenhagen: Antagonism, the Commons, and Solidarity." *Antipode* 45:602–20. https://doi.org/10.1111/j.1467-8330.2012.01025.x.

Chen, Jinlei, Shichang Kang, Qinglong You, Yulan Zhang, and Wentao Du. 2022. "Projected Changes in Sea Ice and the Navigability of the Arctic Passages under Global Warming of 2°C and 3°C." *Anthropocene* 40:100349. https://doi.org/10.1016/j.ancene.2022.100349.

Chinard, Gilbert. 1947. "Eighteenth-Century Theories on America as Human Habitat." *Proceedings of the American Philosophical Society* 91:25–57.

Ciplet, David, and J. Timmons Roberts. 2017. "Climate Change and the Transition to Neoliberal Environmental Governance." *Global Environmental Change* 46:148–56.

Clark, Brett, and Andrew K. Jorgenson. 2012. "The Treadmill of Destruction and the Environmental Impacts of Militaries." *Sociological Compass* 6(7):557–69.

Clark, Brett, Andrew K. Jorgenson, and Jeffrey Kentor. 2010. "Militarization and Energy Consumption: A Test of Treadmill of Destruction Theory in Comparative Perspective." *International Journal of Sociology* 40:23–43.

Clark, Brett, and Richard York. 2005. "Carbon Metabolism: Global Capitalism, Climate Change, and the Biospheric Rift." *Theory and Society* 34:391–428.

Clark, Christopher. 2012. "The Agrarian Context of American Capitalist Development." Pp. 13–37 in *Capitalism Takes Command: The Social Transformation of Nineteenth-Century America*, edited by M. Zakim and G. J. Kornblith. Chicago: Chicago University Press.

Climate and Security Advisory Group (CSAG). 2016. *Briefing Book for a New Administration*. September 14. Accessed December 17, 2017. https:// climateandsecurity.files.wordpress.com/2016/09/climate-and-security -advisory-group_briefing-book-for-a-new-administration_2016_11.pdf.

Climate Security Act. 2008. S. 3036. 110th Congress, 2d Session. Accessed August 4, 2017. https://www.congress.gov/110/bills/s3036/BILLS-110s3036 pcs.xml#toc-id366c58d3-2bbe-440a-97fe-8be325576e2d.

Cloud, John. 2001. "Imagining the World in a Barrel: CORONA and the Clandestine Convergence of the Earth Sciences." *Social Studies of Science* 31(2):231–251.

CNA Military Advisory Board (MAB). 2007. *National Security and the Threat of Climate Change*. Alexandria, VA: CNA Corporation. Accessed April 3, 2013. cna.org/cna_files/pdf/national%20security%20and%20the%20threat %20of%20climate%20change.pdf.

CNA Military Advisory Board (MAB). 2014. *National Security and the Accelerating Risks of Climate Change*. Alexandria, VA: CNA Corporation. Accessed January 9, 2015. cna.org/cna_files/pdf/MAB_5-8-14.pdf.

CNA Military Advisory Board (MAB). 2017. "The Role of Water Stress in Instability and Conflict." Accessed November 7, 2023. https://www.cna.org /archive/CNA_Files/pdf/crm-2017-u-016532-final.pdf.

CNN. 2015. "Democratic Debate: What Is Our Greatest National Security Threat?" Posted October 13. Accessed March 10, 2018. www.youtube.com /watch?v=pGNHxZgb8WQ.

Coen, Deborah. 2018. *Climate in Motion: Science, Empire, and the Problem of Scale*. Chicago: Chicago University Press.

Coen, Deborah R. 2011. "Imperial Climatographies from Tyrol to Turkestan." *Osiris* 26(1):45–65.

Coen, Deborah R. 2016. "Big Is a Thing of the Past: Climate Change and Methodology in the History of Ideas." *Journal of the History of Ideas* 77(2):305–21.

Coen, Deborah R. 2020. "The Advent of Climate Science." In *Oxford Research Encyclopedia of Climate Science*. https://doi.org/10.1093/acrefore/978019 0228620.013.716.

Coen, Deborah R., and Fredrik Albritton Jonsson. 2022. "Between History and Earth System Science." *Isis* 113(2):407–16. https://doi.org/10.1086/719648.

Collin, Nicholas. 1789. "Introduction to Volume the Third: An Essay on Those Inquiries in Natural Philosophy, Which at Present are Most Beneficial to the United States of America." *Transactions of the American Philosophical Society* 3:iii–xxvii.

Collins, H. M. 1974. "The TEA Set: Tacit Knowledge and Scientific Networks." *Science Studies* 4(2):165–85.

Collis, Christy, and Klaus Dodds. 2008. "Assault on the Unknown: The Historical and Political Geographies of the International Geophysical Year (1957-8)." *Science and Geopolitics: The International Geophysical Year, 1957–8* 34(4):555–73. https://doi.org/10.1016/j.jhg.2008.05.016.

Conklin, Alice L. 1997. *A Mission to Civilize: The Republican Idea of Empire in France and West Africa, 1895–1930*. Stanford, CA: Stanford University Press.

Cook, John, Naomi Oreskes, Peter T. Doran, William R. Anderegg, Bart Verheggen, Ed W. Maibach, J. S. Carlton, et al. 2016. "Consensus on Consensus: A Synthesis of Consensus Estimates on Human-Caused Global Warming." *Environmental Research Letters* 11(4):048002.

Corbin, Diana F. M. 1888. *A Life of Matthew Fontaine Maury*. London: Sampson.

Cornwall Alliance. 2024. "Landmark Documents." Accessed May 1, 2024. https://cornwallalliance.org/landmark-documents/.

CO2 Coalition. 2024. "Publications." Accessed May 1, 2024. http://co2coalition.org/.

Cox, Henry J. 1904. "Use of Weather Bureau Records in the Courts." Pp. 303–12 in *Yearbook of the U.S. Department of Agriculture*. Washington, DC: Government Printing Office.

Cox, Henry J. 1991. "Personal View of Henry J. Cox." P. 29 in *The Beginning of the National Weather Service: The Signal Service Years (1870–1891) as Viewed by Early Weather Pioneers*, edited by G. K. Grice. Washington, DC: U.S. National Weather Service.

Cox, John D. 2002. *Storm Watchers: The Turbulent History of Weather Prediction from Franklin's Kite to El Niño*. Hoboken, NJ: Wiley.

Craft, Erik D. 1998. "The Value of Weather Information Services for Nineteenth-Century Great Lakes Shipping." *American Economic Review* 88(5):1059–76.

Craft, Erik D. 1999. "Private Weather Organizations and the Founding of the United States Weather Bureau." *Journal of Economic History* 59(4):1063–71.

Crawford, Elizabeth. 1996. *Arrhenius: From Ionic Theory to the Greenhouse Effect*. Canton, MA: Science History Publications.

Cronon, William. 1991. *Nature's Metropolis: Chicago and the Great West*. New York: Norton.

Cronon, William. 1995. "The Trouble with Wilderness: Or, Getting Back to the Wrong Nature." Pp. 69–90 in *Uncommon Ground: Rethinking the Human Place in Nature*, edited by William Cronon. New York: W. W. Norton. https://www.williamcronon.net/writing/Trouble_with_Wilderness_Main.html.

Cronon, William. 2003. *Changes in the Land: Indians, Colonists, and the Ecology of New England*. New York: Hill and Wang.

Crutzen, Paul. 2006. "Albedo Enhancement by Stratospheric Sulfur Injections: A Contribution to Resolve a Policy Dilemma?" *Climatic Change* 77(3):211–19.

Crutzen, Paul J. 2002. "Geology of Mankind." *Nature* 415:23.

Currie, William. 1792. *An Historical Account of the Climates and Diseases of the United States of America; and of the Remedies and Methods of Treatment, Which have been Found Most Useful and Efficacious, Particularly in Those Diseases Which Depend upon Climate and Situation*. Philadelphia: Dobson.

Cushman, Gregory T. 2005. "Bergen South: The Americanization of the Meteorology Profession in Latin America during World War II." Pp. 197–213 in *From Beaufort to Bjerknes and Beyond: Critical Perspectives on Observing, Analyzing, and Predicting Weather and Climate*, edited by S. Emeis and C. Lüdecke. Munich: Rauner.

Cushman, Gregory T. 2011. "Humboldtian Science, Creole Meteorology, and the Discovery of Human-caused Climate Change in South America." *Osiris* 26:19–44.

Cushman, Gregory T. 2013. "The Imperial Politics of Hurricane Prediction: From Calcutta and Havana to Manila and Galveston, 1839–1900." Pp. 137–62 in *Nation-States and the Global Environment: New Approaches to International Environmental History*, edited by E. Bsumek, D. Kinkela, and M. Lawrence. New York: Oxford University Press.

Dalby, Simon. 2002. *Environmental Security*. New York: Routledge.

Dalby, Simon. 2013. "The Geopolitics of Climate Change." *Political Geography* 37(1):38–47.

Dalezios, Nicolas R., and Panagiotis T. Nastos. 2016. "Milestones of the Diachronic Evolution of Meteorology." *International Global Environmental Issues* 15(1):49–69.

Dalrymple, Dana G. 2009. "The Smithsonian Bequest, Congress, and Nineteenth-Century Efforts to Increase and Diffuse Agricultural Knowledge in the United States." *Agricultural History Review* 57(2):207–35.

Darby, William. 1818. *The Emigrants' Guide to the Western and Southwestern States and Territories*. New York: Kirk and Mercein.

Dauber, Michele Landis. 2013. *The Sympathetic State: Disaster Relief and the Origins of the American Welfare State*. Chicago: University of Chicago Press.

Davenport, Coral, and Eric Lipton. 2017. "How G.O.P. Leaders Came to View Climate Change as Fake Science." *New York Times*, June 3. Accessed October 4, 2018. www.nytimes.com/2017/06/03/us/politics/republican-leaders-climate-change.html.

Davis, Diana K. 2005. "Potential Forests: Degradation Narratives, Science, and Environmental Policy in Protectorate Morocco, 1912–1956." *Environmental History* 10(2):211–38.

Davis, Diana K. 2016. *The Arid Lands: History, Power, Knowledge*. Cambridge, MA: MIT Press.

Davis, Joseph H. 2004. "An Annual Index of U.S. Industrial Production, 1790–1915." *Quarterly Journal of Economics* 119(4):1177–1215.

Dean, Michael. 1994. *Critical and Effective Histories: Foucault's Methods and Historical Sociology*. London: Routledge.

Demeritt, David. 2001. "The Construction of Global Warming and the Politics of Science." *Annals of the Association of American Geographers* 91(2):307–37.

Dennis, Michael. 2004. "Reconstructing Sociotechnical Order: Vannevar Bush and US Science Policy." Pp. 225–53 in *States of Knowledge: The Co-production of Science and Social Order*, edited by S. Jasanoff. New York: Routledge.

de Sherbinin, Alex. 2013. "Climate Change Hotspots Mapping: What Have We Learned?" *Climatic Change* 123(1):23–37.

de Souza Leão, Luciana, and Gil Eyal. 2019. "The Rise of Randomized Controlled Trials in International Development in Historical Perspective." *Theory and Society* 48(3):383–418.

Deutch, John. [1996] 2007. "The Environment on the Intelligence Agenda." Speech delivered to the World Affairs Council in Los Angeles, California, July 25. Accessed January 5, 2019. https://www.cia.gov/news-information /speeches-testimony/1996/dci_speech_072596.html.

Dezalay, Yves, and Bryant Garth. 2002. *The Internationalization of Palace Wars: Lawyers, Economists and the Contest to Transform Latin American States*. Chicago: University of Chicago Press.

Diffenbaugh, Noah S., Deepti Singh, Justin S. Mankin, Daniel E. Horton, Daniel L. Swain, Danielle Touma, Allison Charland, Yunjie Liu, Matz Haugen, Michael Tsiang, and Bala Rajaratnam. 2017. "Quantifying the Influence of Global Warming on Unprecedented Extreme Climate Events." *Proceedings of the National Academy of Sciences* 114(19):4881–86.

Disturnell, John. 1849. *The Emigrant's Guide to New Mexico, California, and Oregon*. New York: Disturnell.

Dobkin, Adin. 2014. "Castellaw and Adams in The Commercial Appeal: If Military Sees Climate Risk, Why Do We Deny?" April 10. Accessed February 20, 2018. https://www.americansecurityproject.org/castellaw -and-adams-in-the-commercial-appeal-if-military-sees-climate-risk-why -do-we-deny/.

Doel, Ronald. 2003. "Constituting the Postwar Earth Sciences: The Military's Influence on the Environmental Sciences in the USA after 1945." *Social Studies of Science* 33(5):635–66.

Doel, Ronald, and Kristine Harper. 2006. "Prometheus Unleashed: Science as a Diplomatic Weapon in the Lyndon B. Johnson Administration." *Osiris* 21(2):66–85.

Donlon, Craig, et al. 2012. "The Global Monitoring for Environment and Security (GMES) Sentinel-3 Mission." *Remote Sensing of Environment* 120:37–57.

Donnelly, Fay. 2013. *Securitization and the Iraq War: The Rules of Engagement in World Politics*. New York: Routledge.

Doose, Katja. 2021. "A Global Problem in a Divided World: Climate Change Research during the Late Cold War, 1972–1991." *Cold War History* 21(4):469–89. https://doi.org/10.1080/14682745.2021.1885377.

Dorn, Michael L. 2001. "(In)temperate Zones: Daniel Drake's Medico-moral Geographies of Urban Life in the Trans-Appalachian American West." *Journal of the History of Medicine and Allied Sciences* 55(3):256–91.

Dörries, Matthias. 2011. "The Politics of Atmospheric Sciences: 'Nuclear Winter' and Global Climate Change." *Osiris* 26(1):198–223.

Dowd, Charles F. 1884. "Origin and Early History of the New System of National Time." *Proceedings of the American Metrological Society* 4:90–101.

Drake, Daniel. 1825. "Geological Account of the Valley of the Ohio." *Transactions of the American Philosophical Society* 2:124–39.

Drake, Daniel. 1850. *A Systematic Treatise, Historical, Etiological, and Practical, on the Principal Diseases of the Interior Valley of North America: As They Appear in the Caucasian, African, Indian, and Esquimaux Varieties of Its Population.* Philadelphia: Lippincott.

Drayton, Richard. 2000. *Nature's Government: Science, Imperial Britain, and the "Improvement" of the World.* New Haven, CT: Yale University Press.

Driver, Felix. 2001. *Geography Militant: Cultures of Exploration and Empire.* Oxford: Blackwell.

Dunbar, James. 1781. *Essays on the History of Mankind in Rude and Cultivated Ages.* London: Strahan.

Duncan, James S. 2007. *In the Shadows of the Tropics: Climate, Race and Biopower in Nineteenth-Century Ceylon.* Hampshire, UK: Ashgate.

Dunlap, Riley, and Robert Brulle. 2015. *Climate Change and Society: Sociological Perspectives.* New York: Oxford University Press.

Dunlap, Riley E., and Aaron M. McCright. 2015. "Challenging Climate Change: The Denial Countermovement." Pp. 300–332 in *Climate Change and Society: Sociological Perspectives*, edited by Riley E. Dunlap and Robert Brulle. New York: Oxford University Press.

Dunne, John P., Jasmin G. John, Alistair J. Adcroft, Stephen M. Griffies, Robert W. Hallberg, Elena Shevliakova, Ronald J. Stouffer, et al. 2012. "GFDL's ESM2 Global Coupled Climate–Carbon Earth System Models. Part I: Physical Formulation and Baseline Simulation Characteristics." *Journal of Climate* 25(19):6646–65.

Dunwoody, Henry H. C., ed. 1883. *Weather Proverbs.* War Department Signal Service Notes 9. Washington, DC: Government Printing Office.

Dunwoody, Henry H. C. 1894. "The Value of Forecasts." Pp. 121–28 in *Yearbook of the U.S. Department of Agriculture.* Washington, DC: Government Printing Office.

Dyer, Gwynne. 2008. *Climate Wars.* London: Random House.

Eckholm, Nils. 1901. "On the Variations of the Climate of the Geological and Historical Past and Their Causes." *Quarterly Journal of the Royal Meteorological Society* 27(117):1–62.

Edson, Carroll E. 1921. "President's Address: The Future for Research in Climatology." *Transactions of the American Climatological and Clinical Association* 37:1–10.

Edwards, Paul N. 2000. "A Brief History of Atmospheric General Circulation Modeling." Pp. 67–90 in *General Circulation Model Development: Past, Present, and Future,* edited by D. Randall. San Diego, CA: Academic Press.

Edwards, Paul N. 2001. "Representing the Global Atmosphere: Computer Models, Data, and Knowledge about Climate Change." Pp. 31–65 in *Changing the Atmosphere: Expert Knowledge and Environmental Governance,* edited by C. Miller and P. Edwards. Cambridge, MA: MIT Press.

Edwards, Paul N. 2006. "Meteorology as Infrastructural Globalism." *Osiris* 21:229–50.

Edwards, Paul N. 2010. *A Vast Machine: Computer Models, Climate Data, and the Politics of Global Warming.* Cambridge, MA: MIT Press.

Ehrlich, Paul, and Anne H. Ehrlich. 2013. "Can a Collapse of Global Civilization Be Avoided?" *Proceedings of the Royal Academy B* 280(1754). https://doi.org /10.1098/rspb.2012.2845.

Eisenhower, Dwight. 1954a. "Executive Order 10521: Administration of Scientific Research by Agencies of the Federal Government." Accessed November 5, 2015. www.presidency.ucsb.edu/ws/?pid=59228.

Eisenhower, Dwight. 1954b. "Statement by the President Upon Signing Executive Order Strengthening the Scientific Programs of the Federal Government." The American Presidency Project. Accessed November 12, 2015. www.presidency.ucsb.edu/ws.

Eisenstadt, Peter. 1998. "Almanacs and the Disenchantment of Early America." *Pennsylvania History* 65(2):143–69.

Elias, Norbert. [1939] 2000. *The Civilizing Process. Sociogenetic and Psychogenetic Investigations.* Oxford: Blackwell.

Engels, Friedrich. [1845] 1887. *Condition of the Working Class in England.* Accessed September 4, 2018. https://www.marxists.org/archive/marx/works /download/pdf/condition-working-class-england.pdf.

England, Merton. 1982. *A Patron for Pure Science: The National Science Foundation's Formative Years, 1945-1957.* Washington, DC: The National Science Foundation.

Environmental Justice Foundation (EJF). 2011. "Climate Change and Migration: Forced Displacement, 'Climate Refugees' and the Need for a New Legal Instrument." Accessed March 4, 2018. ejfoundation.org/resources/downloads /EJF_climate-change-and-migration-2011.pdf.

Environmental Justice Foundation (EJF). 2017. *Beyond Borders: Climate, Migration and Conflict.* Accessed March 3, 2018. ejfoundation.org/films /beyond-borders-film.

Epstein, Alex. 2014. *The Moral Case for Fossil Fuels.* New York: Penguin.

Epstein, Steven. 1996. *Impure Science: AIDS, Activism, and the Politics of Science.* Berkeley: University of California Press.

Espeland, Wendy N. and Mitchell L. Stevens. 1998. "Commensuration as a Social Process." *Annual Review of Sociology* 24:313–43.

Espy, James. 1841. *The Philosophy of Storms.* Boston: Little and Brown.

Espy, James. 1843. *First Report on Meteorology to the Surgeon-General of the United States Army.* Washington, DC: Surgeon General Office.

Evans, Peter B., Dietrich Rueschemeyer, and Theda Skocpol, eds. 1985. *Bringing the State Back In.* Cambridge, UK: Cambridge University Press.

Extinction Rebellion. 2018. "The Emergency." Accessed May 3, 2019. https:// rebellion.earth/the-truth/the-emergency/.

Eyal, Gil. 2013. "Spaces between Fields." Pp. 158–82 in *Bourdieu and Historical Analysis,* edited by P. Gorski. Durham, NC: Duke University Press.

Eyal, Gil, and Larissa Buchholz. 2010. "From the Sociology of Intellectuals to the Sociology of Interventions." *Annual Review of Sociology* 36:117–37.

Fairhead, James, and Melissa Leach. 1996. *Misreading the African Landscape: Society and Ecology in a Forest-Savanna Mosaic.* Cambridge, UK: Cambridge University Press.

Falconer, William. 1781. *Remarks on the Influence of Climate, Situation, Nature of Country, Population, Nature of Food and Way of Life on the Disposition and Temper, Manners and Behaviors, Intellects, Laws and Customs, Form of Government, and Religion, of Mankind.* London: Dilly.

Farbotko, Carol, and Heather Lazrus. 2012. "The First Climate Refugees? Contesting Global Narratives of Climate Change in Tuvalu." *Global Environmental Change* 22(2):382–90.

Farrell, Justin. 2016. "Corporate Funding and Ideological Polarization about Climate Change." *Proceedings of the National Academy of Sciences* 113(1):92–97.

Fassig, Oliver L. 1911. "The Normal Temperature of Porto Rico, West Indies." *Monthly Weather Review* 39(2):299–302.

Fawcett, Edwin B., William E. Hubert, and Albert L. Stickles II. 1956. "Organization and Operation of the Joint Numerical Weather Prediction Unit." *Weatherwise* 9(3):75–106.

Feldman, Theodore S. 1990. "Late Enlightenment Meteorology." Pp. 143–79 in *The Quantifying Spirit in the Eighteenth Century,* edited by T. Frangsmyr, J. L. Heilbron, and R. E. Rider. Berkeley: University of California Press.

Felli, Romain. 2015. "Environment, Not Planning: The Neoliberal Depoliticization of Environmental Policy by Means of Emissions Trading." *Environmental Politics* 24(5):641–60.

Felli, Romain. 2021. *The Great Adaptation: Climate, Capitalism and Catastrophe*. New York: Verso.

Fleming, James R. 1990. *Meteorology in America, 1800–1870*. Baltimore, MD: Johns Hopkins University Press.

Fleming, James R., ed. 1996. *Historical Essays on Meteorology: 1919–1995*. Boston: American Meteorological Society.

Fleming, James R. 1998a. *Historical Perspectives on Climate Change*. Oxford: Oxford University Press.

Fleming, James R. 1998b. "Arrhenius and Current Climate Concerns: Continuity or a 100-Year Gap?" *Eos* 79:406–10.

Fleming, James R. 2000a. "T. C. Chamberlin, Climate Change, and Cosmogony." *Studies in History and Philosophy of Science Part B: Studies in History and Philosophy of Modern Physics* 31(3):293–308.

Fleming, James R. 2000b. "Storms, Strikes, and Surveillance: The U.S. Army Signal Office, 1861–1891." *Historical Studies in the Physical and Biological Sciences* 30(2):315–32.

Fleming, James R. 2007. *The Callendar Effect: The Life and Work of Guy Stewart Callendar*. Boston: American Meteorological Society.

Fleming James R. 2010. *Fixing the Sky: The Checkered History of Weather and Climate Control*. New York: Columbia University Press.

Fleming, James R. 2011. "Planetary-Scale Fieldwork: Harry Wexler on the Possibilities of Ozone and Climate Control." Pp. 190–211 in *Knowing Global Environments: Historical Perspectives on the Field Sciences*, edited by J. Vetter. Piscataway, NJ: Rutgers University Press.

Fleming, James R. 2014. "Climate Physicians and Surgeons." *Environmental History* 19: 338–45.

Fleming, James R., and Vladimir Jankovic. 2011. "Introduction: Revisiting *Klima*." *Osiris* 26(1):1–15.

Fleming, James R., Vladimir Jankovic, and Deborah R. Coen, eds. 2006. *Intimate Universality: Local and Global Themes in the History of Weather and Climate*. Sagamore Beach, MA: Science History Publications.

Flint, Timothy. 1826. *Recollections of the Last Ten Years*. Boston: Cummings.

Forry, Samuel. 1840. *Statistical Report on the Sickness and Mortality of the US Army*. Washington, DC: Gideon.

Forry, Samuel. 1842a. *The Climate of the United States and Its Endemic Influences*. New York: Langley.

Forry, Samuel. 1842b. "Do the Various Races of Man Constitute a Single Species?" *New York Lancet* 2(6):113–33.

Forry, Samuel. 1843a. *Meteorology: Comprising a Description of Atmospheric and Its Phenomena, the Laws of Climate in General*. New York: Winchester.

Forry, Samuel. 1843b. "The Mosaic Account of the Unity of the Human Race, Confirmed by the Natural History of the American Aborigines." *American Biblical Repository* 10(19):29–80.

Forry, Samuel. 1843c. "Bibliographic Notice: Climate of the United States, Meteorology." *New York Journal of Medicine and the Collateral Sciences* 1(1):116.

Forry, Samuel. 1848. "Vital Statistics—The Development of Man's Faculties, and the Laws of His Mortality and Reproduction, Viewed in their Relations to Hygiology or State Medicine." *New York Journal of Medicine* 10(30):289–307.

Forry, Samuel. 1856. "Considerations on the Distinctive Characteristics of the American Aboriginal Tribes." Pp. 354–65 in *Information Respecting the History, Condition, and Prospects of the Indian Tribes of the United States*, edited by Henry Schoolcraft. Philadelphia: Lippincott.

Forster, Thomas. [1817] 1829. *Illustrations of the Atmospherical Origin of Epidemic Disorders of Health: And of Its Relation to the Predisponent Constitutional Causes [. . .]: And of the Twofold Means of Prevention, Mitigation, and Cure, by Change of Air, and by Diet, Regularity, and Simple Medicines*. Chelmsford, UK: Meggy & Chalk.

Forsyth, Tim. 2002. *Critical Political Ecology: The Politics of Environmental Science*. New York: Routledge.

Forsyth, Tim. 2008. "Political Ecology and the Epistemology of Social Justice." *Geoforum* 39(2):756–64.

Foster, John B. 1999. "Marx's Theory of Metabolic Rift: Classical Foundations for Environmental Sociology." *American Journal of Sociology* 105(2):366–405.

Foster, John B. 2000. *Marx's Ecology: Materialism and Nature*. New York: Monthly Review Press.

Foster, John B., and Hannah Holleman. 2012. "Weber and the Environment: Classical Foundations for a Postexemptionalist Sociology." *American Journal of Sociology* 117(6):1625–73.

Foucault, Michel. [1965] 1988. *Madness and Civilization: A History of Insanity in the Age of Reason*. New York: Vintage Books.

Foucault, Michel. 1977. *Discipline and Punish: The Birth of the Prison*. Translated by A. Sheridan. New York: Vintage.

Foucault, Michel. 1978. *The History of Sexuality*. Vol. 1, *An Introduction*. New York: Random House.

Foucault, Michel. 1980. *Power/Knowledge: Selected Interviews and Other Writings, 1972–1977*. Edited by C. Gordon. New York: Pantheon.

Foucault, Michel. 1991. "Governmentality." Pp. 87–104 in *The Foucault Effect: Studies in Governmentality*, edited by G. Burchell, C. Gordon, and P. Miller. Chicago: University of Chicago Press.

Foucault, Michel. 1994. *The Order of Things: An Archaeology of the Human Sciences*. New York: Routledge.

Foucault, Michel. 2004. *Security, Territory, Population: Lectures at the College de France, 1977–1978*. Translated by G. Burchell. New York: Picador.

Foucault, Michel. 2008. *The Birth of Biopolitics: Lectures at the College de France, 1978–1979*. Translated by G. Burchell. New York: Picador.

Fourcade, Marion. 2006. "The Construction of a Global Profession: The Transnationalization of Economics." *American Journal of Sociology* 112(1):145–94.

Fourcade, Marion. 2011. "Cents and Sensibility: Economic Valuation and the Nature of 'Nature.'" *American Journal of Sociology* 116(6):1721–77.

Fourcade, Marion, and Rakesh Khurana. 2013. "From Social Control to Financial Economics: The Linked Ecologies of Economics and Business in Twentieth Century America." *Theory and Society* 42(2):121–59.

Fox News. 2015. "RNC Rips O'Malley for Linking Rise of ISIS to Climate Change." July 25. Accessed February 8, 2016. https://www.foxnews.com /politics/rnc-rips-omalley-for-linking-rise-of-isis-to-climate-change.

Frangsmyr, Tore, J. L. Heilbron, and Robin E. Rider, eds. 1990. *The Quantifying Spirit in the Eighteenth Century*. Berkeley: University of California Press.

Frank, Johann. [1799] 1976. *A System of Complete Medical Police*. Edited by Erna Lesky. Baltimore, MD: Johns Hopkins University Press.

Frank, Neil. 2019. "'Climate-Change Experts Seek Dialogue with Gov. Abbott'— Dr. Neil Frank's Response." Accessed February 3, 2019. http://co2coalition .org/2019/02/01/climate-change-experts-seek-dialogue-with-gov-abbott-dr -neil-franks-response/.

Franklin, Benjamin. 1755. *Observations Concerning the Increase of Mankind, Peopling of Countries, &c.* Boston: Kneeland.

Freedman, Andrew. 2013. "Climate and Conflict: Warmer World May Be More Violent." *Climate Central*, August 1. Accessed May 20, 2019. https://climate central.org/news/a-warmer-world-likely-to-be-a-more-violent-world-study -shows-16299.

Frémont, John C. 1845. *Report of the Exploring Expedition to the Rocky Mountains in the Year 1842, and to Oregon and North California in the Years 1843-44*. Washington, DC: Gales and Seaton.

Frémont, John C., and Charles Preuss. 1846. "Topographical Map of the Road from Missouri to Oregon, Section V." U.S. Congress. Accessed October 20, 2017. www.loc.gov/item/99446202/.

Frickel, Scott, and Kelly Moore, eds. 2006. *The New Political Sociology of Science: Institutions, Networks, and Power*. Madison: University of Wisconsin Press.

Friedman, Lisa. 2019. "White House Tried to Stop Climate Science Testimony, Documents Show." *New York Times*, June 8. Accessed December 5, 2023.

https://www.nytimes.com/2019/06/08/climate/rod-schoonover-testimony
.html.

Friedman, Robert. 1989. *Appropriating the Weather: Vilhelm Bjerknes and the Construction of a Modern Meteorology*. Ithaca, NY: Cornell University Press.

Frisinger, H. Howard. 1983. *The History of Meteorology: To 1800*. Boston: American Meteorological Society.

Frymer, Paul. 2017. *Building an American Empire: The Era of Territorial and Political Expansion*. Princeton, NJ: Princeton University Press.

Fujimura, Joan H. 1992. "Crafting Science: Standardized Packages, Boundary Objects, and 'Translation.'" Pp. 168–211 in *Science as Practice and Culture*, edited by A. Pickering. Chicago: University of Chicago Press.

Garcia, Denise. 2010. "Warming to a Redefinition of International Security: The Formation of a Norm against Climate Change." *International Relations* 24(3):271–92.

Garriott, Edward B. 1903. *Weather Folk-Lore and Local Weather Signs*. U.S. Department of Agriculture Bulletin 33 (Weather Bureau 294). Washington, DC: Government Printing Office.

Garriott, Edward B. 1904. *Long-Range Weather Forecasts*. U.S. Department of Agriculture, Weather Bureau Bulletin 35. Washington, DC: Government Printing Office.

Gauthier, Marine. 2018. *Mai-Ndombe: Will the REDD+ Laboratory Benefit Indigenous Peoples and Local Communities?* Washington, DC: Rights and Resources Initiative.

Geden, Oliver, and Silke Beck. 2014. "Renegotiating the Global Climate Stabilization Target." *Nature Climate Change* 4:747–48.

Gehlbach, Stephen H. 2016. *American Plagues: Lessons from Our Battles with Disease*. Lanham, MD: Rowman & Littlefield.

Geophysical Year 3c 1958 issue U.S. stamp.jpg. n.d. Wikimedia Commons. Accessed June 16, 2016. https://commons.wikimedia.org/wiki/File: Geophysical_Year_3c_1958_issue_U.S._stamp.jpg

Gieryn, Thomas. 1983. "Boundary-Work and the Demarcation of Science from Non-science: Strains and Interests in Professional Ideologies of Scientists." *American Sociological Review* 48(6):781–95.

Gieryn, Thomas F. 1999. *Cultural Boundaries of Science: Credibility on the Line*. Chicago: University of Chicago Press.

Gill, Victoria. 2017. "Warmer Arctic Is the 'New Normal.'" *BBC News*, December 13. Accessed December 14, 2017. bbc.com/news/science-environment -42330771.

Gilles, Nathan. 2017. "Naomi Oreskes—Scientific Consensus, Climate Change, and the Merchants of Doubt." American Association for the Advancement of Science. April 21. Accessed March 12, 2018. www.aaas.org/blog/member -spotlight/naomi-oreskes-scientific-consensus-climate-change-and -merchants-doubt.

Gillett, Mary. 1987. *The Army Medical Department, 1818–1865*. Washington, DC: U.S. Army Center of Military History.

Giorgi, Filippo. 2006. "Climate Change Hot-Spots." *Geophysical Research Letters* 33(8):L08707. https://doi.org/10.1029/2006GL025734.

Giorgi, Filippo, and Linda O. Mearns. 1991. "Approaches to Regional Climate Change Simulation: A Review. *Reviews of Geophysics* 29:191–216.

Giorgi, Filippo, and Linda O. Mearns. 1999. "Introduction to Special Section: Regional Climate Modeling Revisited." *Journal of Geophysical Research* 104(D6):6335–52.

Glidden, A. C. [1895] 1897. "State Weather Service." P. 4 in *Michigan Weather Services 1895 Annual Report*. Lansing, MI: Robert Smith Printing Co.

Go, Julian. 2011. *Patterns of Empire: The British and American Empires, 1688 to the Present*. New York: Cambridge University Press.

Go, Julian. 2016. *Postcolonial Thought and Social Theory*. New York: Oxford University Press.

Goetzmann, William H. 1966. *Exploration and Empire: The Explorer and the Scientist in the Winning of the American West*. New York: Knopf.

Goldberg, Daniel T. 2002. *The Racial State*. Malden, MA: Blackwell.

Goldman, Mara J., Paul Nadasdy, and Matthew D. Turner. 2011. *Knowing Nature: Conversations at the Intersection of Political Ecology and Science Studies*. Chicago: University of Chicago Press.

Goldman, Michael. 2004. "Eco-Governmentality and Other Transnational Practices of a 'Green' World Bank." Pp. 166–92 in *Liberation Ecologies: Environment, Development and Social Movements*, edited by R. Peet and M. Watts. New York: Routledge.

Goldstein, Jesse. 2018. *Planetary Improvement: Cleantech Entrepreneurship and the Contradictions of Green Capitalism*. Cambridge, MA: MIT Press.

Golinski, Jan. 1999. "Barometers of Change: Meteorological Instruments as Machines of Enlightenment." Pp. 63–93 in *The Sciences in Enlightened Europe*, edited by W. Clark, J. Golinski, and S. Schaffer. Chicago: Chicago University Press.

Golinski, Jan. 2008. "American Climate and the Civilization of Nature." Pp. 153–74 in *Science and Empire in the Atlantic World*, edited by J. Delbourgo and N. Dew. New York: Routledge.

Golinski, Jan. 2010. *British Weather and the Climate of Enlightenment*. Chicago: University of Chicago Press.

Goodman, Sherri W. 1996. "DOD Environmental Security: Investing in the Future." *Federal Facilities Environmental Journal* 7(3):97–105.

Gordon, Collin. 1991. "Governmental Rationality: An Introduction." Pp. 1–52 in *The Foucault Effect: Studies in Governmentality*, edited by G. Burchell, C. Gordon, and P. Miller. Chicago: University of Chicago Press.

Gorski, Phillip S., ed. 2013. *Bourdieu and Historical Analysis*. Durham, NC: Duke University Press.

Gosling, F. G. 1999. *The Manhattan Project: Making the Atomic Bomb.* Washington, DC: U.S. Department of Energy.

Gould, Kenneth. 2007. 'The Ecological Costs of Militarization. *Peace Review: A Journal of Social Justice* 19(4):331–34.

Graham, Stephen. 2010. *Cities Under Siege: The New Military Urbanism.* New York: Verso.

Graunt, John. 1662. *Natural and Political Observations [. . .] Made upon the Bills of Mortality, with Reference to the Government, Religion, Trade, Growth, Air, Diseases, and the Several Changes of the Said City.* London: Roycroft.

Greely, Adolphus W. 1888. *American Weather. A Popular Exposition of the Phenomena of the Weather.* New York: Dodd.

Grove, Richard H. 1995. *Green Imperialism: Colonial Expansion, Tropical Island Edens and the Origins of Environmentalism, 1600–1860.* Cambridge, UK: Cambridge University Press.

Grundmann, Reiner. 2016. "Climate Change as a Wicked Social Problem." *Nature Geoscience* 9(8):562–63.

G7. *A New Climate for Peace: Taking Action on Climate and Fragility Risk.* Accessed May 1, 2024. https://climate-diplomacy.org/sites/default/files/2020-11/NewClimateForPeace_FullReport_small_0.pdf.

Gunderson, Ryan, Diana Stuart, and Brian Petersen. 2018. "The Political Economy of Geoengineering as Plan B: Technological Rationality, Moral Hazard, and New Technology." *New Political Economy* 24(5):1–20.

Guttman, Nathaniel B., and Robert G. Quayle. 1996. "A Historical Perspective of U.S. Climate Divisions." *Bulletin of the American Meteorological Society* 77(2):293–304.

Hacker, Barton. 2000. "Military Patronage and the Geophysical Sciences in the United States: An Introduction." *Historical Studies in the Physical and Biological Sciences* (30)2:309–13.

Hacking, Ian. 1990. *The Taming of Chance.* Cambridge, UK: Cambridge University Press.

Hagen, Raymond. 1986. "The Office of Naval Research: Windows to the Origins." Pp. 4–27 in *40 Years of Excellence: 1946–1986,* Special Issue of *Naval Research Reviews* 38(3).

Haigh, T., M. Priestley, and C. Rope. 2016. *ENIAC in Action: Making and Remaking the Modern Computer.* Cambridge, MA: MIT Press.

Hall, John R. 1980. "The Time of History and the History of Times." *History and Theory* 19(2):113–31.

Hall, John R. 1999. *Cultures of Inquiry: From Epistemology to Discourse in Sociohistorical Research.* Cambridge, UK: Cambridge University Press.

Hall, John R. 2016. "Social Futures of Global Climate Change: A Structural Phenomenology." *American Journal of Cultural Sociology* 4(1):1–45.

Hallsworth, Simon, and John Lea. 2011. "Reconstructing Leviathan: Emerging Contours of the Security State." *Theoretical Criminology* 15(2):141–57.

Hamblin, John D. 2005. *Oceanographers and the Cold War: Disciplines of Marine Science.* Seattle: University of Washington Press.

Hamilton, Gary G., and John R. Sutton. 1989. "The Problem of Control in a Weak State: Domination in the United States, 1880–1920." *Theory and Society* 18:1–46.

Hann, Julius von. [1883] 1903. *Handbook of Climatology.* Edited by Robert D. Ward. New York: Macmillan.

Hansen, James, D. Johnson, A. Lacis, S. Lebedeff, P. Lee, D. Rind, and G. Russell. 1981. "Climate Impact of Increasing Atmospheric Carbon Dioxide." *Science* 213(4511):957–66.

Hansson, Anders. 2024. "Making the 1.5°C Aspirational Climate Target Tangible with Carbon Dioxide Removal and Boundary Work." Pp. 268–75 in *Climate, Science and Society: A Primer*, edited by Zeke Baker, Tamar Law, Mark Vardy, and Stephen Zehr. London: Routledge.

Harding, T. Swann. 1940. "Henry L. Ellsworth, Commissioner of Patents." *Journal of Farm Economics* 22(3):621–27.

Harper, Kristine C. 2003. "Research from the Boundary Layer: Civilian Leadership, Military Funding and the Development of Numerical Weather Prediction (1946–55)." *Social Studies of Science* 33(5):667–96.

Harper, Kristine C. 2008. *Weather by the Numbers: The Genesis of Modern Meteorology.* Cambridge, MA: MIT Press.

Harper, Kristine C. 2017. *Make It Rain: State Control of the Atmosphere in Twentieth Century America.* Chicago: University of Chicago Press.

Harrington, Mark. 1895. "What Meteorology Is Doing for the Farmer." Pp. 117–20 in *Yearbook of the US Department of Agriculture*. Washington, DC: Government Printing Office.

Harrison, Mark. 1996. "'The Tender Frame of Man': Disease, Climate, and Racial Difference in India and the West Indies, 1760–1860." *Bulletin of the History of Medicine* 70(1):68–93.

Harrison, Mark. 1999. *Climates and Constitutions: Health, Race, Environment and British Imperialism in India 1600–1850.* New York: Oxford University Press.

Hart, David M., and David G. Victor. 1993. "Scientific Elites and the Making of US Policy for Climate Change Research, 1957–74." *Social Studies of Science* 23(4):643–80.

Harvey, David. 2014. *Seventeen Contradictions and the End of Capitalism.* New York: Oxford University Press.

Hawken, Paul, Hunter Lovins, and Amory Lovins. 1999. *Natural Capitalism: Creating the Next Industrial Revolution.* New York: Little, Brown.

Headrick, Daniel. 1991. *The Invisible Weapon: Telecommunications & International Politics 1851–1945*. London: Oxford University Press.

Hecht, Gabrielle. 2011. *Entangled Geographies: Empire and Technopolitics in the Global Cold War*. Cambridge, MA: MIT Press.

Henderson, Gabriel. 2014. "The Dilemma of Reticence: Helmut Landsberg, Stephen Schneider, and Public Communication of Climate Risk, 1971–1976." *History of Meteorology* 6:53–78.

Henry, Alfred J. 1906. *Climatology of the United States*. Washington, DC: Government Printing Office.

Henry, Joseph. 1858. *Meteorology in Its Connection with Agriculture*. Washington, DC: Government Printing Office.

Hess, David J. 2006. "Antiangiogenesis Research and the Dynamics of Scientific Fields: Historical and Institutional Perspectives in the Sociology of Science." Pp. 122–47 in *The New Political Sociology of Science: Institutions, Networks, and Power*, edited by S. Frickel and K. Moore. Madison: University of Wisconsin Press.

Hess, David J. 2011. "Bourdieu and Science and Technology Studies: Toward a Reflexive Sociology." *Minerva* 49(3):333–48.

Heyck, Hunter, and David Kaiser. 2010. "New Perspectives on Science and the Cold War: Introduction." *Isis* 101(2):362–66.

Hiatt, Alfred. 2007. "The Map of Macrobius before 1100." *Imago Mundi* 59(2):149–76.

Hickey, Donald, and Connie Clark, eds. 2016. *The Routledge Handbook of the War of 1812*. New York: Routledge.

Hirschman, Daniel A. 2016. "Inventing the Economy, Or: How We Learned to Stop Worrying and Love the GDP." PhD dissertation, University of Michigan. Accessed November 24, 2018. https://deepblue.lib.umich.edu/bitstream/handle/2027.42/120713/dandanar_1.pdf?sequence=1&isAllowed=y.

Hirschman, Daniel, and Elizabeth Popp Berman. 2014. "Do Economists Make Policies? On the Political Effects of Economics1." *Socio-Economic Review* 12(4):779–811.

Hlebica, Joe. 2001. "Roger Revelle and the Great Age of Exploration." *Scripps Institution of Oceanography Explorations* 8(1):22–29.

Hoerling, Martin, Jon Eischeid, Judith Perlwitz, Xiaowei Quan, Tao Zhang, and Philip Pegion. 2012. "On the Increased Frequency of Mediterranean Drought." *Journal of Climate* 25(6):2146–61.

Hoffman, Frederick L. 1902. *Windstorm and Tornado Insurance: A Compilation of the Most Trustworthy and Important Data Relating to the Destruction of Life and Property through Violent Storms*. 3rd ed. New York: The Spectator Company.

Holland, Andrew. 2013. "Bay of Bengal: A Hotspot for Climate Insecurity." August 10. Accessed March 16, 2018. sustainablesecurity.org/2013/08/10/bay-of-bengal-a-hotspot-for-climate-insecurity/.

Holmes, Oliver W., Enoch Hale, G. C. Shattuck, Daniel Drake, John Bell, Austin Flint, and W. Selden. 1848. "Report of the Committee on Medical Literature." *Transactions of the American Medical Association* 1:249–88.

Holyoke, Edward. 1793. "An Estimate of the Excels of the Heat and Cold of the American Atmosphere beyond the European, in the Same Parallel of Latitude: To Which Are Added, Some Thoughts on the Causes of This Excels." *Memoirs of the American Academy of Arts and Sciences* 2:65–88.

Homer-Dixon, Thomas. 1991. "On the Threshold: Environmental Changes as Causes of Acute Conflict." *International Security* 16:76–116.

Homer-Dixon, Thomas. 1994. "Environmental Scarcities and Violent Conflict: Evidence from Cases." *International Security* 19:5–40.

Hong, Wei. 2008. "Domination in a Scientific Field: Capital Struggle in a Chinese Isotope Lab." *Social Studies of Science* 38(4):543–70.

Hooks, Gregory, and Chad Smith. 2004. "The Treadmill of Destruction: National Sacrifice Areas and Native Americans." *American Sociological Review* 69(4):558–75.

Horkheimer, Max. [1937] 1975. "Traditional and Critical Theory." Pp. 188–243 in *Critical Theory: Selected Essays*. New York: Continuum.

Horsman, Reginald. 1981. *Race and Manifest Destiny: The Origins of American Racial Anglo-Saxonism*. Cambridge, MA: Harvard University Press.

Hsiang, Solomon M., Marshall Burke, and Edward Miguel. 2013. "Quantifying the Influence of Climate on Human Conflict." *Science* 341(6151). https://doi .org/10.1126/science.1235367.

Hulme, Mike. 2008. "The Conquering of Climate: Discourses of Fear and Their Dissolution." *Geographical Journal* 174(1):5–16.

Hulme, Mike. 2011. "Reducing the Future to Climate: A Story of Climate Determinism and Reductionism." *Osiris* 26(1):245–66.

Hulme, Mike. 2014. *Can Science Fix Climate Change? A Case against Climate Engineering*. Malden, MA: Polity.

Hulme, Mike. 2017. *Weathered: Cultures of Climate*. London: Sage.

Hulme, Mike, Ruth Doherty, Todd Ngara, Mark New, and David Lister. 2001. "African Climate Change: 1900–2100." *Climate Research* 17:145–68.

Humboldt, Alexander von. 1817. *Des Lignes Isothermes et de la Distribution de la Châleur sur le Golbe*. Paris: Perronneau.

Humboldt, Alexander von. [1845] 1849. *Cosmos: Sketch of a Physical Description of the Universe*. Vol. 1. London: Longman, Brown, Green, and Longmans.

Hunt, Gaillard, ed. 1908. *The Writings of James Madison*. Vol. 3, *1809–1819*. New York: Putnam's Sons.

Hunter, John. 1788. *Observations on the Diseases of the Army in Jamaica: And on the Best Means of Preserving the Health of Europeans, in that Climate*. London: Nicol.

Huntington, Ellsworth. 1913. "Changes in Climate and History." *American Historical Review* 18(2):213–32.

Huntington, Ellsworth. 1915. *Civilization and Climate*. New Haven, CT: Yale University Press.

Huntington, Ellsworth. 1919. *The Red Man's Continent: A Chronical of Aboriginal America*. New Haven, CT: Yale University Press.

Indigenous Climate Action., n.d. Accessed August 4, 2018. www.indigenous climateaction.com/who-we-are.

Intergovernmental Panel on Climate Change (IPCC). 1992. *Climate Change: The 1990 and 1992 IPCC Assessments and 1992 IPCC Supplement*. Geneva: World Meteorological Organization.

Intergovernmental Panel on Climate Change (IPCC). 2007a. "Regional Climate Projections." In *Climate Change 2007: The Physical Science Basis; Contribution of Working Group I to the Fourth Assessment Report of the Intergovernmental Panel on Climate Change*, edited by S. Solomon, D. Qin, M. Manning, Z. Chen, M. Marquis, K. B. Averyt, M. Tignor, and H. L. Miller. Cambridge, UK: Cambridge University Press.

Intergovernmental Panel on Climate Change (IPCC). 2007b. *Climate Change 2007: Impacts, Adaptation and Vulnerability*. Contribution of Working Group II to the Fourth Assessment Report of the Intergovernmental Panel on Climate Change, 2007. Edited by M. L. Parry, O. F. Canziani, J. P. Palutikof, P. J. van der Linden and C. E. Hanson. New York: Cambridge University Press.

Intergovernmental Panel on Climate Change (IPCC). 2014. *Climate Change 2013: The Physical Science Basis*. Contribution of Working Group I to the Fifth Assessment Report of the Intergovernmental Panel on Climate Change. Edited by T. F. Stocker, D. Qin, G.-K. Plattner, M. Tignor, S. K. Allen, J. Boschung, A. Nauels, Y. Xia, V. Bex, and P. M. Midgley. Cambridge, UK: Cambridge University Press.

Intergovernmental Panel on Climate Change (IPCC). 2018. *Global Warming of 1.5°C: An IPCC Special Report*. Accessed May 1, 2024. https://www.ipcc.ch /sr15/ www.ipcc.ch/2018/10/08/summary-for-policymakers-of-ipcc-special -report-on-global-warming-of-1-5c-approved-by-governments/.

Intergovernmental Panel on Climate Change (IPCC). 2023. *Climate Change 2023: Synthesis Report*. Geneva: IPCC.

International Climate Justice Network. 2002. *Bali Principles of Climate Justice*. August 29. Accessed January 3, 2016. www.ejnet.org/ej/bali.pdf.

International Meteorological Committee. 1879. *Report of the Proceedings of the Second International Meteorological Congress at Rome, 1879*. London: Her Majesty's Secretary Office.

Jackson, Robert. [1791] 1795. *A Treatise on the Fevers of Jamaica: With Some Observations on the Intermitting Fever of America, and an Appendix, Containing Some Hints on the Means of Preserving the Health of Soldiers in Hot Climates*. Philadelphia: Campbell.

Jackson, Robert. 1804. *A Systematic View of the Formation, Discipline, and Economy of Armies*. London: Stockdale.

Jaffee, David. 1990. "The Village Enlightenment in New England, 1760–1820." *William and Mary Quarterly* 47(3):327–46.

Jamail, Dahr. 2018. "As Antarctic Melting Accelerates, Worst-Case Scenarios May Come True." *Turthout*, April 5. Accessed October 1, 2018. www.truth -out.org/news/item/44069-as-antarctic-melting-accelerates-worst-case -scenarios-may-come-true.

James, Edwin, ed. 1823. *Account of an Expedition from Pittsburgh to the Rocky Mountains*. Philadelphia: Carey and Lea.

Jankovic, Vladimir. 2000. *Reading the Skies: A Cultural History of English Weather, 1650–1820*. Chicago: University of Chicago Press.

Jankovic, Vladimir. 2010. *Confronting the Climate: British Airs and the Making of Environmental Medicine*. New York: Palgrave.

Jasanoff, Sheila. 1996. "Beyond Epistemology: Relativism and Engagement in the Politics of Science." *Social Studies of Science* 26(2):393–418.

Jasanoff, Sheila, ed. 2004. *States of Knowledge: The Co-production of Science and Social Order*. New York: Routledge.

Jasanoff, Sheila, and Marybeth Martello, eds. 2004. *Earthly Politics: Local and Global in Environmental Governance*. Cambridge, MA: MIT Press.

Jasparro, Christopher, and Jonathan Taylor. 2008. "Climate Change and Regional Vulnerability to Transnational Security Threats in Southeast Asia." *Geopolitics* 13(2):232–56.

Jefferson, Thomas. 1780. "Letter to George Rogers Clark. 25 December." US National Archives and Records Administration, Thomas Jefferson Papers. Accessed January 12, 2017. founders.archives.gov/documents/Jefferson/01-04 -02-0295.

Jefferson, Thomas. [1785] 1794. *Notes on the State of Virginia*. Philadelphia: Carey.

Jefferson, Thomas. 1803. "Instructions to Meriwether Lewis." In *Rivers, Edens, Empires: Lewis and Clark and the Revealing of America*. Accessed June 4, 2017. loc.gov/exhibits/lewisandclark/transcript57.html.

Johnson, Chalmers. 2004. *The Sorrows of Empire: Militarism, Secrecy, and the End of the Republic*. New York: Metropolitan Books.

Johnson, James. [1813] 1827. *The Influence of Tropical Climates on European Constitutions, to Which Is Added Tropical Hygiene, or the Preservation of Health in all Hot Climates*. London: Underwood.

Jonsson, Fredrik Albritton. 2012. "The Industrial Revolution in the Anthropocene." *Journal of Modern History* 84(3):679–96.

Jorgensen, David. 2007. "The Evolving Publication Process of the AMS." *Bulletin of the American Meteorological Society* 88(7):1122–34.

Kahn, Brian. 2015. "Climate Change a 'Contributing Factor' in Syrian Conflict." Climate Central, March 2. Accessed April 1, 2018. https://www

.climatecentral.org/news/climate-change-contributing-factor-syrian
-conflict-18718.

Kahn, Brian. 2016. "Obama Just Tied Climate Change to National Security."
Climate Central, September 22. Accessed May 12, 2019. www.climatecentral
.org/news/obama-climate-change-national-security-20723.

Kahn, Herman. 1976. *The Next 200 Years: A Scenario for America and the
World*. New York: Morrow.

Kahn, Herman, and Anthony J. Wiener. 1967. *The Year 2000: A Framework for
Speculation on the Next Thirty-Three Years*. London: Macmillan.

Kallis, Giorgos, Federico Demaria, and Giacomo D'Alisa. 2015. "Introduction:
Degrowth." Pp. 1–18 in *Degrowth: A Vocabulary for a New Era*, edited by
G. D'Alisa, F. Demaria, and G. Kallis. New York: Routledge.

Kaplan, Joseph. 1954. "The Scientific Program of the International Geophysical Year." *Proceedings of the National Academy of Sciences*
40(10):926–31.

Kaufmann, Eric. 1998. "'Naturalizing the Nation': The Rise of Naturalistic
Nationalism in the United States and Canada." *Comparative Studies in
Society and History* 40(4):666–95.Keeling, Charles. 1960. "The Concentration
and Isotopic Abundances of Carbon Dioxide in the Atmosphere." *Tellus*
12:200–203.

Kelley, Colin P., Shahrzad Mohtadi, Mark A. Cane, Richard Seager, and
Yochanan Kushnir. 2015. "Climate Change in the Fertile Crescent and
Implications of the Recent Syrian Drought." *Proceedings of the National
Academy of Sciences* 112(11):3241–46.

Kelley, Terry P. 1990. *Global Climate Change: Implications for the United States
Navy*. Report to the United States Navy War College. Accessed May 6, 2017.
http://documents.theblackvault.com/documents/weather/climatechange
/globalclimatechange-navy.pdf.

Kelly, Andrew S. 2014. "The Political Development of Scientific Capacity in the
United States." *Studies in American Political Development* 28(1):1–25.

Kelves, Daniel J. 1995. *The Physicists: The History of a Scientific Community in
Modern America (Revised Edition)*. New York: Knopf.

Kendi, Ibram X. 2016. *Stamped from the Beginning: The Definitive History of
Racist Ideas in America*. New York: Nation.

Khazan, Olga. 2013. "Hot Weather Actually Makes Us Want to Kill Each Other."
Atlantic, August. Accessed July 6, 2017. www.theatlantic.com/international
/archive/2013/08/hotterweatheractuallymakesuswanttokilleachother
/278282/.

Ki-Moon, Ban. 2007. "A Climate Culprit in Darfur." *Washington Post*, June 16.
Accessed September 8, 2017. washingtonpost.com/wpdyn/content/article
/2007/06/15/ AR2007061501857.

King, Gregory. [1696] 1804. *Natural and Political Observations and Conclusions upon the State and Condition of England*. London: Stockdale.

King, Katrina Quisumbing. 2019. "Recentering U.S. Empire: A Structural Perspective on the Color Line." *Sociology of Race and Ethnicity* 5(1):11–25.

King, Marcus, and Sherri Goodman. 2011. "Defense Community Perspectives on Uncertainty and Confidence Judgments." *Climatic Change* 108:803.

Kitoh, Akio, Akiyo Yatagai, and Pinhas Alpert. 2008. "First Super-High-Resolution Model Projection That the Ancient 'Fertile Crescent' Will Disappear in This Century." *Hydrological Resources Letters* 2:1–4.

Klein, Naomi. 2014. *This Changes Everything: Capitalism vs. the Climate*. New York: Simon & Schuster.

Klinke, Ian, and Mark Bassin. 2018. "Introduction: Lebensraum and Its Discontents." *Journal of Historical Geography* 61:53–58.

Knorr-Cetina, Karin. 2003. *Epistemic Cultures: How the Sciences Make Knowledge*. Cambridge, MA: Harvard University Press.

Koelsch, William. 1996. "From Geo- to Physical Science: Meteorology and the American University, 1919–1945." Pp. 511–40 in *Historical Essays on Meteorology: 1919–1995*, edited by J. Fleming. Boston: American Meteorological Society.

Kolbert, Elizabeth. 2021. *Under a White Sky: The Nature of the Future*. New York: Penguin.

Köppen, Wladimir. [1884] 2011. "The Thermal Zones of the Earth according to the Duration of Hot, Moderate and Cold Periods and of the Impact of Heat on the Organic World." Translated by E. Volker and S. Bronniman. *Meteorologische Zeitschrift*, 20(3):351–60.

Köppen, Wladimir, and Alfred Wegener. [1924] 2015. *The Climates of the Geological Past*. Stuttgart: Schweizerbart.

Kovarsky, Joel. 2014. *The True Geography of Our Country: Jefferson's Cartographic Vision*. Charlottesville: University of Virginia Press.

Krige, John. 2010. "Post-WW2 Transatlantic Science Policies: Building the Arsenal of Knowledge." *Centaurus* 52:280–96.

Kruger, Tobias. 2013. *Discovering the Ice Ages: International Reception and Consequences for a Historical Understanding of Climate*. Translated by A. M. Hentschel. Boston: Brill.

Kuhn, Thomas. [1962] 1996. *The Structure of Scientific Revolutions*. Chicago: University of Chicago Press.

Kupperman, Karen O. 1984. "Fear of Hot Climates in the Anglo-American Colonial Experience." *William and Mary Quarterly* 41(2):213–40.

Kwa, Chunglin. 2001. "The Rise and Fall of Weather Modification: Changes in American Attitudes toward Technology, Nature, and Society." Pp. 135–65 in *Changing the Atmosphere: Expert Knowledge and Environmental Governance*, edited by C. Miller and P. Edwards. Cambridge, MA: MIT Press.

Labe, Zachary, Gudrun Magnusdottir, and Hal Stern. 2018. "Variability of Arctic Sea Ice Thickness Using PIOMAS and the CESM Large Ensemble." *Journal of Climate* 31(8):3233–47. https://doi.org/10.1175/JCLI-D-17-0436.1.

Lahsen, Myanna. 2005. "Seductive Simulations? Uncertainty Distribution around Climate Models." *Social Studies of Science* 35(6):895–922.

Lahsen, Myanna. 2008. "Experiences of Modernity in the Greenhouse: A Cultural Analysis of a Physicist 'Trio' Supporting the Backlash against Global Warming." *Global Environmental Change* 18(1):204–19.

Lahsen, Myanna. 2013. "Anatomy of Dissent: A Cultural Analysis of Climate Skepticism." *American Behavioral Scientist* 57(6):732–53.

Lahsen, Myanna. 2024. "We Cannot Afford Not to Perform Constructionist Studies of Mainstream Climate Science." Pp. 29–38 in *Climate, Science and Society: A Primer*, edited by Z. Baker, T. Law, M. Vardy, and S. Zehr. London: Routledge.

Lahsen, Myanna, and Jesse Ribot. 2022. "Politics of Attributing Extreme Events and Disasters to Climate Change." *WIREs Climate Change* 13(1):e750. https://doi.org/10.1002/wcc.750.

Lamb, Herbert .H. 1959. "Our Changing Climate, Past and Present." *Weather* 14: 299–318.

Lamb, Peter. 2002. "The Climate Revolution: A Perspective." *Climatic Change* 54(1):1–9.

Landsberg, Helmut E. 1941. *Physical Climatology*. DuBois, PA: Gray Printing.

Landsberg, Helmut E. 1955. "Weather 'Normals' and Normal Weather." *Weekly Weather and Crop Bulletin* (January 31):7–8.

Larson, Erik. 1999. *Isaac's Storm: A Man, a Time, and the Deadliest Hurricane in History*. New York: Random House.

Latour, Bruno. 1983. "Give Me a Laboratory and I Will Raise the Word." Pp. 141–70 in *Science Observed: Perspectives on the Social Study of Science*, edited by K. Knorr-Cetina and M. Mulkay. London: Sage.

Latour, Bruno. 1987. *Science in Action: How to Follow Scientists and Engineers through Society*. Cambridge, MA: Harvard University Press.

Latour, Bruno. 1988. *The Pasteurization of France*. Translated by A. Sheridan and J. Law. Cambridge, MA: Harvard University Press.

Latour, Bruno. 2018. *Down to Earth*. Cambridge, UK: Polity.

Latour, Bruno, and Steve Woolgar. 1986. *Laboratory Life: The Construction of Scientific Facts*. Princeton, NJ: Princeton University Press.

Lave, Rebecca. 2012. *Fields and Streams: Field Restoration, Neoliberalism, and the Future of Environmental Science*. Athens: University of Georgia Press.

Lave, Rebecca, Philip Mirowski, and Samuel Randalls. 2010. "Introduction: STS and Neoliberal Science." *Social Studies of Science* 40(5):659–75.

Lavery, Colm. 2016. "Situating Eugenics: Robert DeCourcy Ward and the Immigration Restriction League of Boston." *Journal of Historical Geography* 53:54–62.

Law, John. 1992. "Notes on the Theory of the Actor-Network: Ordering, Strategy and Heterogeneity." *Systems Practice* 5:379–93.

Lawson, Thomas. 1840. *Meteorological Register for the Years 1826-1830*. Philadelphia: Haswell.

Lawson, Thomas. 1844. *Directions for Taking Meteorological Observations*. Army Medical Department. Washington, DC: Surgeon General Office.

Leach, Melissa, and Robin Mearns, eds. 1996. *The Lie of the Land: Challenging Received Wisdom on the African Environment*. London: The International African Institute.

Leffler, Melvyn P. 1992. *A Preponderance of Power: National Security, the Truman Administration, and the Cold War*. Stanford, CA: Stanford University Press.

Lehmann, Philipp. 2022. *Desert Edens: Colonial Climate Engineering in the Age of Anxiety*. Princeton, NJ: Princeton University Press.

Leighly, John. 1954. "Climatology." Pp. 334-61 in *American Geography: Inventory and Prospect*, edited by J. E. Preston and C. F. Jones. Syracuse, NY: Syracuse University Press.

Leiserowitz, A., N. Smith, and J. R. Marlon. 2010. *Americans' Knowledge of Climate Change*. New Haven, CT: Yale Project on Climate Change Communication.

Lemke, Thomas. 2001. "'The Birth of Bio-Politics': Michel Foucault's Lecture at the Collège de France on Neo-Liberal Governmentality." *Economy and Society* 30(2):190-207.

Leslie, Stuart. 1993. *The Cold War and American Science: The Military-Industrial-Academic Complex at MIT and Stanford*. New York: Columbia University Press.

Le Treut, Hervé, R. Somerville, U. Cubasch, et al. 2007. "Historical Overview of Climate Change." Pp. 93-127 in *Climate Change 2007: The Physical Science Basis*. Contribution of Working Group I to the 4th Assessment Report of the Intergovernmental Panel on Climate Change, edited by S. Solomon, D. Qin, M. Manning, et al. Cambridge, UK: Cambridge University Press.

Levene, Mark. 2013. "Climate Blues: Or How Awareness of the Human End Might Reinstall Ethical Purpose to the Writing of History." *Environmental Humanities* 2:147-67.

Levy, Jonathan. 2014. *Freaks of Fortune: The Emerging World of Capitalism and Risk in America*. Cambridge, MA: Harvard University Press.

Levy, Marc. 1995. "Is the Environment a National Security Issue?" *International Security* 20:35-62.

Lewis, John. 2008. "Smagorinsky's GFDL: Building the Team." *Bulletin of the American Meteorological Society* 89(9):1339-53.

Li, Jinbao, Shang-Ping Xie, Edward R. Cook, Mariano S. Morales, Duncan A. Christie, Nathaniel C. Johnson, Fahu Chen, et al. 2013. "El Niño Modulations over the Past Seven Centuries." *Nature Climate Change* 3:822.

Li, Tania M. 2007. *The Will to Improve: Governmentality, Development, and the Practice of Politics*. Durham, NC: Duke University Press.

Libecap, G. D., and Z. K. Hansen. 2002 "'Rain Follows the Plow' and Dryfarming Doctrine: The Climate Information Problem and Homestead Failure in the Upper Great Plains, 1890–1925." *Journal of Economic History* 62(1): 86–120.

Library of Congress. 1923. " Group Outside U.S. Weather Bureau Kiosk." [Photograph]. Accessed May 1, 2024. https://www.loc.gov/item/2016835284/.

Lindzen, Richard, Edward Lorenz, and George Platzman, eds. 1990. *The Atmosphere—A Challenge: The Science of Jule Gregory Charney.* Boston: The American Meteorological Society.

Lionello, P., and F. Giorgi. 2007. "Winter Precipitation and Cyclones in the Mediterranean Region: Future Climate Scenarios in a Regional Simulation." *Advances in Geoscience* 12:153–58.

Liu, Sida, and Mustafa Emirbayer. 2016. "Field and Ecology." *Sociological Theory* 34(1):62–79.

Livingstone, David N. 1991. "The Moral Discourse of Climate: Historical Considerations on Race, Place, and Virtue." *Journal of Historical Geography* 17(4):413–34.

Livingstone, David N. 2002. "Race, Space, and Moral Climatology: Notes toward a Genealogy." *Journal of Historical Geography* 28(2):159–80.

Livingstone, David N. 2003. *Putting Science in Its Place: Geographies of Scientific Knowledge.* Chicago: Chicago University Press.

Livingstone, David N. 2012. "Reflections on the Cultural Spaces of Climate." *Climatic Change* 113(1):91–93.

Livingstone, David N. 2015. "The Climate of War: Violence, Warfare and Climatic Reductionism." *WIREs Climate Change* 6(5):437–44.

Locher, Fabien. 2009. "Les Météores de la Modernité: La Dépression, le Télégraphe et la Prévision Savante du Temps (1850–1914)." Translated by Neil O'Brien and Semma Sarangi. *Revue d'Histoire Moderne et Contemporaine* (56)4:77–103.

Locher, Fabien, and Jean-Baptiste Fressoz. 2012. "The Frail Climate of Modernity: A Climate History of Environmental Reflexivity." *Critical Inquiry* 38(3):579–98.

Lohmann, Larry. 2016. "Toward a Political Economy of Neoliberal Climate Science." Pp. 305–16 in *Routledge Handbook of the Political Economy of Science*, edited by D. Tyfield, R. Lave, S. Randalls, and C. Thorpe. London: Routledge.

Lomborg, Bjorn. 2001. *The Skeptical Environmentalist: Measuring the Real State of the World.* New York: Cambridge University Press.

Longstreth, T. Morris. 1915. *Reading the Weather.* New York: Outing Publishing Company.

Loomis, Elias. 1868. *Treatise on Meteorology.* New York: Harper & Brothers.

Lorenz, Edward N. 1963. "Deterministic Nonperiodic Flow." *Journal of the Atmospheric Sciences* 20(2):130–41.

Lourenço, Tiago C., Rob Swart, Hasse Goosen, and Roger Street. 2016. "The Rise of Demand-Driven Climate Services." *Nature Climate Change* 6:13–14.

Lovell, Joseph. [1817] 1873. "Remarks on the Sick Report of the Northern Division for the Year ending June 30, 1817." Pp. 102–7 in *The Medical Department of the United States Army from 1775 to 1873*, edited by H. E. Brown. Washington, DC: Surgeon General's Office.

Lovell, Joseph. [1818] 1873. "Regulations of the Medical Department of the United States Army." Pp. 110–24 in *The Medical Department of the United States Army from 1775 to 1873*, edited by H. E. Brown. Washington, DC: Surgeon General's Office.

Lovell, Joseph. 1826. *Meteorological Register for the Years 1822–1825*. Washington, DC: Krafft.

Loveman, Mara. 2005. "The Modern State and the Primitive Accumulation of Symbolic Power." *American Journal of Sociology* 110(6):1651–83.

Loveman, Mara. 2012. *National Colors: Racial Classification and the State in Latin America*. London: Oxford University Press.

Lowe, John J., and Michael J. C. Walker. 2015. *Reconstructing Quaternary Environments*. London: Routledge.

Mabey, Nick, Jay Gulledge, Bernard Finel, and Katherine Silverthorne. 2011. *Degrees of Risk: Defining a Risk Management Framework for Climate Security*. London: E3G.

MacDonald, Matt. 2013. "Discourses of Climate Security." *Political Geography* 33:42–51.

MacDonald, Matt. 2021. *Ecological Security: Climate Change and the Construction of Security*. Cambridge, UK: Cambridge University Press.

MacKenzie, Donald. 2009. "Making Things the Same: Gases, Emissions Rights, and the Politics of Carbon Markets." *Accounting, Organizations, and Society* 34:440–55.

MacKenzie, Donald, and Yuval Millo. 2003. "Constructing a Market, Performing a Theory: The Historical Sociology of a Financial Derivatives Exchange." *American Journal of Sociology* 109(1):107–45.

MacKenzie, Donald, Fabian Muniesia, and Lucia Siu, eds. 2007. *Do Economists Make Markets? On the Performativity of Economics*. Princeton, NJ: Princeton University Press.

Madley, Benjamin. 2018. *An American Genocide: The United States and the California Indian Catastrophe, 1846–1873*. New Haven, CT: Yale University Press.

Maguire, William, ed. 1889. *Exploratory Travels through the Western Territories of North America, by Zebulon Montgomery Pike*. Denver, CO: Lawrence and Co.

Mahony, Martin. 2014. "The Predictive State: Science, Territory and the Future of the Indian Climate." *Social Studies of Science* 44(1):109–33.

Mahony, Martin. 2016. "For an Empire of 'All Types of Climate': Meteorology as an Imperial Science." *Journal of Historical Geography* 51:29–39.

Mahony, Martin, and Angelo M. Caglioti. 2017. "Relocating Meteorology." *History of Meteorology* 8:1–14.

Mahony, Martin, and Georgina Endfield. 2018. "Climate and Colonialism." *WIREs Climate Change* 9(2):e510.

Mahony, Martin, and Mike Hulme. 2016. "Epistemic Geographies of Climate Change: Science, Space and Politics." *Progress in Human Geography* 42(3). https://doi.org/10.1177/0309132516681485.

Malm, Andreas. 2016. *Fossil Capital: The Rise of Steam Power and the Roots of Global Warming.* London: Verso.

Malm, Andreas. 2021. *How to Blow Up a Pipeline.* London: Verso.

Malone, Thomas, Edward Goldberg, and Walter Munk. 1998. *Roger Randall Dougan Revelle, 1901–1991: A Biographical Memoir.* National Academy of Sciences Biographical Memoirs. Washington, DC: National Academies Press.

Manabe, Syukuro, and Kirk Bryan. 1969. "Climate Calculations with a Combined Ocean-Atmosphere Model." *Journal of Atmospheric Sciences* 26:786–89.

Mann, Michael. 1993. *The Sources of Social Power.* Vol. 2, *The Rise of Classes and Nation-States, 1760–1914.* New York: Cambridge University Press.

Mann, Michael E., Zhihua Zhang, Scott Rutherford, Raymond S. Bradley, Malcolm K. Hughes, Drew Shindell, Caspar Ammann, Greg Faluvegi, and Fenbiao Ni. 2009. "Global Signatures and Dynamical Origins of the Little Ice Age and Medieval Climate Anomaly." *Science* 326(5957):1256–60.

Maring, D. T. 1909. "Weather Bureau Kiosks." *Monthly Weather Review* 37(3):89–91.

Markusson, Nils, Franklin Ginn, Navraj Singh Ghaleigh, and Vivian Scott. 2014. "'In Case of Emergency Press Here': Framing Geoengineering as a Response to Dangerous Climate Change." *WIREs Climate Change* 5(2):281–90.

Marsh, George Perkins. 1864. *Man and Nature; Or, Physical Geography as Modified by Human Action.* New York: Scribner.

Martin, Craig. 2006. "Experience of the New World and Aristotelian Revisions of the Earth's Climates during the Renaissance." *History of Meteorology* 3:1–16.

Marvin, Charles F. 1888. "Self-Recording Rain Gauge." *Science* 11(264):97–98.

Marvin, Charles F. 1896. "The Weather Bureau and Its Voluntary Observers." Pp. 555–56 in *Yearbook of Agriculture.* U.S. Department of Agriculture. Washington, DC: Government Printing Office.

Marvin, Charles F. 1909. "Instructions for Installing Street Kiosks." *Weather Bureau Bulletin.* U.S. Department of Agriculture. Washington, DC: Government Printing Office.

Marvin, Charles F. 1920. "The Status and Problems of Meteorology." *Proceedings of the National Academy of Sciences* 6(10):561–72.

Marx, Karl. [1867] 2015. *Capital: A Critique of Political Economy.* Vol. 1. Accessed December 2, 2017. www.marxists.org/archive/marx/works /download/pdf/Capital-Volume-I.pdf.

Mathiesen, Karl. 2015. "Climate Change Has Left the US Exposed in the Arctic, Say Military Experts." *Mother Jones*, June 11. Accessed February 1, 2018. http://motherjones.com/environment/2015/06/climate-change-has-left-us -exposed-arctic-say-military-experts/.

Matthew, Richard A. 2002. "In Defense of Environment and Security Research." *The Wilson Center Environmental Change and Security Program* 8:109–24.

Matthew, Richard A., Jon Barnett, Bryan McDonald, and Karen L. O'Brien, eds. 2009. *Global Environmental Change and Human Security.* Cambridge, MA: MIT Press.

Matthews, Jessica T. 1989. "Redefining Security," *Foreign Affairs* 68(2):162–77.

Maury, Matthew F. 1846. *Address Delivered before the Philodemic Society, at the commencement of Georgetown College, Aug. 28, 1846.* Washington, DC: Gedeon.

Maury, Matthew F. 1869. *Physical Survey of Virginia: Her Geographical Position; Its Commercial Advantages and National Importance.* Richmond, VA: Nye.

Maury, T. B. 1880. "The International Weather-Service." *Popular Science Monthly* 16:280–312.

Mazlish, Bruce. 2004. *Civilization and Its Contents.* Stanford, CA: Stanford University Press.

Mazuzan, George. 1988. "Up, Up, and Away: The Reinvigoration of Meteorology in the United States, 1958 to 1962." *Bulletin of the American Meteorological Society* 69(10):1152–63.

McAdie, Alexander. 1891. *Mean Temperatures and Their Corrections in the United States.* Washington, DC: Signal Office.

McCandless, Peter. 2011. *Slavery, Disease, and Suffering in the Southern Lowcountry.* Cambridge, UK: Cambridge University Press.

McCright, Aaron, and Riley E. Dunlap. 2000. "Challenging Global Warming as a Social Problem: An Analysis of the Conservative Movement's Counter-Claims." *Social Problems* 47(4):499–522.

McCright, Aaron M., and Riley E. Dunlap. 2003. "Defeating Kyoto: The Conservative Movement's Impact on U.S. Climate Change Policy." *Social Problems* 50(3):348–73.

McCright, Aaron M., and Riley E. Dunlap. 2011. "The Politicization of Climate Change and Polarization in the American Public's Views of Global Warming, 2001–2010." *Sociological Quarterly* 52(2):155–94.

McGann, Edward W. 1905. "Unreliable Weather Forecasters." *US Department of Agriculture Climate and Crop Service of the United States Weather Bureau* 18(1):3.

McGranahan, Gordon, Deborah Balk, and Bridget Anderson. 2007. "The Rising Tide: Assessing the Risks of Climate Change and Human Settlements in Low Elevation Coastal Zones." *Environment and Urbanization* 19(1):17–37.

McGuire W. 2016. "Cross-field Effects of Science Policy on the Biosciences: Using Bourdieu's Relational Methodology to Understand Change." *Minerva* 54(3):325–51.

McSweeney, Bill. 1996. "Identity and Security: Buzan and the Copenhagen School." *Review of International Studies* 22(1):81–93.

Meadows, Donella H., Dennis L. Meadows, Jørgen Randers, and William Behrens III. 1972. *The Limits to Growth: A Report for the Club of Rome's Project on the Predicament of Mankind.* New York: Universe Books.

MEDEA. 1995. *Scientific Utility of Navy Environmental Data: A Medea Special Task Force Report.* Washington, DC: Central Intelligence Agency.

Medvetz, Thomas. 2012. *Think Tanks in America.* Chicago: The University of Chicago Press.

Meisinger, C. LeRoy. 1921. "Notes on Meteorology and Climatology." *Science* 53(1371):337–39.

Menga, Filippo, and Erik Swyngedouw, eds. 2018. *Water, Technology and the Nation-State.* London: Routledge.

Mercer, Harriet, and Thomas Simpson. 2023. "Imperialism, Colonialism, and Climate Change Science." *WIREs Climate Change* 14(6):e851.

Merchant, Carolyn. 1983. *The Death of Nature: Women, Ecology, and the Scientific Revolution.* New York: Harper & Row.

Merton, Robert K. 1942. *The Normative Structure of Science in the Sociology of Science: Theoretical and Empirical Investigations.* Chicago: University of Chicago Press.

Mesarovic, Mihajlo, and Eduard Pestel. 1974. *Mankind at the Turning Point: The Second Report to the Club of Rome.* New York: Dutton.

Meyer, William B. 2014. *Americans and Their Weather.* 2nd ed. New York: Oxford University Press.

Mialet, Hélène. 2003. "The 'Righteous Wrath' of Pierre Bourdieu." *Social Studies of Science* 33(4):613–21.

Michaels, Pat. 2008. Interview on *Russia Today.* Accessed May 20, 2024. https://www.cato.org/node/14433/embed.

Michigan Weather Service. 1895. "State Weather Service." P. 4 in *Michigan Weather Services 1895 Annual Report.* Lansing, MI: Smith.

Milankovitch, Milutin. 1920. *Mathematical Theory of Heat Phenomena Produced by Solar Radiation.* Paris: Gauthier-Villars.

Milham, Willis I. [1912] 1918. *Meteorology: A Text-book on the Weather, the Causes of Its Changes, and Weather Forecasting, for the Student and General Reader.* New York: Macmillan.

Miller, Clark A. 2004. "Climate Science and the Making of a Global Political Order." Pp. 46–66 in *States of Knowledge: The Co-production of Science and Social Order*, edited by S. Jasanoff. New York: Routledge.

Miller, Eric R. 1931. "New Light on the Beginnings of the Weather Bureau from the Papers of Increase A. Lapham." *Monthly Weather Review* 59(2):65–70.

Miller, Mark J. 2000. "A Durable International Migration and Security Nexus: The Problem of the Islamic Periphery in Transatlantic Ties." Pp. 15–27 in *Redefining Security: International Migration and Global Security*, edited by D. Grahm and N. Poku. London: Praeger.

Miller, Todd. 2017. *Storming the Wall: Climate Change, Migration, and Homeland Security*. San Francisco, CA: City Lights.

Mirowski, Philip. 1989. *More Heat Than Light: Economics as Social Physics, Physics as Nature's Economics*. New York: Cambridge University Press.

Mirowski, Philip. 2002. *Machine Dreams: Economics Becomes a Cyborg Science*. New York: Cambridge University Press.

Mitchell, J. F. B., T. C. Johns, M. Eagles, W. J. Ingram, and R. A. Davis. 1999. "Towards the Construction of Climate Change Scenarios." *Climatic Change* 41:547–81.

Mitchell, Timothy. 1991. "The Limits of the State: Beyond Statist Approaches and Their Critics." *American Political Science Review* 85(1):77–96.

Mitchell Timothy. 1998. "Fixing the Economy." *Cultural Studies* 12(1):82–101.

Mitchell, Timothy. 1999. "Society, Economy, and the State Effect." Pp. 76–97 in *State/Culture: State Formation after the Cultural Turn*, edited by George Steinmetz. Ithaca, NY: Cornell University Press.

Mitchell, Timothy. 2005. "The Work of Economics: How a Discipline Makes Its World." *European Journal of Sociology* 46(2):297–320.

Mitman, Gregg. 2003. "Hay Fever Holiday: Health, Leisure, and Place in Gilded-Age America." *Bulletin of the History of Medicine* 77(3):600–635.

Mitman, Gregg, and Ronald Numbers. 2003. "From Miasma to Asthma: The Changing Fortunes of Medical Geography in America." *History and Philosophy of the Life Sciences* 25(3):391–412.

Monmonier, Mark. 2000. *Air Apparent: How Meteorologists Learned to Map, Predict, and Dramatize Weather*. New York: Oxford University Press.

Monthly Weather Review. 1889. "Chart VII. Departures From Normal Precipitation, 1889." 17(13):c7.

Moore, Barrington, Jr. 1966. *Social Origins of Dictatorship and Democracy: Lord and Peasant in the Making of the Modern World*. Boston: Beacon Press.

Moore, Jason W. 2017. *Anthropocene or Capitalocene? Nature, History, and the Crisis of Capitalism*. New York: PM Press.

Moore, Kelly. 2007. *Disrupting Science: Social Movements, American Scientists, and the Politics of the Military, 1945–1975*. Princeton, NJ: Princeton University Press.

Moore, Kelly, Daniel Lee Kleinman, David Hess, and Scott Frickel. 2011. "Science and Neoliberal Globalization: A Political Sociological Approach." *Theory and Society* 40(5):505–32.

Moore, Willis L. 1903. "Special Instructions." P. 3 in *Instructions for Obtaining and Tabulating Records from Recording Instruments*, 2nd ed., edited by C. F. Marvin. Circular A, Instrument Division. U.S. Department of Agriculture. Washington, DC: Weather Bureau.

Moore, Willis L. 1904. *Climate: Its Physical Basis and Controlling Factors.* Washington, DC: U.S. Weather Bureau.

Moore, Willis L. 1905. "Fake Rain Making: A Letter from the Chief of Bureau." *Monthly Weather Review* 33(4):152–53.

Morgan, Judith, and Neil Morgan. 1996. *Roger: A Biography of Roger Revelle.* La Jolla, CA: Scripps Institution of Oceanography.

Morgan, Kimberly J., and Ann Shola Orloff. 2017. *The Many Hands of the State.* Cambridge, UK: Cambridge University Press.

Morton, Samuel G. 1839. *Crania Americana.* Philadelphia: Dobson.

Moss, R., P. L. Scarlett, M. A. Kenney, H. Kunreuther, R. Lempert, J. Manning, B. K. Williams, et al. 2014. "Decision Support: Connecting Science, Risk Perception, and Decisions." Pp. 620–47 in *Climate Change Impacts in the United States: The Third National Climate Assessment*, edited by J. M. Melillo, T. C. Richmond, and G. W. Yohe. Washington, DC: US Global Change Research Program.

Mount, Harry A. 1921. "Making Weather to Order: The Indoor Meteorology That Makes All the Difference between Industrial Success and Failure." *Scientific American* 124(10):189, 198–99.

MSNBC. 2015. "MSNBC Panel: O'Malley's Right, Global Warming Sparked Rise of ISIS." *Weekends with Alex Witt*, July 25. Accessed March 3, 2018. https://grabien.com/story.php?id=33187.

Mudge, Stephanie. 2018. *Leftism Reinvented: Western Parties from Socialism to Neoliberalism.* Cambridge, MA: Harvard University Press.

Mudge, Stephanie L., and Antoine Vauchez. 2012. "Building Europe on a Weak Field: Law, Economics, and Scholarly Avatars in Transnational Politics." *American Journal of Sociology* 118(2):449–92.

Mukerji, Chandra. 1989. *A Fragile Power: Scientists and the State.* Princeton, NJ: Princeton University Press.

Mukerji, Chandra. 1997. *Territorial Ambitions and the Gardens of Versailles.* New York: Cambridge University Press.

Mukerji, Chandra. 2003. "Intelligent Uses of Engineering and the Legitimacy of State Power." *Technology and Culture* 44(4):655–76.

Mukerji, Chandra. 2009. *Impossible Engineering: Technology and Territoriality on the Canal du Midi.* Princeton, NJ: Princeton University Press.

Mukerji, Chandra. 2011. "Jurisdiction, Inscription, and State Formation: Administrative Modernism and Knowledge Regimes." *Theory and Society* 40:223–45.

Murphy, Raymond. 1995. "Sociology as If Nature Did Not Matter: An Ecological Critique." *British Journal of Sociology* 46(4):688–707.

Myer, Albert J. 1866. *A Manual of Signals, for the Use of Signal Officers in the Field, and for Military and Naval Students, Military Schools, Etc.* New York: Nosrand.

Myers, Norman. 2005. "Environmental Refugees: An Emergent Security Issue." Paper presented at the 13th Economic Forum, Prague, May 23–27, 2005. Accessed March 8, 2015. populationmedia.org/wp-content/uploads/2008/03/norman-myers-environmental-refugees-an-emergent-security-issue.pdf.

Myers, Norman, and Jennifer Kent. 1995. *Environmental Exodus: An Emergent Crisis in the Global Arena.* Washington, DC: Climate Institute.

Namias, Jerome. 1983. "The History of Polar Front and Air Mass Concepts in the United States—An Eyewitness Account." *Bulletin of the American Meteorological Society* 64(7):734–55.

Namias, Jerome. 1991. *Francis W. Reichelderfer 1895–1983: A Biographical Memoir.* National Academy of Sciences Biographical Memoirs. Washington, DC: National Academies Press.

Nansen Initiative. 2015. *Agenda for the Protection of Cross-Border Displaced Persons in the Context of Disasters and Climate Change.* Accessed March 6, 2018. nanseninitiative.org/wp-content/uploads/2015/02/PROTECTION-AGENDA-VOLUME-1.pdf.

Narain, Sunita. 2019. "Equity: The Final Frontier for an Effective Climate Change Agreement." Pp. xxiv–xxix in *Climate Futures: Re-imagining Global Climate Justice*, edited by Kum-Kum Bhavnani, John Foran, Priya A. Kurian, and Debashish Munshi. London: Zed.

National Academy of Sciences (NAS). 1958. *Research and Education in Meteorology: An Interim Report.* Washington, DC: National Academies Press.

National Academy of Sciences (NAS). 2010. *"Climate Stabilization Targets: Emissions, Concentrations, and Impacts over Decades to Millennia.* Washington, DC: National Academies Press. Accessed December 3, 2017. nationalacademies.org/includes/StabilizationTargetsFinal.pdf.

National Academy of Sciences (NAS). 2016. "Assessing Approaches to Updating the Social Cost of Carbon." Project documents. Accessed December 8, 2018. https://sites.nationalacademies.org/DBASSE/BECS/CurrentProjects/DBASSE_167526.

National Research Council (NRC). 1977. *Energy and Climate: Studies in Geophysics.* Washington, DC: The National Academies Press.

National Research Council (NRC). 1983. *Changing Climate: Report of the Carbon Dioxide Assessment Committee.* Washington, DC: The National Academies Press.

National Science Foundation. 1965. *Weather and Climate Modification: Report of the Special Commission on Weather Modification.* Washington, DC: National Science Foundation.

Nebeker, Frederik. 1995. *Calculating the Weather: Meteorology in the 20th Century*. San Diego, CA: Academic Press.

Needell, Allan. 2000. *Science, Cold War, and the American State: Lloyd V. Berkner and the Balance of Professional Ideals*. Newark, NJ: Harwood Academic Publishers.

Nicollet, Joseph N. 1839. *Essay on Meteorological Observations*. Washington, DC: Gideon.

Nordhaus, William D. 2017. "Revisiting the Social Cost of Carbon." *Proceedings of the National Academy of Sciences* 114(7):1518.

Norgaard, Kari M. 2011. *Living in Denial: Climate Change, Emotions, and Everyday Life*. Cambridge, MA: MIT Press.

Northern Pacific Railroad. 1893. *The Official Northern Pacific Railroad Guide, for the Use of Tourists and Travelers over the Lines of the Northern Pacific Railroad, Its Branches and Allied Lines*. St. Paul, MN: Riley.

Norwood, Paul, and Benjamin Jensen. 2016. "Wargaming the Third Offset Strategy." *Joint Forces Quarterly* 83(4):34–39.

Nott, Josiah. 1851. *An Essay on the Natural History of Mankind: Viewed in Connection with Negro Slavery*. Mobile, AL: Dade, Thompson and Co.

Novak, William J. 2008. "The Myth of the 'Weak' American State." *American Historical Review* 113(3):752–72.

Nuccitelli, Dana. 2015. *Climatology versus Pseudoscience: Exposing the Failed Predictions of Global Warming Skeptics*. Santa Barbara, CA: Praeger.

Null, Schuyler, and Lauren Herzer Risi. 2016. *Navigating Complexity: Climate, Migration, and Conflict in a Changing World*. Washington, DC: Wilson Center.

O'Connor, James. 1998. *Natural Causes: Essay in Ecological Marxism*. New York: Guilford Press.

Oels, Angela. 2005. "Rendering Climate Change Governable: From Biopower to Advanced Liberal Government?" *Journal of Environment and Planning* 7(3):185–207.

Oels, Angela. 2012. "From 'Securitization' of Climate Change to 'Climatization' of the Security Field: Comparing Three Theoretical Perspectives." Pp. 185–205 in *Climate Change, Human Security and Violent Conflict*, edited by J. Scheffran, M. Brzoska, H. Brauch, P. Link, and J. Schilling: Berlin: Springer.

Ogle, Vanessa. 2013. "Whose Time Is It? The Pluralization of Time and the Global Condition, 1870s–1940s." *American Historical Review* 118(5): 1376–1402.

Oldfield, Jonathan D. 2016. "Mikhail Budyko's (1920–2001) Contributions to Global Climate Science: From Heat Balances to Climate Change and Global Ecology." *WIREs Climate Change* 7(5):682–92. https://doi.org/10.1002/wcc.412.

Oldfield, Jonathan D. 2018. "Imagining Climates Past, Present and Future: Soviet Contributions to the Science of Anthropogenic Climate Change,

1953–1991." *Journal of Historical Geography* 60:41–51. https://doi.org/10.1016
/j.jhg.2017.12.004.

Ollson, Ola. 2016. "Climate Change and Market Collapse: A Model Applied to
Darfur." *Games* 7(9):1–27.

Omi, Michael, and Howard Winant. 2015. *Racial Formation in the United
States.* 3rd ed. New York: Routledge.

O'Neill, Brian C., F. Landis MacKellar, and Wolfgang Lutz. 2001. *Population
and Climate Change.* New York: Cambridge University Press.

Oreskes, Naomi. 2003. "A Context of Motivation: US Navy Oceanographic
Research and the Discovery of Sea-Floor Hydrothermal Vents." *Social
Studies of Science* 33(5):697–742.

Oreskes, Naomi. 2015. "How Earth Science Has Become a Social Science."
Historical Social Research 40(2):246–70.

Oreskes, Naomi. 2021. *Why Trust Science?* Princeton, NJ: Princeton University
Press.

Oreskes, Naomi, and Erik M. Conway. 2010. *Merchants of Doubt: How a
Handful of Scientists Obscured the Truth on Issues from Tobacco Smoke to
Global Warming.* New York: Bloomsbury Press.

Oreskes, Naomi, Erik M. Conway, and Matthew Shindell. 2008. "From Chicken
Little to Dr. Pangloss: William Nierenberg, Global Warming, and the Social
Deconstruction of Scientific Knowledge." *Historical Studies in the Natural
Sciences* 38(1):109.

Oreskes, Naomi, and John Krige, eds. 2014. *Science and Technology in the
Global Cold War.* Cambridge, MA: MIT Press.

Osborne, Michael A. 2000. "Acclimatizing the World: A History of the Paradigmatic Colonial Science." *Osiris* 15:135–51.

Otto, Friederike E. L., Luke Harrington, Katharina Schmitt, Sjoukje Philip,
Sarah Kew, Geert Jan van Oldenborgh, Roop Singh, Joyce Kimutai, and
Piotr Wolski. 2020. "Challenges to Understanding Extreme Weather Changes
in Lower Income Countries." *Bulletin of the American Meteorological Society*
101(10):E1851–60.

Oxford Geoengineering Programme. 2018. "Why Consider Geoengineering?"
Accessed January 5, 2019. http://www.geoengineering.ox.ac.uk/www
.geoengineering.ox.ac.uk/what-is-geoengineering/why-consider
-geoengineering/.

PAGES 2k Consortium, Moinuddin Ahmed, et al. 2013. "Continental-Scale
Temperature Variability during the Past Two Millennia." *Nature Geoscience*
6:339.

Panofsky, Aaron. 2011. "Field Analysis and Interdisciplinary Science: Scientific
Capital Exchange in Behavior Genetics." *Minerva* 49:295–316.

Panofsky, Aaron. 2014. *Misbehaving Science: Controversy and the Development
of Behavior Genetics.* Chicago: Chicago University Press.

Parenti, Christian. 2015. "The Environment Making State: Territory, Nature, and Value." *Antipode* 47(4):829–48.

Parr, Adrian. 2013. *The Wrath of Capital: Neoliberalism and Climate Change Politics*. New York: Columbia University Press.

Parsons, Meg. 2014. "Destabilizing Narratives of the 'Triumph of the White Man over the Tropics': Scientific Knowledge and the Management of Race in Queensland, 1900–1940." Pp. 213–32 in *Climate, Science, and Colonization: Histories from Australia and New Zealand*, edited by James Beattie, Emily O'Gorman, and Matthew Henry. New York: Palgrave.

Pearse, Rebecca, and Steffen Böhm. 2014. "Ten Reasons Why Carbon Markets Will Not Bring about Radical Emissions Reduction." *Carbon Management* 5(4):325–37.

Pelling, Mark. 2011. *Adaptation to Climate Change: From Resilience to Transformation*. New York: Routledge.

Perch-Nielsen, Sabine L., Michèle B. Bättig, and Dieter Imboden. 2008. "Exploring the Link between Climate Change and Migration." *Climatic Change* 91:375–93.

Peron, François. [1806] 1983. "Experiments on the Physical Strength of the Savage People of Van Diemen's Land and New Holland and of the Inhabitants of Timor." Pp. 145–59 in *The Baudin Expedition and the Tasmanian Aborigines, 1802*, edited by B. Plomley. Hobart: Blubber Head Press.

Peron, François. [1809] 2012. *A Voyage of Discovery to the Southern Hemisphere, Performed by Order of the Emperor Napoleon, During the Years 1801, 1802, 1803, and 1804*. London: Phillips. Accessed December 21, 2018. http://gutenberg.net.au/ebooks12/1203691h.html.

Peterson, Nicole, and Kenneth Broad. 2009. "Climate and Weather Discourse in Anthropology: From Determinism to Uncertain Futures." Pp. 70–86 in *Anthropology and Climate Change*, edited by S. A. Crate and M. Nuttal. Walnut Creek, CA: Left Coast Press.

Petty, William. [1687] 1890. "Five Essays on Political Arithmetic." Pp. 103–31 in *Essays on Mankind and Political Arithmetic*. New York: Mershon.

Phillips, Norman. 1956. "The General Circulation of the Atmosphere: A Numerical Experiment." *Quarterly Journal of the Royal Meteorological Society* 82:123–64.

Phillips, Norman. 1995. *Jule Gregory Charney 1917–1981: A Biographical Memoir*. National Academy of Sciences Biographical Memoirs. Washington, DC: National Academy of Sciences.

Phillips, Norman. 1998. "Carl-Gustaf Rossby: His Times, Personality, and Actions." *Bulletin of the American Meteorological Society* 79(6):1097–2012.

Pickering, Andrew. 1993. "The Mangle of Practice: Agency and Emergence in the Sociology of Science." *American Journal of Sociology* 99(3):559–89.

Pickett, Neil. 1992. *A History of the Hudson Institute*. Indianapolis, IN: Hudson Institute.

Pickford, James. 1858. *Hygiene, Or, Health as Depending Upon the Conditions of the Atmosphere, Foods and Drinks, Motion and Rest, Sleep and Wakefulness, Secretions, Excretions, and Retentions, Mental Emotions, Clothing, Bathing, &c.* London: Churchill.

Pielke, Roger A., Jr., Joel Gratz, Christopher W. Landsea, Douglas Collins, Mark A. Saunders, and Rade Musulin. 2008. "Normalized Hurricane Damages in the United States: 1900–2005." *Natural Hazards Review* 9(1):29–42.

Pietruska, Jamie L. 2011. "US Weather Bureau Chief Willis Moore and the Reimagination of Uncertainty in Long-Range Forecasting." *Environment and History* 17(1):79–105.

Pietruska, Jamie L. 2017. *Looking Forward: Prediction and Uncertainty in Modern America.* Chicago: Chicago University Press.

Pike, Zebulon. 1805–1807. "Zebulon Pike's Notebook of Maps, Traverse tables, and Meteorological Observations." National Archives and Records Administration, Records of the Adjutant General's Office, Record Group 94. Accessed June 5, 2017. https://catalog.archives.gov/id/5928242.

Platt, Isaac Hull. 1886. "The Problem of Acclimatization." *Transactions of the American Climatological Association* 2:104–14.

Podesta, John, and Pete Ogden. 2007. "The Security Implications of Climate Change." CAP 20. November 26. Accessed March 12, 2024. www.american progress.org/issues/green/news/2007/11/26/3657/the-security-implications -of-climate-change/.

Pointer, John. 1738. *A Rational Account of the Weather, Showing the Signs of Its Several Changes, with the Philosophical Reasons of Them.* 2nd ed. London: Ward.

Polanyi, Karl. [1944] 2001. *The Great Transformation: The Economic and Political Origins of Our Time.* Boston: Beacon Press.

Pols, Hans. 2012. "Notes from Batavia, the Europeans' Graveyard: The Nineteenth-Century Debate on Acclimatization in the Dutch East Indies." *Journal of the History of Medicine and Allied Sciences* 67(1):120–48.

Ponko, Vincent. 1997. "The Military Explorers of the American West, 1838–1860." Pp. 332–411 in *North American Exploration*, vol. 3, *A Continent Comprehended*, edited by John Allen. Lincoln: University of Nebraska Press.

Priest, Dana, and William M. Arkin. 2011. *Top Secret America: The Rise of the New American Security State.* New York: Little, Brown.

Priestly, Joseph. 1774. *Experiments and Observations of Different Kinds of Air.* London: Johnson.

Pringle, John. 1753. *Observations on the Diseases of the Army in Camp and Garrison.* London: Millar.

Puckrein, Gary. 1973. "Climate, Health and Black Labor in the English Americas." *Journal of American Studies* 13(2):179–93.

Pulver, Simone, and Stacy D. VanDeveer. 2009. "'Thinking about Tomorrows': Scenarios, Global Environmental Politics, and Social Science Scholarship." *Global Environmental Politics* 9(2):1–13.

Pumphrey, Carolyn, ed. 2008. *Global Climate Change: National Security Implications.* Carlisle, PA: Strategic Studies Institute.

Quarles, Donald A. 1954. Letter to Joseph Dodge. Eisenhower Presidential Library, National Archives. Accessed November 5, 2015. https://www.eisenhower .archives.gov/.

Raines, Rebecca R. 2011. *Getting the Message through A Branch History of the U.S. Army Signal Corps.* Washington, DC: Center of Military History.

Rainger, Ronald. 2000. "Science at the Crossroads: The Navy, Bikini Atoll, and American Oceanography in the 1940s." *Historical Studies in the Physical and Biological Sciences* 30(2):349–71.

Rainger, Ronald. 2001. "Constructing a Landscape for Postwar Science: Roger Revelle, the Scripps Institution of Oceanography and the University of California, San Diego." *Minerva* 39(3):327–52.

Raleigh, Clionadh, Lisa Jordan, and Idean Salehyan. 2008. *Assessing the Impact of Climate Change on Migration and Conflict.* Report to the World Bank. Accessed April 5, 2014. http://siteresources.worldbank.org/EXTSOCIAL DEVELOPMENT/Resources/SDCCWorkingPaper_MigrationandConflict.pdf.

Ramsay, David. 1790. *A Dissertation on the Means of Preserving Health, in Charleston, and the Adjacent Low Country.* Charleston, SC: Markland & M'Iver.

Randalls, Samuel. 2010. "Weather Profits: Weather Derivatives and the Commercialization of Meteorology." *Social Studies of Science* 40(5):705–30.

Randalls, Samuel, and James Kneale. 2021. "A Fragile Network: Effecting Hail Insurance in Britain, 1840–1900." *Enterprise & Society* 22(3):739–69.

Randers, Jorgen. 2012. *2052: A Global Scenario for the Next Forty Years.* White River Junction, VT: Chelsea Green.

Raskin, Paul D. 2005. "Global Scenarios: Background Review for the Millennium Ecosystem Assessment." *Ecosystems* 8:133–42.

Ratzel, Friedrich. 1896. *History of Mankind.* Vol. 1. London: Macmillan.

Ratzel, Friedrich. [1901] 2018. "Lebensraum: A Biogeographical Study." *Journal of Historical Geography* 61:59–80.

Reed, Isaac A. 2011. *Interpretation and Social Knowledge: On the Use of Theory in the Human Sciences.* Chicago: Chicago University Press.

Reed, William G. 1916. "Weather Insurance." *Monthly Weather Review* 44(10):575–80.

Reuveny, Rafael, and Will H. Moore. 2009. "Does Environmental Degradation Influence Migration? Emigration to Developed Countries in the Late 1980s and 1990s." *Social Science Quarterly* 90(3):461–79.

Revelle, Roger. 1958. "Sun, Sea, and Air: IGY Studies of the Heat and Water Budget of the Earth." *Geophysical Monographs Series* 2:147–53.

Revelle, Roger. 1975. "Transcript: 25th Reunion of People Who Participated in Midpac and Capricorn Expeditions by Scripps Institution of Oceanography." Accessed June 10, 2016. libraries.ucsd.edu/speccoll/siooralhistories/2011-49 -revelle.pdf.

Revelle, Roger, Wallace Broecker, Charles D. Keeling, and Joseph Smagorinsky. 1965. "Atmospheric Carbon Dioxide." Pp. 111–33 in *Restoring the Quality of Our Environment: Report of the President's Science Advisory Council*, edited by J. Tukey. Washington, DC: Government Printing Office.

Revelle, Roger, and Hans Suess. 1957. "Carbon Dioxide Exchange between Atmosphere and Ocean and the Question of an Increase of Atmospheric CO_2 during the Past Decades." *Tellus* 9(1):18–27.

Rich, Nathaniel. 2018. "Losing Earth: The Decade We Almost Stopped Climate Change." *New York Times Magazine*, August 1. Accessed August 4, 2018. https://www.nytimes.com/interactive/2018/08/01/magazine/climate-change -losing-earth.html.

Ripple, William J., Christopher Wolf, Thomas M. Newsome, Phoebe Barnard, and William R. Moomaw. 2020. "World Scientists' Warning of a Climate Emergency." *BioScience* 70(1):8–12. https://doi.org/10.1093/biosci/biz088.

Ripple, William J., et al. 2017. "World Scientists' Warning to Humanity: A Second Notice." *BioScience* 67(12):1026–28. Accessed January 4, 2019. https://doi.org /10.1093/biosci/bix125 (list of signatories).

Rising Tide North America. 2018. "Climate Justice." Accessed May 2, 2018. https:// risingtidenorthamerica.org/about-rising-tide-north-america/climate-justice/.

Rising Tide North America. n.d. *The Climate Movement Is Dead: Long Live the Climate Movement*. Accessed December 18, 2018. https://risingtidenorth america.org/publications/the-climate-movement-is-dead-long-live-the -climate-movement/.

Robbins, Paul. 2012. *Political Ecology: A Critical Introduction*. Cambridge, UK: Blackwell.

Roberts, J. Timmons, and Bradley Park. 2006. *A Climate of Injustice: Global Inequality, North-South Politics, and Climate Policy*. Cambridge, MA: MIT Press.

Robertson, Morgan. 2012. "Measurement and Alienation: Making a World of Ecosystem Services." *Transactions of the Institute of British Geographers* 37(3):386–401.

Robertson, William. [1777] 1817. *The History of America*. Vol. 2 of *The Works of William Robertson*, edited by Dugald Stewart. London: Cadell and Davies.

Rockstrom, Johan, Will Steffen, Kevin Noone, Åsa Persson, Stuart Chapin III, Eric Lambin, Timothy M. Lenton, et al. 2009. "Planetary Boundaries: Exploring the Safe Operating Space for Humanity." *Ecology and Society* 14(2). https://doi.org/10.5751/ES-03180-140232.

Rockwell, Stephen J. 2010. *Indian Affairs and the Administrative State in the Nineteenth Century*. Cambridge, UK: Cambridge University Press.

Rogers, Will, and Jay Gulledge. 2010. *Lost in Translation: Closing the Gap between Climate Science and National Security Policy.* Washington, DC: Center for a New American Security.

Rohli, Robert V., and Gregory D. Bierly. 2011. "The Lost Legacy of Robert DeCourcy Ward in American Geographical Climatology." *Progress in Physical Geography* 35:547–64.

Romm, Joe. 2015. "The Link between Climate Change and ISIS Is Real." July 23. Accessed December 1, 2017. https://thinkprogress.org/the-link-between -climate-change-and-isis-is-real-399789412b41/.

Rosa, Eugene, and Thomas Dietz. 1998. "Climate Change and Society: Speculation, Construction, and Scientific Investigation." *International Sociology* 13(4):421–55.

Rose, Nikolas, Pat O'Malley, and Mariana Valverde. 2006. "Governmentality." *Annual Review of Law and Social Science* 2:83–104.

Rose, Nikolas, and Peter Miller. 1992. "Political Power beyond the State: Problematics of Government." *British Journal of Sociology* 43(2):173–205.

Rosenberg, Charles E. 1997. "Rationalization and Reality in the Shaping of American Agricultural Research, 1875–1914." *Social Studies of Science* 7:401–22.

Rossby, Carl-Gustav. 1934. "Comments on Meteorological Research." *Journal of Aeronautical Science* 1:32–34.

Rossby, Carl-Gustav. 1959. "Current Problems in Meteorology." Pp. 9–50 in *The Atmosphere and the Sea in Motion,* edited by Bert Bolin. New York: Rockefeller Institute Press.

Rothchild, Emma. 1995. "What Is Security?" *Deadelus* 124(3):53–98.

Rousseau, Jean-Jacques. [1762] 1913. *The Social Contract.* New York: Dutton.

Rubin, Morton. 1958. "Synoptic Meteorology and the IGY." *Geophysical Monographs Series* 2:154–60.

Rueschemeyer, Dietrich, and Theda Skocpol, eds. 1996. *States, Social Knowledge, and the Origins of Modern Social Policies.* Princeton, NJ: Princeton University Press.

Ruget, Vanessa. 2003. "Scientific Capital in American Political Science: Who Possesses What, When and How?" *New Political Science* 24(3):469–78.

Rupke, Nicolaas, and Karen Wonders. 2000. "Humboldtian Representations in Medical Cartography." Pp. 163–75 in *Medical Geography in Historical Perspective,* edited by Nicolaas Rupke. London: Wellcome Trust Centre.

Rush, Benjamin. 1774. *An Oration [. . .] before the American Philosophical Society Containing, an Enquiry into the Natural History of Medicine among the Indians in North-America, and a Comparative View of Their Diseases and Remedies, with Those of Civilized Nations.* Philadelphia: Crukshank.

Rush, Benjamin. 1786. "An Enquiry into the Cause of the Increase of Bilious and Intermitting Fevers in Pennsylvania, with Hints for Preventing Them." *Transactions of the American Philosophical Society* 2:206–12.

Rush, Benjamin. 1790. *An Inquiry into the Effects of Spirituous Liquors on the Human Body and the Mind, to Which Is Added, a Moral and Physical Thermometer*. Boston: Thomas.

Rush, Benjamin. 1794. *An Account of the Bilious Remitting Yellow Fever, as It Appeared in the City of Philadelphia, in the Year 1793*. Philadelphia: Dobson.

Rush, Benjamin. 1799. *Observations upon the Origin of the Malignant Bilious, or Yellow Fever in Philadelphia, and upon the Means of Preventing It*. Philadelphia: Budd and Bartram.

Rusnock, Andrea A. 2002. *Vital Accounts: Quantifying Health and Population in Eighteenth-Century England and France*. New York: Cambridge University Press.

Sabin Center. 2020. "Silencing Science Tracker." Columbia University, Sabin Center for Climate Change Law. Accessed May 1, 2024. columbiaclimatelaw.com/resources/silencing-science-tracker.

Sagendorph, Robb. 1970. *America and Her Almanacs: Wit, Wisdom and Weather, 1639–1970*. Boston: Little.

Sapiro, Gisèle. 2003. "Forms of Politicization in the French Literary Field." *Theory and Society* 32(5):633–52.

Sapolsky, Harvey. 1990. *Science and the Navy: The History of the Office of Naval Research*. Princeton, NJ: Princeton University Press.

Schapiro, Mark. 2016. *The End of Stationarity: Searching for the New Normal in the Age of Carbon Shock*. White River Junction, VT: Chelsea Green.

Scheffran, Jürgen. 2008. "Climate Change and Security." *Bulletin of the Atomic Scientists* 64(2):19–25.

Scheffran, Jürgen, and Antonella Battaglini. 2011. "Climate and Conflicts: The Security Risks of Global Warming." *Regional Environmental Change* 11(Supplement 1):S27–S39.

Schlosberg, David, and Lisette B. Collins. 2014. "From Environmental to Climate Justice: Climate Change and the Discourse of Environmental Justice." *WIREs Climate Change* 5(3):359–74.

Schnaiberg, Allan. 1980. *The Environment: From Surplus to Scarcity*. New York: Oxford University Press.

Schneider, Stephen H. 1977. "Climate Change and the World Predicament: A Case Study for Interdisciplinary Research." *Climatic Change* 1(1):21–43.

Schott, Charles A. 1876. *Tables, Distribution, and Variations of the Atmospheric Temperature in the United States*. Washington, DC: Smithsonian Institution.

Schoonover, Rod. 2019. "The White House Blocked My Report on Climate Change and National Security." *New York Times*, July 30. Accessed December 5, 2023. https://www.nytimes.com/2019/07/30/opinion/trump-climate-change.html.

Schubert, Julia. 2021. *Engineering the Climate: Science, Politics, and Visions of Control*. Manchester: Mattering Press.

Schulten, Susan. 2012. *Mapping the Nation: History and Cartography in Nineteenth-Century America*. Chicago: University of Chicago Press.

Schwartz, Peter, and Doug Randall. 2003. *An Abrupt Climate Change Scenario and Its Implications for United States National Security*. Washington, DC: U.S. Department of Defense.

Scott, James C. 1998. *Seeing Like a State: How Certain Schemes to Improve the Human Condition Have Failed*. New Haven, CT: Yale University Press.

Seager, Richard, Haibo Liu, Naomi Henderson, Isla Simpson, Colin Kelley, Tiffany Shaw, Yochanan Kushnir, and Mingfang Ting. 2014. "Causes of Increasing Aridification of the Mediterranean Region in Response to Rising Greenhouse Gases." *Journal of Climate* 27(12):4655–76.

Sellers, Bev. 2017. "Preface." Pp. 6–8 in *Whose Land Is It Anyway? A Decolonization Handbook*, edited by P. McFarlane and N. Schabus. Vancouver: Federation of Post-Secondary Educators of BC.

Seth, Suman. 2018. *Difference and Disease: Medicine, Race, and the Eighteenth-Century British Empire*. New York: Cambridge University Press.

Shabecoff, Philip. 1988. "Global Warming Has Begun, Expert Tells Senate." *New York Times*, June 24. Accessed January 15, 2019. https://www.nytimes.com/1988/06/24/us/global-warming-has-begun-expert-tells-senate.html.

Shackley, Simon, and Brian Wynne. 1996: "Representing Uncertainty in Global Climate Change Science and Policy: Boundary Ordering Devices and Authority." *Science, Technology and Human Values* 21:275–302.

Shaman, Jeffrey, Susan Solomon, Rita Colwell, and Christopher Field. 2013. "Fostering Advances in Interdisciplinary Climate Science." *Proceedings of the National Academy of Sciences* 110:3653–56.

Shapin, Steven, and Simon Schaffer. 1985. *The Leviathan and the Air-Pump: Hobbes, Boyle, and the Experimental Life*. Princeton, NJ: Princeton University Press.

Shattuck, Lemuel. 1841. *The Vital Statistics of Boston*. Philadelphia: Lea and Blanchard.

Shattuck, Lemuel. 1850. *Report of the Sanitary Commission of Massachusetts, 1850*. Cambridge, MA: Harvard University Press.

Sherry Consulting. 2009. *Climate Change and National Security: A Field Map and Analysis of Funding Opportunities*. Accessed April 7, 2016. static1.squarespace.com/static/57fe7c73cd0f685a4c0dacd3/t/582cf93eb3db2bc03f81748b/1479342399529/SC_ClimateChangeNationalSecurity.pdf.

Short, John R. 2001. *Representing the Republic: Mapping the United States 1600–1900*. London: Reaktion.

Short, John R. 2009. *Cartographic Encounters: Indigenous Peoples and the Exploration of the New World*. Chicago: University of Chicago Press.

Short, Thomas. 1750. *New Observations, Natural, Moral, Civil, Political and Medical on City, Town, and Country Bills of Mortality [. . .] with an Appendix on the Weather and Meteors*. London: Longman.

Simon, Julian L., ed. 1996. *The State of Humanity*. New York: Wiley-Blackwell.

Simpson, Charles R. 1992. "The Wilderness in American Capitalism: The Sacralization of Nature." *International Journal of Politics, Culture, and Society* 5(4):555–76.

Singer, Fred, ed. 2008. *Summary for Policymakers of the Report of the Nongovernmental International Panel on Climate Change*. The Heartland Institute. Accessed November 24, 2017. annkal.org/downloads/environment /naturenothumanactivityrulesclimate.pdf.

Sismondo, Sergio. 2011. "Bourdieu's Rationalist Science of Science: Some Promises and Limitations." *Cultural Sociology* 5(1):83–97.

Skocpol, Theda. 1995. *Protecting Soldiers and Mothers: The Political Origins of Social Policy in United States*. Cambridge, MA: Harvard University Press.

Skowronek, Stephen. 1982. *Building a New American State: The Expansion of National Administrative Capacities, 1877–1920*. Cambridge, UK: Cambridge University Press.

Smagorinsky Joseph. 1991. "Climatology's Scientific Maturity." Pp. 29–36 in *Climate in Human Perspective: A Tribute to Helmut E. Landsberg*, edited by F. Baer, N. L. Canfield, and J. M. Mitchell. Boston: Kluwer.

Smagorinsky, Jospeh. 1983. Interview with Hessam Taba. *World Meteorological Organization Bulletin* 32(4):277–90.

Smith, Dan, and Janani Vivekananda. 2007. *A Climate of Conflict: The Links between Climate Change, Peace and War*. London: International Alert.

Smith, Neil. 1984. *Uneven Development: Nature, Capital, and the Production of Space*. Athens: Georgia University Press.

Smith, Paul. 2007. "Climate Change, Weak States, and the War on Terror in South and Southeast Asia." *Contemporary Southeast Asia* 29(2):264–85.

Smithsonian Institution. 1859. *Annual Report of the Board of Regents of the Smithsonian Institution, 1858*. Washington, DC: Government Printing Office.

Smithsonian Institution. 1864. *Annual Report of the Board of Regents of the Smithsonian Institution, 1863*. Washington, DC: Government Printing Office.

Solow, Andrew R. 2013. "Global Warming: A Call for Peace on Climate and Conflict." *Nature* 497:179–80.

Somers, Margaret R. 1995. "What's Political or Cultural about Political Culture and The Public Sphere? Toward an Historical Sociology of Concept Formation." *Sociological Theory* 13(2):113–44.

Somerville, Scott. 1979. "A Vennorable Weather Prophet." *Chinook* 1(3):36–37.

Spash, Clive L. 2010. "The Brave New World of Carbon Trading." *New Political Economy* 15(2):169–95.

Spratt, David, and Philip Sutton. 2008. *Code Red: The Case for Climate Emergency*. Melbourne, VIC: Scribe.

Stampnitzky, Lisa. 2011. "Disciplining an Unruly Field: Terrorism Experts and Theories of Scientific/Intellectual Production." *Qualitative Sociology* 31:1–19.

Stampnitzky, Lisa. 2013. *Disciplining Terror: How Experts Invented "Terrorism"*. Cambridge, UK: Cambridge University Press.

Star, Susan, and James Griesemer. 1989. "Institutional Ecology, 'Translations' and Boundary Objects: Amateurs and Professionals in Berkeley's Museum of Vertebrate Zoology, 1907–39." *Social Studies of Science* 19(3):387–420.

Steelman John R, ed. 1947. *Science and Public Policy: A Report to the President*. 5 vols. Washington, DC: Government Printing Office.

Stehr, Nico, and Hans V. Storch, eds. 2000. *Eduard Bruckner: The Sources and Consequences of Climate Change and Climate Variability in Historical Times*. Dordrecht: Springer.

Stein, Jeff. 2017. "Michael Flynn, Russia and a Grand Scheme to Build Nuclear Power Plants in Saudi Arabia and the Arab World." *Newsweek*, June 23. Accessed February 7, 2018. newsweek.com/2017/06/23/flynn-russia-nuclear -energy-middle-east-iran-saudi-arabia-qatar-israel-donald-623396.html.

Steinberg, Ted. 1991. *Acts of God: The Unnatural History of Natural Disaster in America*. London: Oxford University Press.

Steinmetz, George. 2008. "The Colonial State as a Social Field: Ethnographic Capital and Native Policy in the German Overseas Empire before 1914." *American Sociological Review* 73(4):589–612.

Stern, Nicholas. 2007. *The Economics of Climate Change: The Stern Review*. Cambridge, UK: Cambridge University Press.

Sterne, Jonathan. 2003. "Bourdieu, Technique and Technology." *Cultural Studies* 17(3–4):367–89.

Stewart, Irvin. 1948. *Organizing Scientific Research for War: The Administrative History of the Office of Scientific Research and Development*. Boston: Little, Brown.

Stewart, Mark G. 2020. *Climate Change and National Security: Balancing the Costs and Benefits*. Cato Institute. Accessed December 15, 2023. https://www.cato.org/publications/climate-change-national-security -balancing-costs-benefits.

Strawa, Anthony W., Gary Latshaw, Stanley Farkas, Philip Russell, and Steven Zornetzer. 2020. "Arctic Ice Loss Threatens National Security: A Path Forward." *Orbis* 64(4):622–36. https://doi.org/10.1016/j.orbis.2020.08.010.

Stuart, Douglas T. 2008. *Creating the National Security State: A History of the Law that Transformed America*. Princeton, NJ: Princeton University Press.

Sullivan, Walter. 1961. *Assault on the Unknown: The International Geophysical Year*. New York: McGraw-Hill.

Sweezy, Kevin Z. 2016. *Prelude to the Dust Bowl: Drought in the Nineteenth-Century Southern Plains*. Norman: University of Oklahoma Press.

Swift, Jeremy. 1996. "Desertification: Narratives, Winners and Losers." Pp. 73–90 in *The Lie of the Land: Challenging Received Wisdom on the African Environment*, edited by M. Leach and R. Mearns. London: The International African Institute.

Swyngedouw, Erik. 2010. "Apocalypse Forever? Post-political Populism and the Spectre of Climate Change." *Theory, Culture & Society* 27(3):213–32.

Sydenham, Thomas. [1686] 1742. *Schedula Monitoria, or an Essay on the Rise of a New Fever.* Pp. 495–521 in *The Entire Works of Dr. Thomas Sydenham: The Second Edition,* edited by John Swan. London: Cave.

Szasz, Andrew. 2016. "Novel Framings Create New, Unexpected Allies for Climate Activism." Pp. 150–70 in *Reframing Climate Change: Constructing Ecological Geopolitics,* edited by S. O'Lear and S. Dalby. New York: Routledge.

Tamboukou, Maria. 2006. "Writing Genealogies: An Exploration of Foucault's Strategies for Doing Research." *Discourse: Studies in the Cultural Politics of Education* 20(2):201–17.

Taylor, Marcus. 2014. *The Political Ecology of Climate Change Adaptation: Livelihoods, Agrarian Change and the Conflicts of Development.* New York: Routledge.

Thew, Harriet, Lucie Middlemiss, and Jouni Paavola. 2020. "'Youth Is Not a Political Position': Exploring Justice Claims-Making in the UN Climate Change Negotiations." *Global Environmental Change* 61:102036.

Thomas, Michael D. 2015. "Climate Securitization in the Australian Political–Military Establishment." *Global Change, Peace & Security* 27(1):97–118.

Thompson, Kenneth. 1981. "The Question of Climatic Stability in America before 1900." *Climatic Change* 3(3):227–41.

Tilly, Charles, ed. 1975. *The Formation of National States in Western Europe.* Princeton, NJ: Princeton University Press.

Tilly, Charles. 1985. "War Making and State Making as Organized Crime." Pp. 169–91 in *Bringing the State Back In,* edited by P. B. Evans, D. Rueschemeyer, and T. Skocpol. Cambridge, UK: Cambridge University Press.

Tilly, Charles. 1990. *Coercion, Capital, and European States, AD 990–1990.* Cambridge, MA: Basil Blackwell.

Tilton, James. [1781] 1813. *Economical Observations on Military Hospitals; and the Prevention and Cure of Diseases Incident to an Army.* Wilmington, DE: Wilson.

Titley, David. 2011. "Climate Change and National Security." TEDxPentagon. Posted January 24. Accessed April 5, 2017 www.youtube.com/watch?v=7udNMqRmqV8.

Titley, David. 2017. "How the Military Fights Climate Change." TED talk. Center for Solutions to Weather and Climate Risk, Penn State University. Accessed May 12, 2019. https://www.ted.com/talks/david_titley_how_the_military_fights_climate_change.

Torres, Roger R., David M. Lapola, Jose A. Marengo, and Magda A. Lombardo. 2012. "Socio-climatic Hotspots in Brazil." *Climatic Change* 115(3–4):597–609.

Trattner, Walter I. 1999. *From Poor Laws to Welfare State: A History of Social Welfare in America.* 6th ed. New York: Free Press.

Trombetta, Maria J. 2011: "Rethinking the Securitization of the Environment: Old Beliefs, New Insights." Pp. 135–49 in *Securitization Theory: How Security Problems Emerge and Dissolve*, edited by in T. Balzacq. London: Routledge.

Trombetta, Maria J. 2014. "Linking Climate-Induced Migration and Security within the EU: Insights from the Securitization Debate." *Critical Studies on Security* 2(2):131–47.

Tuathail, Gearóid Ó. 1996. *Critical Geopolitics: The Politics of Writing Global Space*. Minneapolis: University of Minnesota Press.

Tyndall, John. 1861. "On the Absorption and Radiation of Heat by Gases and Vapors, and on the Physical Connexion of Radiation, Absorption, and Conduction." *London, Edinburgh, and Dublin Philosophical Magazine and Journal of Science* 22(146):169–94, 273–85.

UK Ministry of Defense. 2007. *Development, Concepts and Doctrine Centre Global Strategic Trends Programme: 2007–2036*. Accessed January 12, 2017. www.cuttingthroughthematrix.com/articles/strat_trends_23jan07.pdf.

Ullman, Richard. 1983. "Redefining Security." *International Security* 8(1):129–53.

Ullrich, J. K. 2015. "Climate Fiction: Can Books Save the Planet?" *Atlantic*, August 13. Accessed May 12, 2018. https://www.theatlantic.com /entertainment/archive/2015/08/climate-fiction-margaret-atwood-literature /400112/.

Ungar, Sheldon. 1992. "The Rise and (Relative) Decline of Global Warming as a Social Problem." *Sociological Quarterly* 33(4):483–501.

United Nations (UN). 1992. *United Nations Framework Convention on Climate Change*. Accessed November 8, 2017. unfccc.int/resource/docs/convkp /conveng.pdf.

United Nations (UN). 1993. *Report of the United Nations Conference on Environment and Development*. New York: United Nations.

United Nations (UN) Brundtland Commission. 1987. *World Commission on Environment and Development: Our Common Future*. www.un-documents .net/our-common-future.pdf.

United Nations Development Program (UNDP). 1994. *Human Development Report 1994*. New York: Oxford University Press.

United Nations Development Program (UNDP). 2021. "Climate Finance for Sustaining Peace: Making Climate Finance Work for Conflict-Affected and Fragile Contexts." Accessed November 18, 2023. https://www.undp.org/sites /g/files/zskgke326/files/2021-12/UNDP-Climate-Finance-for-Sustaining -Peace.pdf.

United Nations (UN) General Assembly. 2009. *Climate Change and Its Possible Security Implications: Report of the Secretary-General*. September 11. Accessed January 24, 2018. refworld.org/docid/4ad5e6380.html.

University Corporation for Atmospheric Research (UCAR) Digital Archives. 1958–1965. Meeting Minutes, Personal Correspondence, Resolutions and Policy Drafts. Accessed March 6, 2016. archives.ucar.edu.

U.S. Advisory Committee on Weather Control. 1957. *Report of the President's Advisory Committee on Weather Control*. Washington, DC: Government Printing Office.

U.S. Agency for International Development (USAID). 2012. "USAID Climate Change Adaptation Plan." Accessed May 1, 2019. https://www.usaid.gov/sites /default/files/documents/1865/Agency%20Climate%20Change%20 Adaptation%20Plan%202012.pdf.

U.S. Agency for International Development (USAID). 2023. "Climate Change: Advancing Global Climate Action." Accessed December 12, 2023. https:// www.usaid.gov/climate.

U.S. Agricultural Society. 1858. "Agricultural Operations of the Patent Office." *Monthly Bulletin of the United States Agricultural Society* 1(5):37–38.

U.S. Atomic Energy Commission. 1954. *In the Matter of J. Robert Oppenheimer: Transcript of Hearing before Personnel Security Board*, Washington, DC, April 12–May 6. Accessed January 28, 2019. https://archive.org/details /OppenheimerHearings.

U.S. Congress. House of Representatives. 2008. *National Intelligence Assessment on the National Security Implications of Global Climate Change to 2030: Hearing before the Permanent Select Committee on Intelligence, House Select Committee on Energy Independence and Global Warming*. 110th Congress., 2d Session. Testimony of Thomas Fingar.

U.S. Congress. House of Representatives. 2017. *National Defense Authorization Act for Fiscal Year 2018: Conference Report to Accompany H.R. 2810*. November 7. Washington, DC: Government Printing Office.

U.S. Congress. House of Representatives. 2019. *The National Security Implications of Climate Change: Hearing before the Committee on Intelligence*. 116th Congress, 2d Session. Statement for the Record by Rod Schoonover. Accessed May 4, 2024. https://int.nyt.com/data/documenthelper/1103-rod -schoonover-testimony/9ea6b07179b17035421f/optimized/full.pdf#page=1.

U.S. Congress. Senate. 1965–1966. *Weather Modification, Parts 1 and 2: Hearings on S. 23 and S. 2916 before the Committee on Commerce*. Washington, DC: Government Printing Office.

U.S. Congress. Senate. 1987–1988. *Greenhouse Effect and Global Climate Change: Hearings before the Committee on Energy and Natural Resources*. 100th Congress, 1st and 2d Sessions, November 9–10, 1987 and June 23, 1988. Washington, DC: Government Printing Office.

U.S. Department of Agriculture (USDA). 1892. "List of Voluntary Observers of the Weather Bureau, Who Furnish Meteorological Reports for the Monthly Weather Review." *Monthly Weather Review* 20(3):80–83.

U.S. Department of Agriculture (USDA). 1896. "The Weather in 1895." In *Yearbook of the United States Department of Agriculture for 1895.* Washington, DC: Government Printing Office.

U.S. Department of Agriculture (USDA). 1906. "Crop Bulletins." In *Yearbook of the United States Department of Agriculture for 1905.* Washington, DC: Government Printing Office.

U.S. Department of Agriculture (USDA). 1919. "Condensed Climatological Summary." *Monthly Weather Review* 47(3):193–99.

U.S. Department of Defense (DOD). 2006. *Quadrennial Defense Review Report.* Accessed April 2, 2016. https://archive.defense.gov/pubs/pdfs/QDR2006 0203.pdf.

U.S. Department of Defense (DOD). 2010. "Quadrennial Defense Review Report." Accessed April 2, 2016. https://archive.defense.gov/qdr/QDR%20as %20of%2029JAN10%201600.pdf.

U.S. Department of Defense (DOD). 2014a. *Department of Defense Climate Change Adaptation Roadmap.* Alexandria, VA: Office of the Assistant Secretary of Defense (Energy, Installations, and Environment).

U.S. Department of Defense (DOD). 2014b. *Quadrennial Defense Review Report.* Accessed December 10, 2023. https://history.defense.gov/Portals/70/Documents /quadrennial/QDR2014.pdf?ver=tXH94SVvSQLVw-ENZ-a2pQ%3d%3d.

U.S. Department of Defense (DOD). 2015. *National Security Implications of Climate-Related Risks and a Changing Climate.* Accessed June 20, 2016. archive.defense.gov/pubs/150724-congressional-report-on-national -implications-of-climate-change.pdf?source=govdelivery.

U.S. Department of Defense (DOD). 2016. "Directive 4715.21: Climate Change Adaptation and Resilience." Accessed January 12, 2017. http://www.dtic.mil /whs/directives.

U.S. Department of Defense (DOD). 2021. *Department of Defense Climate Risk Analysis: A Report to the National Security Council.* Accessed December 12, 2023. https://media.defense.gov/2021/Oct/21/2002877353/-1/-1/0/DOD -CLIMATE-RISK-ANALYSIS-FINAL.PDF.

U.S. Department of Defense (DOD). 2023. "DOD Climate Resilience Portal." Accessed December 12, 2023. climate.mil.

U.S. Department of Homeland Security (DHS). 2010. *Climate Change Adaptation Report.* Accessed July 19, 2018. https://www.hsdl.org/?view&did=4324.

U.S. Global Change Research Program (USGCRP). 2000. *Climate Change Impacts on the United States: The Potential Consequences of Climate Variability and Change.* Cambridge, UK: Cambridge University Press.

U.S. Interagency Climate Change Adaptation Task Force. 2011. *Federal Actions for a Climate Resilient Nation.* Accessed May 1, 2024. https://www.white house.gov/wp-content/uploads/legacy_drupal_files/ceq/2011_adaptation _progress_report.pdf.

U.S. National Intelligence Council (NIC). 2008. *National Security Implications of Global Climate Change through 2030*. Washington, DC: National Intelligence Council.

U.S. National Intelligence Council (NIC). 2012. *Global Trends 2030: Alternative Worlds*. Accessed July 1, 2023. https://www.dni.gov/files/documents/Global Trends_2030.pdf.

U.S. National Intelligence Council (NIC). 2021. *Global Trends 2040: A More Contested World*. Accessed December 5, 2023. https://www.dni.gov/files /ODNI/documents/assessments/GlobalTrends_2040.pdf.

U.S. National Security Council. 2006. *National Strategy for Combating Terrorism*. Washington, DC: National Security Council.

U.S. Patent Office. 1861. *Results of Meteorological Observations, 1854–1859*. Washington, DC: Government Printing Office.

U.S. Signal Service. 1872. *Instructions to Observer Sergeants, Signal Service, U S.A. on Duty at Stations*. Washington, DC: Government Printing Office.

U.S. Signal Service. 1873. *Annual Report of the Chief Signal-Officer to the Secretary of War for the Year 1872*. Washington, DC: Government Printing Office.

U.S. Signal Service. 1874. *Annual Report of the Chief Signal-Officer to the Secretary of War for the Year 1873*. Washington, DC: Government Printing Office.

U.S. Signal Service. 1875a. "Introductory." *Monthly Weather Review* 3(4):1.

U.S. Signal Service. 1875b. "Atmospheric Temperature." *Monthly Weather Review* 3(4):4-5.

U.S. Signal Service. 1876. "Precipitation." *Monthly Weather Review* 4(8):5-6.

U.S. Signal Service. 1878. *Annual Report of the Chief Signal Officer to the Secretary of War for the Year 1877*. Washington, DC: Government Printing Office.

U.S. Signal Service. 1883. *Regulations for the Signal Office*. Washington, DC: Government Printing Office.

U.S. Signal Service. 1884. "Catalogue of Signal Service Instruments." Pp. 29–33 in *History of the Signal Service, with a Catalogue of Publications, Instruments, and Stations*. Washington, DC: War Department.

U.S. Signal Service. 1887. *Annual Report of the Chief Signal Officer to the Secretary of War*. Washington, DC: Government Printing Office.

U.S. Signal Service. 1889. *Annual Report of the Chief Signal Officer to the Secretary of War*. Washington, DC: Government Printing Office.

U.S. Signal Service. 1890. "Appendix 18: Report of Assistant Professor in Charge of The Instrument Division." Pp. 650–62 in *Annual Report of the Chief Signal Officer, United States Army, to the Secretary of War*. Washington, DC: Government Printing Office.

U.S. Weather Bureau. 1897. *Instructions for Voluntary Observers: Prepared under Direction of Willis L. Moore*. Washington, DC: Government Printing Office.

U.S. Weather Bureau. 1908. "Foreign Meteorologists Study American Weather Services." Pp. xvii–xviii in *Report of the Chief of the Weather Bureau, 1906–1907.* Washington, DC: Government Printing Office.

U.S. Weather Bureau. 1911."106 Climatological Sections in the United States." *Monthly Weather Review* 39(1):unpaginated supplement.

U.S. Weather Bureau. 1912. *Summaries of Climatological Data by Sections.* Weather Bureau Bulletin W, 1–2. Washington, DC: Government Printing Office.

U.S. White House. 1991. *National Security Strategy of the United States.* Washington, DC: Government Printing Office.

U.S. White House. 1992. *National Security Strategy of the United States.* Washington, DC: Government Printing Office.

U.S. White House. 1993. *National Security Strategy of the United States.* Washington, DC: Government Printing Office.

U.S. White House. 2010. *National Security Strategy of the United States.* Washington, DC: Government Printing Office.

U.S. White House. 2015. *Findings from Select Federal Reports: The National Security Implications of a Changing Climate.* Accessed August 1, 2017. https://obamawhitehouse.archives.gov/sites/default/files/docs/National _Security_Implications_of_Changing_Climate_Final_051915.pdf.

U.S. White House. 2016. "Presidential Memorandum: Climate Change and National Security." Office of the Press Secretary, September 21. Accessed May 12, 2019. https://obamawhitehouse.archives.gov/the-press-office/2016 /09/21/presidential-memorandum-climate-change-and-national-security.

U.S. White House. 2017. *National Security Strategy of the United States.* Washington, DC: Government Printing Office.

Valencius, Conevery Bolton. 2002. *The Health of the Country: How American Settlers Understood Themselves and Their Land.* New York: Basic Books.

Valencius, Conevery Bolton, David I. Spanagel, Emily Pawley, Sara Stidstone Gronim, and Paul Lucier. 2016. "Science in Early America: Print Culture and the Sciences of Territoriality." *Journal of the Early Republic* 36(1):73–123.

Vandergeest, Peter, and Nancy Lee Peluso. 2011. "Political Violence and Scientific Forestry: Emergencies, Insurgencies, and Counterinsurgencies in Southeast Asia" Pp. 152–66 in *Knowing Nature: Conversations at the Intersection of Political Ecology and Science Studies*, edited by M. J. Goldman, P. Nadasdy, and M. Turner. New York: Routledge.

Vennor, Henry G. 1877. *Vennor's Winter Almanac and Weather Record, 1877–78.* Montreal: Dougall.

Vetta, James A. 2011. "Climate Change and National Security: Implications for Space Systems." *Crosslink* 12(2). Accessed February 23, 2018. www.aerospace .org/crosslinkmag/summer2011/crosslink-summer-2011/.

Victor, David G., and Charles F. Kennel. 2014. "Climate Policy: Ditch the 2°C Warming Goal." *Nature* 514(7520):30–31.

Vidal, John. 2009. "Global Warming Could Create 150 Million 'Climate Refugees' by 2050." *Guardian*, November 3. Accessed December 14, 2017. https://www.the guardian.com/environment/2009/nov/03/global-warming-climate-refugees.

Visher, Stephen. 1924. *Climatic Laws: Ninety Generalizations with Numerous Corollaries as to the Geographic Distribution of Temperature, Wind, Moisture, Etc.; A Summary of Climate*. New York: J. Wiley & Sons.

Volney, Constantine. 1804. *View of the Climate and Soil of the United States of America*. London: Johnson.

Voosen, Paul. 2018. "Trump White House Quietly Cancels NASA Research Verifying Greenhouse Gas Cuts." *Science Magazine*, May 9. Accessed June 1, 2018. www.sciencemag.org/news/2018/05/trump-white-house-quietly -cancels-nasa-research-verifying-greenhouse-gas-cuts.

Wacquant, Loïc. 2004. "Pointers on Pierre Bourdieu and Democratic Politics." *Constellations* 11(1):3–15.

Wacquant, Loïc, and Aksu Akçaoğlu. 2016. "Practice and Symbolic Power in Bourdieu: The View from Berkeley." *Journal of Classical Sociology* 17(1):55–69.

Wæver, Ole. 1995. "Securitization and Desecuritization." Pp. 46–86 in *On Security*, edited by R. Lipschutz. New York: Columbia University Press.

Waldo, Frank. 1901. "The Blue Hill Meteorological Observatory." *Popular Science Monthly* 59(7):290–307.

Walker, Francis. 1874. *Statistical Atlas of the United States: Based on the Results of the Ninth Census, 1870*. New York: Bien.

Wallace-Wells, David. 2017. "The Uninhabitable Earth." *New York Magazine*, July 10. Accessed January 9, 2018. http://nymag.com/intelligencer/2017/07 /climate-change-earth-too-hot-for-humans.html.

Wallace-Wells, David. 2019. *The Uninhabitable Earth: Life after Warming*. New York: Duggan.

Walz, F. J. 1905. "Fake Weather Forecasts." *Popular Science Monthly* 67(October):503–13.

Wang, Jessica. 1999. *American Science in an Age of Anxiety: Scientists, Anti-communism, and the Cold War*. Chapel Hill: University of North Carolina Press.

Wang, Zuoyue. 2008. *In Sputnik's Shadow: The President's Science Advisory Committee and Cold War America*. New Brunswick, NJ: Rutgers University Press.

Ward, Robert D. 1908. *Climate: Considered Especially in Relation to Man*. New York: Putnam.

Ward, Robert D. 1914. "Lorin Blodget's 'Climatology of the United States': An Appreciation." *Monthly Weather Review* 42(1):33–37.

Ward, Robert D. 1915. "Climatic Subdivisions of the United States." *Bulletin of the American Geographical Society* 47(9):672–80.

Ward, Robert D. 1929. "The Acclimatization of the White Races in the Tropics." *New England Journal of Medicine* 201(13):617–27.

Warren, Louis S. 2002. "The Nature of Conquest: Indians, Americans, and Environmental History." In *A Companion to American Indian History*, edited by P. J. Deloria and N. Salisbury. New York: Wiley.

Water Utility Climate Alliance (WUCA). 2009. *Options for Improving Climate Modeling to Assist Water Utility Planning for Climate Change.* Accessed August 4, 2016. www.wucaonline.org/assets/pdf/pubs-whitepaper-120909.pdf.

Waterman Alan T. 1954. *The International Geophysical Year.* Washington, DC: National Science Foundation.

Watts, Isaac. 2017. "Twin Megastorms Have Scientists Fearing This May Be the New Normal." *Guardian*, September 6. Accessed December 13, 2017. https://www.theguardian.com/world/2017/sep/06/twin-megastorms-irma-harvey-scientists-fear-new-normal.

Watts, Michael J. 2015. "Now and Then: The Origins of Political Ecology and the Rebirth of Adaptation as a Form of Thought." Pp. 19–50 in *Handbook of Political Ecology*, edited by J. McCarthy and T. Perrault. London: Routledge.

WBGU. 2008. *World in Transition: Climate Change as a Security Risk.* Report of the German Advisory Council on Global Change. London: Earthscan.

Weart Spencer R. 1997. "Global Warming, Cold War, and the Evolution of Research Plans." *Historical Studies in the Physical and Biological Sciences* 27(2):319–56.

Weart, Spencer R. 2008. *The Discovery of Global Warming.* 2nd ed. Cambridge, MA: Harvard University Press.

Weart, Spencer. 2013. "Rise of Interdisciplinary Research on Climate." *Proceedings of the National Academies of Science* 110:3657–64.

Weber, Max. [1919] 1946. "Politics as a Vocation." Pp. 77–128 in *From Max Weber: Essays in Sociology*, edited and translated by H. H. Gerth and C. W. Mills. New York: Oxford University Press.

Weber, Max. 1978. *Economy and Society: An Outline of Interpretive Sociology.* Edited by G. Roth and C. Wittich. Berkeley: University of California Press.

Webersik, Christian. 2010. *Climate Change and Security: A Gathering Storm of Global Challenges.* Santa Barbara, CA: Praeger.

Weeks, Sinclair. 1954. Letter to Detlov Bronk, March 23. Accessed November 5, 2015. Eisenhower Presidential Library, National Archives. https://www.eisenhower.archives.gov/research/online_documents/igy/1954_3_23.pdf.

Werrell, Caitlin E., and Francesco Femia. 2013. *The Arab Spring and Climate Change: A Climate and Security Correlations Series.* Center for Climate & Security, Center for American Progress, and Stimpson. February 28. Accessed June 28, 2015. www.americanprogress.org/issues/security/reports/2013/02/28/54579/the-arab-spring-and-climate-change/.

Werrell, Caitlin, and Francesco Femia. 2017. "Reaction: The New National Security Strategy and Climate Change." Center for Climate & Security.

December 18. Accessed January 3, 2018. https://climateandsecurity.org/2017/12/18/reaction-the-new-national-security-strategy-and-climate-change/.

Whiston, William. 1716. *An Account of a Surprizing Meteor, Seen in the Air, March the 6th, 1715/16, at Night [. . .]*. London: Senex.

White, Gregory. 2011. *Climate Change and Migration: Security and Borders in a Warming World*. New York: Oxford University Press.

White, Sam. 2015. "Unpuzzling American Climate: New World Experience and the Foundations of a New Science." *Isis* 106(3):544–66.

White, Sam. 2020. *A Cold Welcome: The Little Ice Age and Europe's Encounter with North America*. Cambridge, MA: Harvard University Press.

Whitelaw, Kevin. 2008. "Intelligence Report Assesses Impact of Climate Change." *U.S. News and World Report*, June 24. Accessed August 23, 2014. usnews.com/news/national/articles/2008/06/24/intelligence-report-assesses-impact-of-climate-change.

Whyte, Kyle Powys. 2013: "Justice Forward: Tribes, Climate Adaptation and Responsibility." *Climatic Change* 120:517–30.

Whyte, Kyle Powys. 2017. "Is It Colonial *Deja Vu*? Indigenous Peoples and Climate Injustice." Pp. 88–105 in *Humanities for the Environment: Integrating Knowledges, Forging New Constellations of Practice*, edited by J. Adamson and M. Davis. New York: Routledge.

Wille, Robert-Jan. 2017. "Colonizing the Free Atmosphere: Wladimir Köppen's 'Aerology', the German Maritime Observatory, and the Emergence of a Trans-Imperial Network of Weather Balloons and Kites, 1873–1906." *History of Meteorology* 8:95–123.

Williams, Samuel. [1794] 1809. *Natural and Civil History of Vermont*. Walpole, NH: Thomas and Carlisle.

Williams, Stephan. 1836. "Medical and Physical Topography of the Town of Deer Field, Franklin Co., Massachusetts." *Boston Medical and Surgical Journal* 15:197–203.

Williamson, Hugh. 1771. "An Attempt to Account for the Change of Climate, Which Has Been Observed in the Middle Colonies in North-America." *Transactions of the American Philosophical Society* 1:272–80.

Wilson, Job. 1815. *An Inquiry into the Nature and Treatment of the Prevailing Epidemic, called Spotted Fever*. Boston: Bradford and Read.

Wilson Center. 2016. "Beyond the Headlines: Climate, Migration, and Conflict (Report Launch)." Accessed August 3, 2018. https://www.wilsoncenter.org/event/beyond-the-headlines-climate-migration-and-conflict-report-launch.

Wolfe, Patrick. 2006. "Settler Colonialism and the Elimination of the Native." *Journal of Genocide Research* 8(4):387–409.

Women's Earth and Climate Action Network (WECAN). 2016. *Women's Climate Action Agenda*. Edited by O. Lake, C. Greensfelder, and E. Colligan. Accessed March 30, 2018. www.wecaninternational.org/uploads/cke_documents/WECAN-Agenda-update-2016-webd.pdf.

Wood, William [1634] 1764. *New England's Prospects [. . .]*. Boston: Fleet.

World Bank. 2014. *Turn Down the Heat: Confronting the New Climate Normal*. Washington, DC: World Bank.

World People's Conference on Climate Change and the Rights of Mother Earth. 2010. Accessed June 4, 2016. pwccc.wordpress.com/support/.

Wu, Yutian, Mingfang Ting, Richard Seager, Huei-Ping Huang, and Mark A. Cane. 2011. "Changes in Storm Tracks and Energy Transports in a Warmer Climate Simulated by the GFDL CM2.1 Model." *Climate Dynamics* 37(1):53–72.

Wynne, Brian. 2010. "Strange Weather, Again: Climate Science as Political Art." *Theory, Culture, and Society* 27(2):289–305.

Zabarenko, Deborah. 2007. "Scientists Charge White House Pressure on Warming." *Washington Post*, January 30. Accessed November 1, 2017. washingtonpost.com/wp-dyn/content/article/2007/01/30/AR2007013000 985.html.

Zelizer, Viviana A. 1978. "Human Values and the Market: The Case of Life Insurance and Death in 19th-Century America." *American Journal of Sociology* 84(3):591–610.

Zerubavel, Eviatar. 1982. "The Standardization of Time: A Sociohistorical Perspective." *American Journal of Sociology* 8(1):1–23.

Zilberstein, Anya. 2016. *A Temperate Empire: Making Climate Change in Early America*. New York: Oxford University Press.

Zizek, Slavoj. 2010. *Living in the End Times*. London: Verso.

Index

Abbe, Cleveland, 128, 138, 277n1
acclimatization, 17, 72–73, 78, 93, 96, 111; and
  health, 27, 36–37, 50, 53–54, 57–60, 62,
  69, 74, 78–79, 82, 90, 92, 122, 151, 245–46,
  255, 272n11; and migration, 10, 23–24,
  201–2, 213, 237, 259, 263, 270n5; and
  race, 19, 73, 87–88, 90–97, 118, 121, 176,
  271n5, 272n5, 275n11, 276n15, 282n2
"actionable" science, 31, 230, 236–37
agriculture: capitalism and commercial,
  108–9, 116–17, 246; climate zones and
  development of, 116–18, 119; dry-farming,
  118; Joseph Henry on, 110–11; role of
  meteorology in, 110–11, 142, 152; scientific
  approaches to, 65, 110–11, 112, 115
almanacs: and American print culture, 40;
  Farmers Almanac, 40; and prophecy, 143,
  257; and weather prediction, 9, 30, 112,
  124–26, 132, 140–41, 143–46, 148, 167, 169,
  171, 182–83, 186, 201–2, 246, 257, 278n6
American Medical Association, 61
American Meteorological Society, 96, 281n10
American Philosophical Society, 50–52, 54,
  56–57, 61, 63, 68, 82, 83, 110, 272n7, 274n7
American Revolution, 48, 54, 73, 79. *See
  also* war
American Security Project (ASP), 220–23, 232

antebellum period, 70–71, 94
Anthropocene, 14, 253–54, 260
apocalypse, 5–6, 26, 200, 214, 242, 256, 265,
  266, 268
Arab Spring, 206, 210
Aristotelian meteorology, 37, 107, 272n12
Arrhenius, Svante, 276n2
atomic bomb, 169

Barton, Edward, 61
Bergen School (meteorology), 165–69, 175
blackness: and African Americans, 71, 90, 272;
  and labor, 91; and tropical climates, 93–94
Blodget, Lorin, 103, 105, 111, 121, 148, 250
Bolin, Bert, 178, 279n5
boundary work, 10, 70. *See also* credibility
Bourdieu, Pierre, 14, 22, 70, 120, 162, 194,
  229, 242, 267, 274n4, 279n2
British Empire, 16, 73, 78, 107
bureaucratic field, 70, 176, 179, 190, 219, 226,
  231, 234–35, 261
Bush, George W., 203, 219–20, 224
Bush, Vannevar, 170
Byers, Horace, 165, 181

Calhoun, John C., 74, 85
Callendar, Guy S, 276n2

Founded in 1893,
UNIVERSITY OF CALIFORNIA PRESS
publishes bold, progressive books and journals
on topics in the arts, humanities, social sciences,
and natural sciences—with a focus on social
justice issues—that inspire thought and action
among readers worldwide.

The UC PRESS FOUNDATION
raises funds to uphold the press's vital role
as an independent, nonprofit publisher, and
receives philanthropic support from a wide
range of individuals and institutions—and from
committed readers like you. To learn more, visit
ucpress.edu/supportus.